"十三五"普通高等教育本科系列教材

水利水电专业英语

主　编　李继清　朱永强
副主编　周　婷　宋晓漓
参　编　姜旭新　张验科　王世玉　杨　卓
主　审　张建海

中国电力出版社
CHINA ELECTRIC POWER PRESS

内 容 提 要

本书为“十三五”普通高等教育本科系列教材，涉及水利水电行业的多个领域，包括水资源、水文学、河流工程、工程结构、环境工程、灌溉、工程施工、水电站设施设备、特殊水电站功能、工程招投标、学术论文写作和翻译等。全书分29章，分别是水资源、水资源开发规划、水资源可持续发展、水文学与水循环、水文学原理——单位线、泥沙、河流、洪水、水污染、坝、堤防、水电站、水环境、气候变化、水利用率与节水、灌溉文化、灌溉方法、灌溉运行评价、大坝的截流和导流、土石坝填筑、水电站主要设备、水电及其特点、抽水蓄能电站、潮汐发电站、小水电、航运和娱乐、招投标、怎样写好你的第一篇学术论文、翻译。

本书可作为高等学校水利水电专业的教学用书，也可作为水利水电企业工程技术人员和管理人员学习专业英语的培训教材。

图书在版编目（CIP）数据

水利水电专业英语/李继清，朱永强主编. —北京：中国电力出版社，2020.8（2024.2重印）
“十三五”普通高等教育本科规划教材
ISBN 978-7-5198-2830-1

Ⅰ.①水… Ⅱ.①李… ②朱… Ⅲ.①水利工程–英语–高等学校–教材 Ⅳ.①TV

中国版本图书馆CIP数据核字（2018）第296390号

出版发行：中国电力出版社
地　　址：北京市东城区北京站西街19号（邮政编码100005）
网　　址：http://www.cepp.sgcc.com.cn
责任编辑：孙　静（010-63412542）曹　慧
责任校对：王晓鹏
装帧设计：郝晓燕
责任印制：吴　迪

印　　刷：北京锦鸿盛世印刷科技有限公司
版　　次：2020年8月第一版
印　　次：2024年2月北京第三次印刷
开　　本：787毫米×1092毫米　16开本
印　　张：22
字　　数：553千字
定　　价：66.00元

前言

本书是作者结合专业英语课程的多年教学和工作经验，悉心设计、细选精编而成。其特点是内容丰富、覆盖面宽，并配有大量的教学辅导环节。本书以服务本科教学为主，但同时也可为水利水电类研究生、大专生以及从事水利水电行业的工作人员提供参考或借鉴。

本书涉及水利水电行业的多个领域，包括水资源（Water Resources）、水文学（Hydrology）、河流工程（River Engineering）、工程结构（Engineering Structure）、环境工程（Environmental Engineering）、灌溉（Irrigation）、工程施工及设备（Engineering Construction and Equipment）、水电站及综合利用（Hydroelectric Power Station and Multifunction）、实践应用（Practical Application）等单元和附录。

全书分29章，分别是：水资源（Water Resources）、水资源开发规划（Planning for Water Resources Development）、水资源可持续发展（Water Resources for Sustainable Development）、水文学与水循环（Hydrology and Water Cycle）、水文学原理——单位线（Principle of Hydrograph—Unit Hydrographs）、泥沙（Sediments）、河流（River）、洪水（Flood）、水污染（Water Pollution）、坝（Dams）、堤防（Levee）、水电站（Hydropower Plants）、水环境（Water Environment）、气候变化（Climate Change）、水利用率与节水（Water Use Efficiency and Water Conservation:Definitions）、灌溉文化（Irrigation Culture）、灌溉方法（Irrigation Methods）、灌溉运行评价（Irrigation Performance Evaluation）、大坝的截流和导流（River Closure and Diversion in Dam Construction）、土石坝填筑（Embankment and Fills）、水电站主要设备（Main Equipment in Hydropower Plants）、水电及其特点（Hydroelectricity and Its Characteristics）、抽水蓄能电站（Pumped-Storage Plants）、潮汐发电站（Tidal Power Station）、小水电（Small Hydropower）、航运和娱乐（Navigation and Recreation）、招投标（Bids）、怎样写好你的第一篇学术论文（How to Write Your First Research Paper）、翻译（EST Translation）。各章（Chapter）分为若干部分（Part），各部分有独立的主题，每一部分就是一篇短文。每一章的各部分既有关联，又可拆分。读者可以根据需要，选择适当的章节自学或组织教学活动。

书中提供了丰富的学习辅导内容。正文中采用下划线对复杂词句进行了重点标注；正文之后，有生词表、重点词组、难句解说，便于读者阅读理解；此外还有英文缩写形式列举、相关专业术语总结，帮助读者加深记忆、重点掌握。形式多样的练习题，有助于读者检验对本章内容的掌握程度，以便改进提高。各章后面，根据正文中出现的有特点的专业词汇，有针对性地配备了构词方法的总结。读者现学现总结，马上扩充词汇，效果更佳。

书后附录给出了全书的习题答案、词汇表、英文缩写、常用的水利协会及组织网址的总结，同时配有模拟考试样卷及参考文献列表，便于读者跨章节查阅及使用。

本书由华北电力大学李继清、朱永强、宋晓漓、张验科和安徽农业大学周婷、新疆额河建管局姜旭新、王世玉以及北京市水文总站杨卓执笔。四川大学张建海教授审阅了全书。编写过程中，纪昌明、王丽萍、吕爱钟等老师提出了宝贵的意见和建议，浙江省水利河口研究院席锐超、国网山东省泰安供电公司田军、北京良乡蓝鑫水利工程设计有限公司梅艳艳、南方电网科学研究院有限责任公司丁泽俊、青岛市水文局王勇、国网山西省长治供电公司许郁、国网河南省安阳供电公司陈彩虹、国网北京市电力公司陈沐乐参与了书稿的翻译和整理工作。李建昌、黄婧、王爽等研究生在素材收集、格式整理、部分课文翻译等方面做了大量的工作，在此表示衷心的感谢！

本书的出版得到国家自然科学基金（项目编号：51641901）、国家重点研发计划项目专题（2017YFC0405906-3、2016YFC0402208-5）、华北电力大学名师培育项目资助，特此致谢！

编者

2020 年 6 月

目 录

Chapter 1 Water Resources

Part 1 Water

Water is one of the Earth's natural resources. It is a finite resource, which means that the total amount of water is limited. Most of the world's water supply is saltwater stored in the oceans. Converting saltwater to freshwater is generally too expensive to be used for industrial, agricultural or household purpose. [1]

The type of water we generally use in human activities is fresh water. Only 3% of the world's water supply is fresh water and 79% of that is frozen, forming the polar icecaps, glaciers, and icebergs (see Fig.1.1). The remaining 21% of the fresh water supply is freshwater available as either surface (1%) or ground water (20%), ground water accounts for 95% of this amount; Easily accessible surface fresh water includes 5 kinds of water, such as lakes (52%), river (1%), soil moisture (38%), atmospheric water vapor (8%) and water within living organisms (1%).Ground water is water that either fills the spaces between soil particles or penetrates the cracks and spaces within rocks.[2]

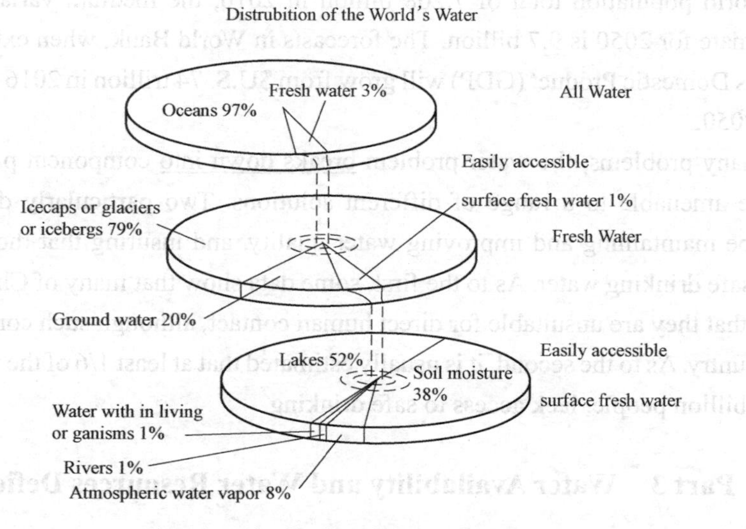

Fig.1.1 Distribution of the World's Water

Most people get their water from ground water sources. Roughly 9 out of every 10 Public Water Systems (PWSs) operate wells to tap ground water and just over half of the total population served by PWSs drinks water from a ground water source. Millions more get their water from private wells which also tap ground water sources. Ground water systems tend to serve smaller numbers of people, while surface water systems generally serve large populations. The quality and quantity of the world's water supply depends on how we choose to use water. Our use can be consumptive, which means that the water is not returned to nature (such as drinking water), or nonconsumptive.[3] Nonconsumptive use returns water, polluted or not, to the system. Creating hydroelectric power is an example of non-consumptive use.

Part 2 World Water Resources

About 119000 km^3 of fresh water precipitates on the continents each year. Much of this evaporates or is absorbed by plants. About 43000 km^3 flows through the world's rivers. Of this renewable fresh water, it is thought that about 9000 km^3 is readily available for human use, and that somewhat more than 6000 km^3 of fresh water is withdrawn from all sources (World Meteorological Organization, 1997).[4]

About 69% of world fresh water are for agriculture, 23% for industry and 8% for direct human use[5] (World Resources Institute, 1998). At first glance there seems to be adequate renewable freshwater, but when uses for ecosystem preservation, large differences in the pattern of regional availability and use, and the expense of developing additional supplies are considered, there is less water for human use than might be supposed. Moreover, taking into account population and economic growth, both of which contribute to increased demands on and pollution of water supplies, world fresh water is expected to become substantially more stressed in future decades. [6]

From a world population total of 7.208 billion in 2016, the medium variant United Nations population estimate for 2050 is 9.7 billion. The forecasts in World Bank, when extrapolated, suggest that world Gross Domestic Product (GDP) will grow from $U.S. 74 trillion in 2016 to more than $U.S. 105 trillion in 2050.

As with many problems, the water problem breaks down into component parts which, though interrelated, are amenable to a range of different solutions. Two particularly difficult and costly problems will be maintaining and improving water quality, and insuring that those on the margins have access to safe drinking water. As to the first, some data show that many of China's river reaches are so polluted that they are unsuitable for direct human contact, although such contact is regular part of life in that country. As to the second, it is usually estimated that at least 1/6 of the world's population, or more than 1 billion people, lack access to safe drinking.

Part 3 Water Availability and Water Resources Deficit

Water resource distribution over the territory of the Earth is uneven. Also, they disagree with population spread and economic development. These are very clearly revealed by analyzing and comparing the specific water availability for a single period of time for different regions and countries. The specific water availability represents the value of actual per capita renewable water resource.

For every design level the specific water availability is determined by dividing water resources without water consumption by the population number. In this case, water resources are assumed to be the river runoff formed in the territory of the given region and summed up with half the river water inflow from outside. So, the specific water availability is meant the residual (after use) per capita quantity of fresh water. Obviously, as population and water consumption grow, the value of specific water availability decreases.

For instance, the greatest water availability of 170-180 thousand m^3 per capita for 1995 is in the regions of Canada with Alaska and in Oceania. At the same time, in densely populated regions of Asia, Central, South Europe and Africa, the modern water availability is within 1.2-5 thousand m^3 per year. In the north of Africa and the Arabian Peninsula, it is as much as 0.2-0.3 thousand m^3 per year. It is worth mentioning that water availability of less than 2000 m^3 per year per capita is considered to be very low, and less than 1000 m^3 per year catastrophically low. With these values of water availability, very serious problems arise unavoidably with population life-support, industry and agriculture development.

Part 4 Water Resources and Water Use

Of particular interest is comparison of water use with renewable water resources of surface waters. These data by all regions of the world for1950, 1995 and 2025 are presented in Fig.1.2. This figure shows a comparison between the total water withdrawal and the values of local water resources summed up with half the inflow from outside. Thus, it is conventionally anticipated that every region can have at its disposal half the fresh water inflow from neighbor regions.

In accordance with the data presented in Fig.1.2, the modern water withdrawal in the world as a whole is not great in total amounting to 8.4% of global water resources.

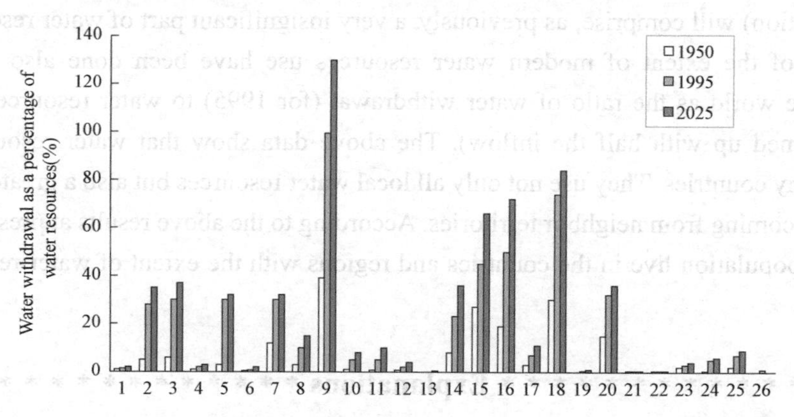

Fig.1.2 Water withdrawal by the natural-economic regions in percentage of water resources for 1950-1995-2025

1—North; 2—Central; 3—South; 4—North part of ETS SU; 5—South part of ETS SU; 6—North; 7—Centeral; 8—South;9—North; 10—South; 11—East; 12—West; 13—Central; 14—North China

By 2025 this figure is expected to increase up to 12.2%. However, water resources in the world are distributed very unevenly, which is seen even when comparing water withdrawal and river runoff by the continents on the average. Even at the present time in Europe and Asia, water withdrawal comprises 15%-17% of water resources, and in the future, it will reach 21%-23%. At the same time in South America and Oceania only 1.2%-1.3% of river runoff are used, and even in the future it is unlikely that this value will be above 1.6%-2.1%.

This distribution of river runoff and water use is especially uneven in the natural-economic regions of the world. Within every continent (except for South America), on the one hand, there are regions with a large extent of using water resources. On the other hand, there are regions with an insignificant water use (especially water consumption) as compared to water resources (see Fig.1.2).[7]

For instance, in Southern and Central parts of Europe, modern water withdrawal amounts already to as much as 24%-30% of water resources. At the same time in the northern part of the continent these values are not above 1.5%-3.0%. In the northern part of North America water withdrawal is not above 1% of water resources, and for the US territory this value is 28%. Even greater contrast takes place for Africa and Asia. In the northern part of America even at the present time renewable resources are almost totally withdrawn (water withdrawal is 95% of water resources). In other regions (especially in Central Africa), water withdrawal is negligibly small as compared with the value of water resources. In Asia, including the regions of Southern, Western, and Central Asia &Kazakhstan the use of water resources is very great (42%-84%). At the same time in the region of Siberia and the Far East, this use is not above 1%. Only in South America, in all regions, the value of using water resources is insignificant being not more than 2%-4%.

Up to 2025, the unevenness in the distribution of water resources and water use will be preserved and even much more increase. At the present time in many regions the use of water resources is already quite great. In the future this use will grow much and reach critical value. By contrast, in northern regions and in the regions with excessive moisture on all the continents water use (especially water consumption) will comprise, as previously, a very insignificant part of water resources.

Analyses of the extent of modern water resources use have been done also for individual countries of the world as the ratio of water withdrawal (for 1995) to water resources (local water resources summed up with half the inflow). The above data show that water resources are fully depleted in many countries. They use not only all local water resources but also a greater part of fresh water inflow incoming from neighbor territories. According to the above results at present about 75% of the Earth's population live in the countries and regions with the extent of water resources use of more than 20%

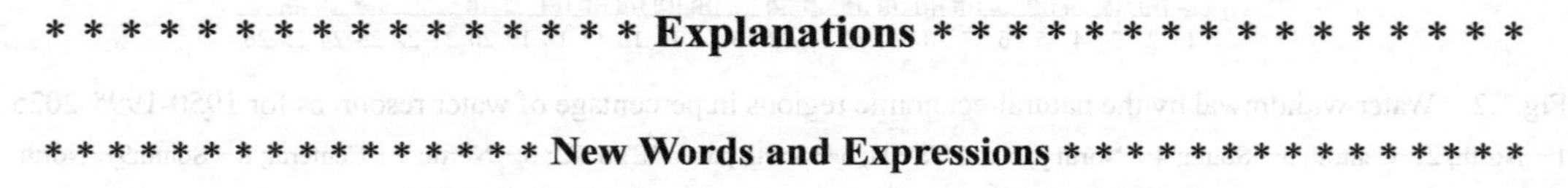

* * * * * * * * * * * * * * * * * **Explanations** * * * * * * * * * * * * * * * * *

* * * * * * * * * * * * * * * * * **New Words and Expressions** * * * * * * * * * * * * * * * *

1. finite ['fainait] adj. 有限的，有限的
2. saltwater ['sɔːlt'wɔːtə] adj. 盐水的，海产的
3. freshwater ['freʃ,wɔːtə(r)] n.淡水，湖水

4. polar [ˈpəulə] adj. [天]两极的，极性的
n. 极性
5. icecap [ˈaisˌkæp] n. 冰盖，冰帽
6. glacier [ˈglæsjəˌˈgleiʃə] n. 冰川
7. iceberg [ˈaisbəg] n. 冰山，冷冰冰的人
8. particle [ˈpa: tikl] n. 微粒，颗粒，粒子
9. tap [tæp] n. 轻打，活栓，水龙头
vt. 轻敲，开发，分接，使流出
vi. 轻叩，轻拍，轻声走
10. hydroelectric [ˈhaidrəuiˈlektrik] adj. 水电的，水力发电的
11. renewable [riˈnju: əbl] adj. 可更新的，可恢复的
12. withdraw [wiðˈdrɔ:] vt. 收回，撤回
vi. 离开，脱离
13. ecosystem [ˈekəusistəm] n. 生态系统
14. regional [ˈri: dʒənəl] adj. 整个地区的，地方的，地域性的
15. variant [ˈvɛəriənt] adj. 不同的
n. 变量
16. uneven [ˈʌnˈi: vən] adj. 不平坦的，不平均的，不均匀的
17. runoff [ˈrʌnwei] n. 径流，水流
18. inflow [ˈinfləu] n. 入流，流入物
19. densely [ˈdensli] adv. 浓密地，浓厚地
20. moisture [ˈmɔistʃə] n. 潮湿，水分，湿气
21. deplete [diˈpli:t] vt. 耗尽，使衰竭

* * * * * * * * * * * * * * * * * Complicated Sentences * * * * * * * * * * * * * * * * *

1. Converting saltwater to freshwater is generally too expensive to be used for industrial, agricultural or household purpose.

【译文】仅仅为了工业、农业或家庭用水而采取咸水淡化措施，通常代价高昂、无法实行。

【说明】too…to 表示“太……以致不能”。

2. Ground water is water that either fills the spaces between soil particles or penetrates the cracks and spaces within rocks.

【译文】地下水是指聚积在土壤或岩层空隙中的水。

【说明】That 引导定语从句解释前面的 water。

3. Our use can be consumptive, which means that the water is not returned to nature (such as drinking water), or nonconsumptive.

【译文】我们的用水可以是消耗性的，意味着水不能回到自然状态（如饮用水），也可以是非消耗性的。

【说明】which 引导定语从句，解释 consumptive 的具体内容。

4. Of this renewable fresh water, it is though that about 9000 km^3 is readily available for human use, and that somewhat more than 6000 km^3 of fresh water is withdrawn from all sources.

【译文】这些可再生的淡水中约有 9000 km^3 能为人类所用，各种来源的淡水总量在 6000 km^3 以上。

【说明】句中 that about 9000 km^3 is readily available for human use 和 that somewhat more than 6000 km^3 of freshwater is withdrawn from all resources 由 and 连接，是句中真正的主语。

5. About 69% of world fresh water are for agriculture, 23% for industry and 8%for direct human use.

【译文】地球上约有69%的淡水用于农业，23%用于工业，8%为人类直接所用。

【说明】句中23%和8%后面都省略了 of world freshwater withdrawals are。

6. Moreover, taking into account population and economic growth, both of which contribute to increased demands on and pollution of water supplies, world fresh water is expected to become substantially more stressed in future decades.

【译文】此外，若考虑人口和经济增长这两个增加供水需求并对水造成污染的因素，地球上淡水在未来几十年中可能会相当紧张。

【说明】句中 which 指代 population and economic growth，increased demands on and pollution of water supplies 作其前面 to 的宾语。

7. On the other hand, there are regions with an insignificant water use (especially water consumption) as compared to water resources.

【译文】另一方面，有些地区相对于其水资源来说，其使用的水（尤其是用水量）非常少。

*** * * * * * * * * * * * * * * * * Summary of Glossary * * * * * * * * * * * * * * * ***

1. natural resource	自然资源
2. water supply	供水，水源
3. surface water	地表水
4. ground water	地下水
5. drinking water	饮用水
6. water resource distribution	水资源分布
7. water availability	水可用性
8. water consumption	耗水量，用水量
9. river reach	河段
10. river runoff	河流径流量

*** * * * * * * * * * * * * * * * * * Abbreviations (Abbr.) * * * * * * * * * * * * * * * * * ***

1. GDP	Gross Domestic Product	国内生产总值
2. PWS	Public Water System	公共用水系统

*** Exercises ***

(1) Converting ________ to freshwater is generally too ________ to be used for ________, agricultural or ________ purpose.

(2) Only 2.5% of the world's water ________ is fresh water and 68.7% of that is ________, forming the ________ icecaps, ________, and icebergs.

(3) Surface water is ________ above the ground ________, such as ________, river, ________ and lakes.

(4) Ground water is water that either ________ the spaces between soil ________ or ________ the

________ and spaces within rocks.

*** * * * * * * * * * * * Word Building (1) -al(ial);-ous * * * * * * * * * * * ***

1. -al; ial [形容词后缀]，表示“……的”，“有……属性的”

Nature	n.	natural	adj.	自然的，自然界的
Part	n.	partial	adj.	部分的，局部的
Essence	n.	essential	adj.	本质的，基本的，精华的
Function	n.	functional	adj.	功能的

2. -al; -ial [名词后缀]，表示：状态、行为或其结果

propose	vt.	proposal	n.	提议，建议
remove	vt.	removal	n.	移动，免职，切除
renew	v.	renewal	n.	更新，恢复，续借，重申
reverse	v.	reversal	n.	颠倒，反转，反向
try	v	trial	n.	试验，考验，审讯，审判

3. -ous [形容词后缀]，表示：……的

danger	n.	dangerous	adj.	危险的
nerve	n.v.	nervous	adj.	紧张的，不安的
disaster	n.	disastrous	adj.	损失惨重的
vary	v.	various	adj.	不同的，多样的
continue	v.	continuous	adj.	连续的，持续的

*** * * * * * * * * * * * * * * * * * Text Translation * * * * * * * * * * * * * * * ***

第一章 水 资 源

第一节 水

水是地球上的一种自然资源。水是有限资源，也就是说，水的总量是有限的。世界上大部分的水源是储存在海洋中的咸水。为了工业、农业或家庭用水，将咸水转变成淡水通常代价高昂、无法实行。（单词个数：52）

人类活动所使用的水通常为淡水。世界上只有3%的水资源为淡水，其中79%为极地冻结的冰冠、冰川和冰山（见图1.1）。淡水总量中剩下的21%为容易利用的淡水或地下淡水，地下水占其中的95%。容易利用的淡水分为5类，如湖泊（52%）、河流（1%）、空气中的水汽（1%）、土壤水（38%）和动植物活体内的水（8%）。地下水是指聚积在土壤或岩层空隙中的水。（单词个数：100）

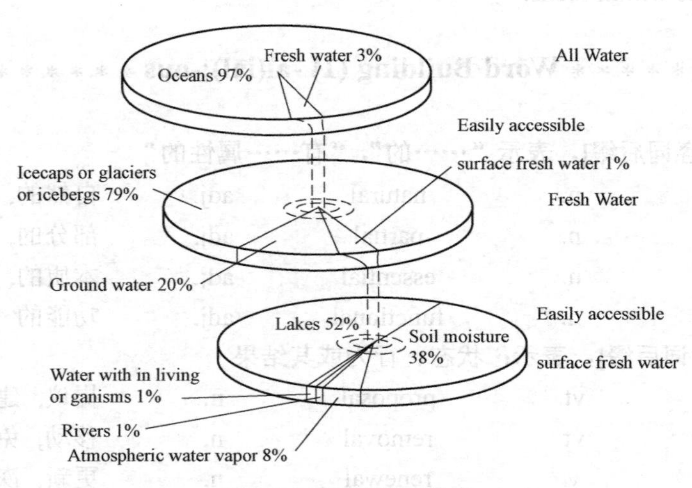

图 1.1　世界水资源分布

大多数人从地下水源取水，大约有 2/3 的公共用水系统打水井来开采地下水。其中，超过半数的人口饮用水是依靠公共用水系统提供的地下水。数百万人则从自家的井中取水，当然也是开采地下水源而获得的。依靠地下水系统提供生活用水的人数量少些，而依靠地面水系统提供生活用水的人则一般多些。世界水源的数量和质量取决于我们选择如何利用水。我们的用水可以是消耗性的，即水不能回到天然状态（如饮用水）；也可以是非消耗性的。非消耗性的使用可使水无论是污染过的，还是未污染的都可以再回到水资源系统。水力发电就是非消耗性使用的一个例子。（单词个数：135）

第二节　世界水资源介绍

每年，陆地上约有 11.9 万 km^3 的淡水凝结成雨或露等，其中大部分都蒸发掉或被植物吸收。全球约有 4.3 万 km^3 的降水回归河流。这些可再生的淡水中约有 9000 km^3 可为人类所用，估计来自各种渠道的淡水约有 6000 km^3 以上。（世界气象组织，1997）（单词个数：65）

地球上可用的淡水中约有 69%用于农业，23%用于工业，8%为人类直接所用（世界资源研究所，1998）。乍一看来，可再生的淡水似乎很充分，但若考虑用于生态系统保护、区域水有效性与水使用格局的巨大差异以及开发更多供水的费用时，人类所用之水就比想象的要少了。此外，若考虑人口和经济增长这两个增加供水需求并对水造成污染的因素，地球淡水在未来的几十年中可能会相当紧张。（单词个数：102）

2016 年世界总人口为 72.08 亿，联合国估计到 2050 年世界人口将达到 97 亿。据世界银行的预测，世界 GDP 将从 2016 年的 74 万亿美元增加到 2050 年的 105 万亿美元。（单词个数：52）

问题虽多，但水问题可归结为要素问题。这些要素尽管内部相关，但应该用于制订一系列不同的解决问题的方案。两个尤其棘手并且代价高昂的问题就是：保持并改善水质和确保那些用水紧缺的人们可饮用安全水。至于前者，有资料表明，中国许多河段污染如此严重，以致于不适合人类直接接触，尽管这种接触是日常生活的一部分。至于后者，通常估计，至少 1/6 的世界人口或 10 亿以上的人口缺乏安全饮用水。（单词个数：96）

第三节 水可用性与水资源匮乏

地球上水资源的地区分布不均匀。水资源与人口分布及经济发展状况也不一致。这些情况可通过对一定时期内不同国家和地区具体水可用性的分析和比较得以清楚地说明。水可用性表示实际人均可再生水资源的值。（单词个数：58）

对于每一设定的值，具体水可用性取决于对水资源的划分，而不考虑按人口数量所消耗的水。在此情况下，水资源即被假定为特定地区所形成的河水净流量。很明显，随着人口和水消费的增长，具体水可用性的数值会下降。（单词个数：82）

例如，1995 年，加拿大、美国阿拉斯加州或大洋洲地区的人均最大可用水量为 170000～180000m³ /(人 · 年)。而与此同时，在亚洲、森特拉尔、欧洲南部和非洲等人口密集的地区，年人均可用水量为 1200～5000m³ /(人 · 年)。在非洲北部与阿拉伯半岛地区，年人均可用水量仅为 200～300m³ /(人 · 年)。值得一提的是，人均用水量不到 2000m³ /(人 · 年)，属于人均可用水资源短缺；人均用水量小于 1000m³ /(人 · 年)，属于人均可用水量严重短缺。水资源的稀缺，必将限制人类日常生活和工农业的发展。（单词个数：119）

第四节 水资源及水利用量

将水利用量与地表水的可再生资源进行比较非常有益处，全球七大洲不同地区分别在 1950 年、1995 年和 2025 年的水资源及利用量数据如图 1.2 所示。该图对总水资源和区域水资源利用的百分数进行比较（一半入流水量来自于外部）。因此，一般认为，各地区都有一半的淡水入流水量来自于临近地区。（单词个数：76）

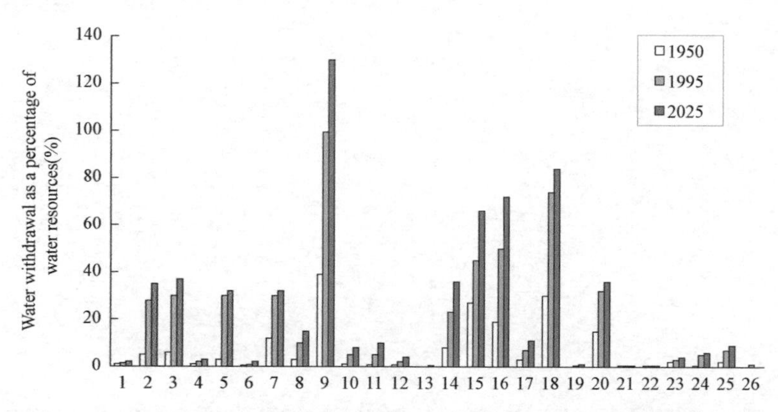

图 1.2 1950-1995-2025 全球所选定的自然-经济区的具体水资源可用性

1—North; 2—Central; 3—South; 4—North part of ETS SU; 5—South part of ETS SU; 6—North; 7—Centeral; 8—South;9—North; 10—South; 11—East; 12—West; 13—Central; 14—North China; 15—South Asia; 16—West Asia; 17—South-East Asia; 18—Middle Asia; 19—Siberia, For East of Russia; 20—Caucasus; 21—East; 23—West; 24—Central; 25—Australia; 26—Oceania

从图 1.2 可以看出，全世界现在的水利用量总体上不多，为全球水资源的 8.4%。

至2025年可望增加到12.2%。然而，世界水资源分布相当不均匀，这种不均匀即便将各大洲的平均水利用量和河流径流量进行比较也能看出。目前的欧洲和亚洲，水利用量占水资源的15%～17%，将来会达到21%～23%。同样，在南美洲和大洋洲，只有1.2%～1.3%的河流径流量得到利用，即便是将来，也不可能超过1.6%～2.1%。(单词个数：123)

世界各地的河川径流和用水分布特别不均衡。除南美洲外，其他各大洲：一方面，有的地区水资源利用量很大；另一方面，也有的地区相对于其水资源储量来说使用的水（尤其是消耗性用水量）微乎其微（见图1.2)。(单词个数：60)

例如：在南欧和中欧地区，现在水利用量已多达水资源的24%～30%，而欧洲的北部地区不超过1.5%～3.0%。北美北部的水利用不超过水资源的1%，而对于美国来说，却达到28%。非洲和亚洲相比差异更大。非洲北部目前可再生水资源几乎全部取尽（取用水量为水资源的95%)。其他地区尤其是中非，与其水资源相比，可取水量小到忽略不计。在亚洲，包括南亚、西亚和中亚以及哈萨克，水资源的使用量很大（42%～84%)。而在西伯利亚和远东地区，其使用率不超过1%。只有在南美洲，所有地区水资源使用量不超过2%～4%。(单词个数：174)

未来到2025年，水资源与水利用量的分布不均会仍然持续，甚至更加严重。目前，许多地区的水资源的利用量已相当大。未来，水利用量将更多，并达到临界值。相比而言，在北半球及所有丰水地区，水利用量（尤其是水消耗量）将和以前类似，对水资源的量不会有太大影响。(单词个数：84)

通过水利用量（1995）与水资源（一半入流水量的地区水资源）的比率对世界各国的现在水资源的利用程度进行分析表明：许多国家水资源将全部耗尽。他们不仅会用完当地水资源，而且还会用去多半来自邻近地区的淡水入流水量。以上情况表明，目前，全球约有75%的人口居聚集在水资源使用量比例高于20%的国家和地区。(单词个数：103)

Chapter 2 Planning for Water Resources Development

Part 1 Introduction of Planning

Planning can be defined as the orderly consideration of a project from the original statement of purpose through the evaluation of alternatives to the final decision on a course of action.[1] It includes all the work associated with the design of a project except the detailed engineering of the structures. It is the basis for the decision to proceed with (or to abandon) a proposed project and is the most important aspect of the engineering for the project. Because each water-development project is unique in its physical and economic setting, it is impossible to describe a simple process that will inevitably lead to the best decision. There is no substitute for "engineering judgment" in the selection of the method of approach to project planning, but each individual step toward the final decision should be supported by quantitative analysis rather than estimates or judgment whenever possible. One often hears the phrase "river-basin planning", but the planning phase is no less important in the case of the smallest project. The planning for an entire river basin involves a much more complex planning effort than the single project, but the difficulties in arriving at the correct decision may be just as great for the individual project.[2]

The term "planning" carries another connotation which is different from the meaning described above. This is the concept of the regional master plan which attempts to define the most desirable future growth pattern for an area.[3] If the master plan is in reality the most desirable pattern of development, then future growth should be guided toward this pattern. Unfortunately, the concept of "most desirable" is subjective, and it is difficult to assure that any master plan meets this high standard when first developed. Subsequent changes in technology, economic development, and public attitude often make a master plan obsolete in a relatively short time. Any plan is based on assumptions regarding the future, and if these assumptions are not realized the plan must be revised. Plan generally must be revised periodically.

An overall regional water management plan, developed with care and closely coordinated with other regional plans, may be a useful tool in determining which of many possible actions should be taken.[4] But it must always be considered subjecting to modification as the technological, economic and social environment change or as new factual data that are developed. A master plan that is no more than a catalog of all physically feasible actions is likely to prove of little value.

Part 2 Levels of Planning

Planning occurs at many levels within each country with the purpose and nature of the planning effort differing at each level. Many countries have a national planning organization with the goal of enhancing the economic growth and social conditions within the country. Even if no such organization exists, the national goals remain, and some form of national planning occurs in the legislative or executive branches of a government. The national planning organization will rarely deal with water problems directly, but in setting goals for production of food, energy, industrial goods, housing, etc, it may effectively specify targets for water management.

To allow for differences between the various regions of the country, regional planning groups may exist and equivalent regional water planning may occur. However, a natural "region" for water planning is the river basin and the tendency is to create river basin commissions, or river basin management units. These groups must assure coordination between the various activities within the river basin. Each specific action in water management is likely to have consequences downstream (and sometimes upstream), and thus these specific actions should not be planned in isolation but must be coordinated.

Planning of specific actions is the lowest level of planning, but it is at this level that important decisions which determine the effectiveness of water management are made. [5] This level is often called project planning although a physical may not necessarily result. For example, a study leading to a plan for flood-plain management is a legitimate project plan.

Part 3 Process of Planning

Project planning usually passes through several phases before the final plan emerges. In each country there is a specified sequence with specific names. The first phase or reconnaissance study is usually a coarse screen intended to eliminate those projects or actions which are clearly infeasible without extensive study, and hence to identify those activities which deserve further study. Following the reconnaissance phase may be one or more phases intended to thoroughly evaluate the feasibility of the proposed activity and in the process to formulate a description of the most desirable actions, i.e. the plan. In many cases a single feasibility study is adequate because the nature of the action is relatively easily evaluated. In other cases, one or more pre-feasibility studies may be undertaken to examine various aspects of the proposal. The idea of several sequential studies is to reduce planning costs by testing the weakest aspects of the project first. If the project is eliminated because of some aspect, the expense of studying all other aspects will have been avoided. If the series of studies is allowed to become a series of increasingly more thorough reviews of all aspects of the project, the cost may be increased because many things will have been redone two or more times.

The feasibility study usually requires that the structural details of a project be specified in sufficient detail to permit an accurate cost estimate.[6] In the final phase of the effort, the details of design must be examined carefully, and the construction drawings and specifications produced. Although, in principle, the issue of feasibility was decided on the basis of the feasibility study, the possibility always exists that the more thorough study may develop information that alters feasibility. Consequently, the decision to proceed should not be made irrevocable until the final design is completed.

* * * * * * * * * * * * * * * * * **Explanations** * * * * * * * * * * * * * * * *

* * * * * * * * * * * * * * * * * **New Words and Expressions** * * * * * * * * * * * * * * * *

1. planning [ˈplænɪŋ] n. 规划
2. evaluation [iˌvæljuˈeiʃən] n. 估价；评估
3. engineering [ˈendʒiˌniəriŋ] n. 工程技术，工程
4. proceed [prəˈsi: d] v. 着手进行，继续进行
5. basin [ˈbeisn] n.（河川的）流域
6. connotation [ˌkɑnəˈteɪʃ(ə)n] n. 含义，含意
7. master [ˈmæstər] adj.精通的；主要的
   n. 主人；专家；硕士；主管；少爷；原件；…桅船
   v. 精通，掌握；控制；战胜，克服；制作…母版
8. assumption [əˈsʌmpʃən] n. 假定，臆断；担任，承担
9. revise [riˈvaiz] v. 校订，修正，校正
   n. 校订，修正，再校稿
10. management [ˈmænidʒmənt] n.管理，处理
11. modification [ˌmɔdifiˈkeiʃən] n. 调整，修正，修饰，修改
12. legislative [ˌledʒisˈleitiv] adj. 有立法权的
    n. 立法机构
13. executive [igˈzekjutiv] adj. 行政的
    n. 执行者，主管
14. tendency [ˈtendənsi] n. 趋势，潮流；倾向；癖性；天分
15. coordination [kəʊˌɔ:diˈneiʃn] n.协调，和谐
16. upstream [ˈʌpˈstri:m] adj. 向上游，逆流地
17. downstream [ˈdaʊnˌstri:m] n. 下游
    adv. 下游地
18. reconnaissance [riˈkɔnəsəns] n. 侦察，搜察，勘察队，勘测
19. feasibility [ˌfi: zəˈbiləti] n. 可行性
20. irrevocable [iˈrevəkəbəl] adj. 不能取消的，不能唤回的，不能变更的

* * * * * * * * * * * * * * * * * **Complicated Sentences** * * * * * * * * * * * * * * * * *

1. Planning can be defined as the orderly consideration of a project from the original statement of purpose through the evaluation of alternatives to the final decision on a course of action.

【译文】规划就是对一个工程从目的的初始陈述、方案评价到行动过程乃至最终决策的有序研究。

【说明】该句是被动结构，其中 from the original statement of purpose through the evaluation of alternatives to the final decision on a course of action 作为后置定语修饰 consideration。

2. The planning for an entire river basin involves a much more complex planning effort than the single project, but the difficulties in arriving at the correct decision may be just as great for the individual project.

【译文】整个流域的规划比单个工程涉及更加复杂的规划程序，而单项工程在取得正确决定时所面临的困难，也可能与整个流域规划一样大。

【说明】本句是由 but 连接的两个简单句，可以作为两个句子翻译。前半句中 for an entire river basin 修饰 The planning，后半句 in arriving at the correct decision 修饰 the difficulties。

3. This is the concept of the regional master plan which attempts to define the most desirable future growth pattern for an area.

【译文】这是一个区域总体规划的概念，它试图为一个地区确定最适宜的未来发展模式。

【说明】句子主体结构为一简单句，但其中由 which 引导的定语从句较长，翻译时可以灵活处理为两个分句。

4. An overall regional water-management plan, developed with care and closely coordinated with other regional plans, may be a useful tool in determining which of many possible actions should be taken.

【译文】一个经过细心研究，并与其他区域规划紧密协调的区域水管理总体规划，可能是一个有用的工具，在决定许多可能方案时应当采用这一有用工具。

【说明】句子主体结构为一个简单句，其中 developed with care and closely coordinated with other regional plans 作定语修饰主语中心词 plan；而 in determining which of many possible actions should be taken 是表语 a useful tool 的定语。

5. Planning of specific actions is the lowest level of planning, but it is at this level that important decisions which determine the effectiveness of water management are made.

【译文】具体项目计划是最低一级的规划，但决定水管理效率的重要决策都是在这一级确定的。

【说明】本句是由 but 连接的两个简单句。It 指代前句的主语 Planning of specific actions，后半句中 which 引导的定语从句修饰的主体是 decisions。

6. The feasibility study usually requires that the structural details of a project be specified in sufficient detail to permit an accurate cost estimate.

【译文】可行性研究通常要求一个工程的结构细节应该规定得充分具体，以便给出一个准确的费用估计。

【说明】句中 that 引导一个宾语从句，该宾语从句为被动结构，to permit an accurate cost estimate 修饰 detail。

****************** Summary of Glossary ******************

1. feasibility study 可行性研究
2. regional master plan 区域总体规划
3. engineering judgment 工程判断

********************** Exercises **********************

(1) Planning can be defined as the orderly consideration of a project from the original _________ of purpose through the evaluation of _________ to the final ______ on a course of action.

(2) An overall regional water-management plan, ________with care and closely coordinated ________ other regional plans, may be a useful tool in determining which of many possible actions should be________。

(3) There is no substitute for "__________ judgment" in the selection of the method of approach __________ project planning.

(4) It is the basis for the decision to proceed ________ (or to abandon) a ________project and is the most important aspect of the engineering for the project.

*** * * * * * * * * * * * * Word Building (2) –able; aqu- * * * * * * * * * * * * ***

1. -able [形容词后缀]，表示：……的，能……的

accept v. acceptable adj. 可接受的
cap- (完成) vt. capable adj. 有能力的，可以……的
exchange v./n. exchangeable adj. 可交换的, 可替换的
move v./n. movable adj. 可移动的
reason v./n. reasonable adj. 合理的, 有道理的

2. aqu- [词根]，表示：水

-atic adj. aquatic adj. 水生的
educt n. aqueduct n. 水道
-ous adj. aqueous adj. 含水的
terr-(陆地) n. terraqueous adj. 有水有地的

*** * * * * * * * * * * * * * * * Text Translation * * * * * * * * * * * * * * * ***

第二章 水资源规划

第一节 规划概述

规划就是对一项工程从目的的初始陈述、方案评价到行动过程，乃至最终决定的有序研究。除了建筑物的细部结构外，它还包括与工程设计有关的全部工作。它是决定某项建设项目继续进行（或放弃）的依据，也是这个工程项目最重要的部分。因为每个水资源开发项目对经济和自然的调节作用都是唯一的，所以，简单的描述方法不可能获得最好的决定。在项目规划方法的选择过程中，没有什么能够代替"工程经验判断"。但迈向最终决策过程的每个独立步骤，无论何时都应采用定量分析方法，而尽可能不用估计和判断。人们经常听到"流域规划"这个词，规划阶段对最小工程来说可能不太重要，整个流域的规划比单个工程涉及更加复杂的规划程序。但某一单项工程在取得正确决定时所面临的困难可能与整个流域规划

一样大。（单词个数：204）

"规划"这个词包含有不同于上面提到过的之外的含义，这是一个区域总体规划的概念。它试图为某一地区确定最适宜的未来发展模式。如果这个总体规划就是实际上的最适宜的发展模式，那么未来的发展就应该向这个模式方面引导。遗憾的是"最适宜的"概念是主观的，它很难保证任何总体规划初次提出就能满足这一高标准。其次，技术的变化、经济的发展和公众态度常常会使一个总体规划在相当短的时间内就变得过时。任何一个规划都是基于对未来的假设，因此，如果这些假设没有实现，那么这个规划就必须修改。规划也必须定期进行全面的修改。（单词个数：131）

一个经过细心研究并与其他区域规划紧密协调的区域水管理总体规划，可能是一个有用的工具，在决定许多可能方案时应当采用这一有用工具。但必须考虑经济和社会环境变化，或当研究出新的有关数据时，随时对它进行调整。一个总体规划，只不过是所有可行性研究中的一种，也很有可能被证明是没有价值的。（单词个数：77）

第二节 规 划 分 级

每个国家许多领域都有规划（都有不同级别的规划），而每一领域（级别）中规划工作的目的和性质又不相同。许多国家有国家规划机构，它们的目标是加强国内的经济发展、改善国内的社会状况。即使没有这样的组织存在，也有国家目标存在，国家规划的某些形式由立法机关或政府行政部门提出。国家规划机构很少直接处理水问题，但在调整粮食、能源、工业商品和住房等生产指标时，它可能有效地为水管理规定目标。（单词个数：100）

考虑到一个国家各地区之间存在差异，因此，一个国家可以有地区规划分支机构，并且还可以存在相应的地区水利规划。然而，由于水利规划的一个自然"区域"是流域，因此有建立流域委员会或流域管理局（官方）的趋势。这些规划组织必须确保流域内各项工作之间的协调。水管理的每个具体行动可能对下游（并且有时对上游）产生影响。因此，这些具体行动不应单独进行规划，而必须进行相互之间的协调。（单词个数：100）

具体行动计划是规划的最低级，它决定了水管理的有效性的实现。决定水管理效率的重要决定就是由这一级制定的。尽管一个具体工程不一定产生，但这一级别的规划常称为工程规划。例如，因某一项研究而产生的一个洪泛区管理方案就是一个合法的工程规划。（单词个数：58）

第三节 规 划 过 程

工程规划通常要经过几个阶段才能成为最终的方案。每一国家的规划都有一个具体名称的规定顺序。第一阶段或初步勘测阶段通常是一个粗略过程，这一阶段是用来淘汰那些具有明显不必广泛研究特征的工程或方案，从而确定哪些是值得进一步研究的工程或方案。初步勘测阶段后可能有一个或更多的阶段要对建议方案的可行性进行详细的论证和评价。在这个过程中，要对最适宜方案的作用进行系统说明，这就是规划。在许多情况下，某一单独方案的可行性分析是足以满足要求的，因为方案的性质比较容易评价。在其他情况下，一个或更多的初步可行性分析可以对建议方案的各个方面进行论证。分几个阶段研究的目的，首要任务是通过考察工程的最薄弱方面来减少规划的费用。如果这个工程由于某方面的原因被淘汰，那么这个工程所有其他方面的研究费用就不用再花了。如果这一系列论证越来越多地成为这个项目所有各方面的一系列的更详细的论据，那么规划费用就会因许多事情将会重做两次或更多次而增加。（单词个数：209）

可行性研究通常要求一个工程的结构细节应该规定得充分具体，以便给出一个准确的费用估计。在规划工作的最终阶段，设计的细节必须进行认真的审查，并且要给出施工图和设计说明书。尽管在原则上可行性分析的结果取决于可行性研究，但更详细的研究可能开发出使可行性改变的资料的可能性总是存在的。所以，不到最终设计完成就不能取消继续做下去的决定。（单词个数：92）

Chapter 3 Water Resources for Sustainable Development

Part 1 Introduction

The adjective "sustainable" stems from a Latin verb "sustinere (to uphold). The corresponding English verb, "to sustain", being in use since the late Middle Ages, has meanings such as: to maintain, to keep going, to keep in being, to keep from falling, to carry on, to withstand, to bear, to support life, to provide for life or bodily needs, to furnish with the necessities of life (Little, 1972).[1] Many of these meanings are encapsulated in term "sustainable development" which is being broadly used nowadays. In fact, "sustainable development" is an old concept that has been used in the management of renewable natural resources to ensure that the rate of harvesting a resource is smaller than the rate of its renewal. [2] As mentioned in the Brundtland Report (World Commission on Environment and Development, WCED,1987), "humanity has the ability to make development sustainable — to ensure that it meets the needs of the present without compromising the ability of future generations to meet their own needs". This aim should be achieved while minimizing the losses (maximizing the gains) to economic, social and environmental systems.

The availability of water in adequate quantity and quality is a necessary condition for sustainable development. Water, the basic element of the life support system of the planet, is indispensable to sustain any form of life and virtually every human activity. The annual runoff into the oceans exceeds 40 000 km^3. Withdrawals currently reaching 3800 km^3 constitute only a small portion (about 9%) of the average annual runoff. At first sight this may look like a relative abundance of water. However, these apparently comforting global figures are largely misleading as water availability at smaller scales is concerned.

Global water consumption has increased about sevenfold since the beginning of the 20th century. This has been caused both by population growth and by increase of the per capita water use. The continuing populating growth with consequences for food production and justified aspirations of nations and individuals towards better living conditions will undoubtedly cause the demand for water to grow further. An adequate and reliable supply of water of proper quality for the entire population of the Globe and for preserving the hydrological, biological and chemical functions of ecosystem is still a remote goal. The increased demand already cannot be met in a number of locations and at all times at present under the natural variabilities of temperature and precipitation.

Water shortage is therefore likely to be the most dominant water problem in the forthcoming century, jeopardizing sustainable development. The number of countries subject to water scarcity, defined as water availability below 500m^3 per capita per year, reaches 12 at present and is likely to grow to 19 by 2025 (Gleick,1993). Some assessments foresee that the portion of the global human population subject to water scarcity may grow to 35% around the year 2025. In a number of countries subject to a dynamic population growth, a dramatic drop of the per capita availability of water has already taken place and aggravation of this process is foreseen to substantially below the level recognized now as the threshold of scarcity. The UN Water Conference in 1977 that agreed that "all peoples, whatever their stage of development and their social and economic condition, have the right to have access to drinking water in quantities and of a quality equal to their basic needs". Access to safe water has therefore become a kind of human right. The UN International Drinking Water Supply and Sanitation Decade (1981—1990) had the goal of arranging for access to safe drinking water and sanitation for the whole population of the Globe. Yet, at the end of the Decade, despite all the unquestionable achievements, a large number of human beings (of the order of one billion) still lacked clean and safe water, largely because population growth has outweighed all the progress achieved in water supply. The number of people without safe water supply has been growing up to present.

According to World Health Organization (WHO), about 80% of all diseases and one third of all deaths in developing countries are related to water-related disease, such as diarrhea, malaria, schistosomiasis, river blindness, Guinea worm, and others which kill globally perhaps 25 000 human beings a day, over 17 people one minute.

Part 2 Concluding Remarks

Sustainable development requires an integrated approach and a holistic perspective, in which a structure of inter-linked components is taken into account.[3] This structure contains not only hydrological or water resources components but also a number of other components, such as environmental, economic, demographic, socio-cultural and institutional subsystems.

Institutional issues play an extremely important part in striving towards sustainable development. Typically, water management is fragmented among several institutions. At the national level, the existence of a central water office, such as a Ministry of Water Resources, that deals with all aspects of water management is very rare. Usually several ministries (e.g. environment, agriculture, forestry, industry, navigation, construction, interior, etc.) hold responsibilities for portions of water problems. Frequently the coordination between these national bodies is very limited or non-existent. The advent of strong water agencies with a clear mandate, adequate resources and technical skills is very welcome. Due to the multi-faceted nature of water issues, it is necessary, though difficult, for an agency to operate across disciplinary and jurisdictional lines.

In the WMO's survey INFOHYDRO (Hydrological Information Referral Service) (World Meteorological Organization, WMO, 1995), 175 countries reported on hydrological data collection activities. However, among the respondents there were 480 agencies, that is on average nearly three agencies per country. This illustrates the fragmentation of hydrology at the national level.

A similar fragmentation of water affairs can also be observed at the international level. There exists no powerful intergovernmental water agency. There are two dozen or so agencies of the United Nations family dealing with water. However, water units or projects in these agencies are usually outsiders, clearly beyond the mainstream.

It is proposed (United Nations Conference on Environment and Development ,UNCED, 1993) to base sustainable development on the principles of: (1) decentralization and devolution of responsibility in water and environmental matters to the parties involved at the lowest level in society (subsidiarity principle); (2) local and private sector participation; and (3) a demand-driven cost recovery approach and equitable charging to enhance sustainability and enforceable legislation at all levels. Further, it is essential to reach significant portions of communities and involve them in the process of consultation to make them understand, accept and support plans. The principles of decentralization and involvement of communities pose ambitious challenges for education and training.

Agenda 21 (UNCED, 1993) proposes that sustainability be built into national accounting. Nations should set priorities and construct implementation plans for sustainable development.

Water is not a free goods any more but rather an economic goods. A change of philosophy is needed; rather than trying to fulfill increasing water demands and devise new costly supply sources, one should strive towards increasing the efficiency of water use, trying to "do more with less".

There is still much to achieve in the area of water demand management. Water pricing is likely to be increasingly important, covering not only the cost of development and water supply but also the cost of resources, in the sense of foregone opportunities. Tradeable water permits offer another mechanism.

There are several cases when the recent behaviour of time series of hydrological variables differs significantly from the historical means.[4] To a large extent this has been caused by direct anthropogenic reason-human activities like deforestation, urbanization, etc. However, many scientists attribute a part of these changes to man induced nonstationarity of the natural climatic system; to climate variability and change (greenhouse effect). For example, decreases of precipitation and water levels in a number of rivers and lakes in Africa have been observed. An increasing severity of extreme events has been reported, with such examples as the recent floods on the Mississippi and on the Rhine. Some experts explain the above observations by such physical mechanisms as changes of prevailing atmospheric circulation patterns and of directions of atmospheric moisture advection. The possibility of climate change may add another dimension to the context of sustainable development.

* * * * * * * * * * * * * * * * * * Explanations * * * * * * * * * * * * * * * * * *

* * * * * * * * * * * * * * * * * New Words and Expressions * * * * * * * * * * * * * * *

1. furnish ['fɜ·niʃ] vt. 提供；陈设，布置
2. encapsulate [in'kæpsə,let] vt. 概括，压缩；封装；把……包于胶囊
3. sevenfold ['sevn'fold] adj. 七倍的 adv. 由七部分组成地
4. precipitation [pri,sipi'teiʃən] n. 降（雨、雪、雹）；降落；急躁，仓促；沉淀，沉淀作用
5. forthcoming [,forθ'kʌmiŋ] adj. 即将到来的现成的，唾手可得的
6. jeopardize ['dʒɛpɚd,aiz] vt. 危及，损害

7. dynamic [daiˈnæmik] adj. 动力的，动态的；有活力的，强有力的；不断变化的

8. dramatic [drəˈmætik] adj. 激动人心的，引人注目的；戏剧的

9. aggravation [ˌægrəˈveʃən] n. [口]激怒，惹恼；加重，加剧，恶化

10. threshold [ˈθrɛʃhold] n. 门槛；开始

11. sanitation [ˌsænəˈteʃən] n. 卫生系统

12. integrated [ˈintəˌgretid] adj. 无种族界限的；和谐的，完整的，完全的，综合的

13. holistic [hoˈlistik] adj. 从整体着眼的，全面的

14. demographic [ˌdiməˈgræfik] adj. 人口的，人口统计学的

15. fragment [ˈfrægmənt] n. 碎片，片段

16. navigation [ˌnævəˈgeʃən] n. 航行（学）；导航，领航

17. advent [ˈædvənt] n. 出现，到来

18. mandate [ˈmændet] n. 授权，正式命令

19. multi-faceted [ˈmʌltiˌfæsitid] adj. 多方面的；多才多艺的

20. fragmentation [ˌfrægmənˈteʃən] n. 分裂

21. decentralization [ˌdisɛntrələˈzeʃən] n. 疏散化

22. devolution [ˌdɛvəˈluʃən] n. 移交，转让

23. equitable [ˈekwitəbl] adj. 公正的，合理的

24. legislation [ˌledʒisˈleʃən] n. 法律，法规；立法，制定法律

25. anthropogenic [ˌænθrəpəuˈdʒenik] adj. 人为的

26. deforestation [ˌdifɔrəsˈteʃən] n. 滥伐森林

27. nonstationarity [ˈnɔnsteiʃəˈnæriti] n. 非平稳性

28. dimension [diˈmenʃən] n. 方面，部分；规模，程度；尺寸，度量

*** * * * * * * * * * * * * * * * * * Complicated Sentences * * * * * * * * * * * * * * * * ***

1. The corresponding English verb, "to sustain", being in use since the late Middle Ages, has meanings such as: to maintain, to keep going, to keep in being, to keep from falling, to carry on, to withstand, to bear, to support life, to provide for life or bodily needs, to furnish with the necessities of life (Little, 1972).

【译文】对应的英文动词"持续"自从中世纪晚期就被使用了，有如下几种含义：维持，继续前进，继续生存；停止堕落；继续进行；承受，支持生活；提供生活或身体所需物质，提供生活必需品（雷特，1972 年）。

【说明】"to sustain"是 the corresponding English verb 的同位语。动名词短语 being in use since the late Middle Ages 是 English verb 的定语，翻译时可变换句式结构，以使句子更通顺。

2. In fact, "sustainable development" is an old concept that has been used in the management of renewable natural resources to ensure that the rate of harvesting a resource is smaller than the rate of its renewal.

【译文】事实上，"可持续发展"是一个古老的概念，早已用于可再生自然资源的管理，以确保资源的利用率小于它的更新率。

【说明】第一个 that 引导定语从句修饰 concept，第二个 that 引导宾语从句。

3. Sustainable development requires an integrated approach and a holistic perspective, in which a structure of inter-linked components is taken into account.

【译文】可持续发展需要一个全面和综合的方法，其内部结构中一些内在联系的因素也要考虑到。

【说明】in which 相当于 where，引导定语从句。

4. There are several cases when the recent behaviour of time series of hydrological variables differs significantly from the historical means.

【译文】在几种情况下，描述水文变量的时间序列最近的情况明显不同于历史资料。

【说明】when 引导的从句是 several cases 的同位语，不是定语。

********************* Summary of Glossary *******************

1. sustainable development	可持续发展
2. per capita	人均
3. private sector	私营部门

********************* Abbreviations (Abbr.) ********************

1. UN	United Nations	联合国
2. WHO	World Health Organization	世界卫生组织
3. WMO	World Meteorological Organization	世界气象组织
4. WCED	World Commission on Environment and Development	世界环境与发展委员会
5. UNCED	United Nations Conference on Environment and Development	联合国环境与发展大会
6. INFOHYDRO	Hydrological Information Referral Service	水文信息采集与获取

******************** Exercises *********************

(1) Many of these meanings are __________ in term “__________ development” which is being __________ used nowadays.

(2) In fact, “sustainable development” is an old ________ that has been used in the management of __________ natural resources to ensure that the rate of ________ a resource is smaller than the rate of its __________.

(3) The__________ of water in adequate quantity and quality is a necessary condition for sustainable development. Water, the basic __________ of the life support system of the planet, is __________ to sustain any form of life and __________ every human activity.

(4) At first sight this may look like a relative__________ of water. However, these apparently comforting global figures are largely __________ as water __________ at smaller scales is concerned.

****************Word Building (3) bi-; de-****************

1. bi- [前缀]，表示：二，两，双

bilateral	adj.	双向的，双面的，双边的
bidirectional	adj.	双向的
bisexual	adj	两性的
bivalent	adj	二价的

2. de- [前缀]，表示相反动作，剥夺，分离：分，解

destruction	n.	破坏，毁灭
detach	vt.	脱开
deform	v.	变形
dewater	vt.	排水
demagnetization	n.	去磁
decaffeinated	adj.	去咖啡因的

****************** Text Translation ******************

第三章 水资源可持续发展

第一节 简 介

形容词“可持续”源于从中世纪晚期就开始使用的拉丁文动词“ sustinere（维护）”。相应的英文动词为“保持”。有如下几种含义：保持，继续前进，继续生存，停止堕落，继续进行，承受，拥护生活，提供生活或身体所需物质，提供生活必需品（雷特，1972 年）。许多这些近似意思都可以用现在已经广泛运用的“可持续发展”来概括了。事实上，“可持续发展”是一个古老的概念，早已用于可再生自然资源的管理，以确保资源的利用率小于它的更新率。正如布伦特兰报告（世界环境和发展委员会，WCED，1987 年）所说：“人类有能力保持可持续发展——以确保满足目前的需要，而又不损害子孙后代满足他们需要的能力。”这一目标的实现，同时应尽量减少经济、社会和环境的损失（收益最大化）。（单词个数：174）

提供充足的水资源的数量和质量是可持续发展的一个必要条件，促进可持续发展。水，是地球上生命支持系统最基本的元素，是维持任何形式的生命和几乎所有人类的活动所必不可少的。流入海洋的年径流量超过 40 000km^3。取用水量一般为 3800km^3，只占年平均径流量的一小部分（约 9%）。乍一看，可能感觉有相当丰富的水资源。但是，这些表面令人欣慰的数据是一个大误导，因为水资源可利用量占比很小。（单词个数：97）

自 20 世纪初以来，全球用水量已增加了 7 倍。这是人口增长和人均用水量增加两方面原因导致的结果。人口持续增长的后果将使粮食生产和国家及个人向更好的生活条件发展，从而导致对水的需求进一步增长。提供充足和可靠的符合质量要求的水，以供给地球上的所有人，并且维护生态系统的水文、生物和化学功能仍然是一个久远的目标。目前，一部分地区已经不能满足日益增长的需求了，并从现在开始，或者说任何时候，都被异变的温度和降水所控制。（单词个数：120）

水资源短缺，在即将到来的世纪中很有可能是最主要的水资源问题，危及可持续发展。水的可利用量人均每年在 $500m^3$ 以下就可以定义为水荒，目前遭受水荒的国家数已经达到12个，并有可能到2025年增长至19个（格雷克，1993年）。一些评估专家预测，到2025年全球水资源短缺的人口比例可能会增长到35%左右。在一些国家人口动态增长，人均可用水量的急剧下降已成事实。这一过程的加剧警示着：水资源短缺的阈值将比现在认定的标准更低。联合国水资源会议于1977年达成共识："所有的人类，不管他们的国家处于哪个发展阶段及其他们的社会和经济条件，都有权获得满足其基本需要的饮用水的数量和质量。"因此，获得安全饮用水成为了一种人权。联合国国际饮水供应和卫生十年（1981—1990年）的目标是全球的人口都能获得安全卫生的饮用水。然而，在十年活动结束的时候，尽管取得了不容置疑的成就，但是大量的人（大约有10亿）仍然缺乏洁净和安全用水，大部分是因为人口增长已超过水供给能力。缺乏安全饮用水的人口数量还在继续增加。（单词个数：259）

根据世界卫生组织的统计，发展中国家约80%的疾病和三分之一的死亡疾病与水有关，如腹泻、疟疾、血吸虫病、盘尾丝虫病、几内亚蠕虫和其他的疾病，造成全球范围内每天大约25 000人死亡，每天每分钟超过17人死亡。（单词个数：46）

第二节 结 语

可持续发展需要一个全面而综合的方法，其内部结构中的一些内在联系的因素也要考虑到。这种结构不仅仅是水文和水资源部分，也有一些其他因素，如环境、经济、人口、社会、文化和体制子系统等因素。（单词个数：46）

制度问题在努力实现可持续发展方面起着极其重要的作用。通常情况下，水的管理是由若干机构分担的。在国家层面上，有中央办公室，如水利部，一般不会涉及各个方面的水的管理。通常有几个部委（如环境、农业、林业、工业、航运、建筑、内务部等）担负有关水的部分问题的职责。通常情况下，这些国家机构之间的协调是非常少的，或者根本没有协调。如果能建立大的水管机构，又同时担负明确的任务、拥有充足的资源和技术技能，将是非常可喜的。鉴于水问题的复杂性，需要建立统一的机构来管理，虽然困难，但还是必要的。（单词个数：121）

1995年，世界气象组织（WMO）在水文信息采集获取（INFOHYDRO）调查中，有175个国家报告了水文数据采集工作，共有480个机构，也就是说，平均每个国家有3个机构。这也说明了国家层面的水文工作不成体系。（单词个数：42）

类似地，也可发现世界范围的水文工作的混乱。没有一个强有力的政府间水资源机构。联合国范围内大约有24个处理水资源问题的相关机构。然而，这些机构中的部门和方案通常都是外行，显然超出了主流行业范畴。（单词个数：51）

1993年，联合国环境与发展大会（环发大会）建议可持续发展要遵循以下原则：①对于社会最底层的团体的环境问题，相关责任和关系要适当地分割和移交权力；②地方和私人要部分合作；③在各个层次上，都由需求驱动的成本回收方式和公平的收费，以提高可持续性和可强制执行的立法。更重要的是，让全社会相关团体参与到商议的过程中来，使他们理解、接受并且支持这些计划。分权和团体选择的原则给教育和培训提出了很大的挑战。（单词个数：102）

环发大会的21世纪日程提出：可持续发展要纳入国家预算。国家应该为可持续发展设置优先权，并且切实履行计划。（单词个数：23）

水资源不再是免费的资源，而是经济商品。观念的转变是必须的。与其努力实现水需求的增长并且规划新的高成本的供给资源，还不如努力增加对水的有效利用，“尽量少花钱多办事”。（单词个数：49）

需水管理上仍有很多工作要做。水价越来越重要，从预先确定的机会这个意义来说，涉及不止供水和发展的成本还有水资源的成本。水许可交易提供了另一种机制。（单词个数：49）

很多情况下的水文变量的时间序列发生了变化，即最近的情况明显不同于历史资料。很大程度上是由于直接的人为原因，诸如森林砍伐、城市化造成的。但是，大部分科学家把一部分变化归因于自然气候系统人为的不稳定。对气候的变异（温室效应），已经观察到非洲的降雨明显减少，许多河流和湖泊的水位也在降低。经常出现日益严重的自然灾害的报道，最近的例子有密西西比河和莱茵河的洪水，一些专家把上述通过物理机制观察到的情况，解释为大气环流模式和大气水分平流方向的变化，气候变化的可能性要提高到可持续发展的层面。（单词个数：143）

Chapter 4 Hydrology and Water Cycle

Part 1 Introduction of Hydrology

Hydrology is the study of occurrence, distribution, and movement of water on, in, and above the earth.[1] As such, it is an earth science. The cycle of movement of water between atmosphere, hydrosphere, lithosphere and biosphere is termed the hydrologic cycle.

There are two broad sub-disciplines within the science of hydrology. The first is surface water hydrology which focuses on water on and above the surface of the earth. Examples of applications of surface water hydrology are flooding and droughts. The second sub-discipline is groundwater hydrology or geohydrology, which focuses on the distribution and movement of water beneath the earth's surface. Groundwater hydrology is important for applications in water supply, irrigation and environmental engineering. Note that in the oceans is separate discipline known as oceanography. And water in the atmosphere is mostly studied in meteorology.

Also included in hydrology is the study of motion of water and water-borne constituents-material carried either as dissolved quantities or in separate phases.[2] A related facet of hydrology is the determination of statistical flow prediction in rivers and streams. This information is essential to design and evaluation of natural and man-made channels, bridge openings and dams.

Hydrology is principally concerned with the part of the cycle after the precipitation of water onto the land and before its return to the oceans; thus, meteorology and oceanography are closely related to hydrology. Hydrologists study the cycle by measuring such variables as the amount and intensity of precipitation, the amount of water stored as snow or in glaciers, the advance and retreat of glaciers, the rate of flow in streams, and the soil-water balance. Hydrology also includes the study of the amount and flow of groundwater. Though the flow of water cannot be seen under the surface, hydrologists can reduce the flow by understanding the characteristics, including permeability, of the soil and bedrock; how water behaves near other sources of water, such as rivers and oceans; and fluid flow models based on water movements on the earth's surface. Hydrology is also important to the study of water pollution, especially of groundwater and other potable water supplies. Knowledge of hydrology is extensively used to determine the movement and extent contamination from landfills, mine runoff; and other potentially contaminated sites to surface and subsurface water.

Part 2 What Is Hydrology?

Hydrology is the science that encompasses the occurrence, distribution, movement and properties of the waters of the earth and their relationship with the environment within each phase of the hydrology cycle. The hydrology cycle is a continuous process by which water is purified by evaporation and transported from the earth's surface (including the oceans) to the atmosphere and back to the land and oceans. All of the physical, chemical and biological processes involving water as it travels its various path in the atmosphere, over and beneath the earth's surface and through growing plants, are of the interest to those who study the hydrologic cycle.[3] There are many pathways the water may take in its continuous cycle of falling as rainfall or snowfall and returning to the atmosphere. It may be captured for millions of years in polar ice caps. It may flow to rivers and finally to the sea. It may soak into the soil to be evaporated directly from the soil surface as dries or be transpired by growing plants.[4] It may percolate through the soil to groundwater reservoirs (aquifers) to be stored or it may flow to wells or springs or back to streams by seepage. The cycle for water may be short, or it may take millions of years.

People tap the water cycle for their own uses. Water is diverted temporarily from one part of the cycle by pumping it from the ground or drawing it from a river or lake. It is used for a variety of activities such as households, businesses and industries; for transporting wastes through sewers; for irrigation of farms and parklands; and for production of electric power.

After use, water is returned to another part of cycle: perhaps discharged downstream or allowed to soak into the ground. Used water normally is lower in quality, even after treatment, which often poses a problem for downstream users.

The hydrologist studies the fundamental transport processes to be able to describe the quantity and quality of water as it moves through the cycle (evaporation, precipitation, stream flow, infiltration, groundwater flow, and other components). The engineering hydrologist, or water resources engineer, is involved in the planning, analysis, design, construction and operation of projects for the control, utilization and management of water resources. Water resources problems are also the concern of meteorologists, oceanographers, geologists, chemists, physicists, biologists, economists, political scientists, specialists in applied mathematics and computer science, and engineers in several fields.

Part 3 Introduction of Water Cycle

Water on Earth is always changing. Its repeating changes make a cycle. As water goes through its cycle, it can be a solid (ice), a liquid (water), or a gas (water vapor). Ice can change to become water or water vapor. Water can change to become ice or water vapor. Water vapor can change to become ice or water.

How do these changes happen? Adding or subtracting heat makes the cycle work. If heat is added to ice, it melts. If heat is added to water, it evaporates. Evaporation turns liquid water into a gas called water vapor. If heat is taken away from water vapor, it condenses. Condensation turns water vapor into a liquid. If heat is taken away from liquid water, it freezes to become ice.

The water cycle is called the hydrologic cycle. In the hydrologic cycle, water from oceans, lakes, swamps, rivers, plants, and even you, can turn into water vapor. Water vapor condenses into millions of tiny droplets that form clouds. Clouds lose their water as rain or snow, which is called precipitation.

Precipitation is either absorbed into the ground or runs off into rivers.[5] Water absorbed into the ground is taken up by plants.Plants lose water from their surfaces as vapor back into the atmosphere. Water that runs off into rivers flows into ponds, lakes, or oceans where it evaporates back into the atmosphere.[6] The cycle continues.

Part 4 Processes of Water Cycle

In nature, water is constantly changing from one state to another. The heat of the sun evaporates water from land and water surfaces, this water vapor (a gas), being lighter than air, rises until it reaches the cold upper air where it condenses into clouds. Clouds drift around according to the direction of the wind until they strike a colder atmosphere. At this point the water further condenses and falls to the earth as rain, sleet, or snow, thus completing the hydrologic cycle.

The hydrologic cycle consists of the passage of water from the oceans into the atmosphere by evaporation and transpiration (or evapotranspiration), onto the lands, over and under the lands as runoff and infiltration, and back to the oceans. There are six important processes that make up the water cycle. These are as follows:

1. Evaporation

Evaporation is the process where a liquid, in this case water, changes from its liquid state to a gaseous state. Liquid water becomes water vapor. Although lower air pressure helps promote evaporation, temperature is the primary factor. For example, all of the water in a pot left on a table will eventually evaporate. It may take several weeks. But, if that same pot of water is put on a stove and brought to a boiling temperature, the water will evaporate more quickly. During the water cycle some of the water in the oceans and freshwater bodies, such as lakes and rivers, is warmed by the sun and evaporates. During the process of evaporation, impurities in the water are left behind. As a result, the water that goes into the atmosphere is cleaner than it was on earth.

2. Condensation

Condensation is the opposite of evaporation. Condensation occurs when a gas is changed into a liquid. Condensation occurs when the temperature of the vapor decreases. When the water droplets formed from condensation are very small, they remain suspended in the atmosphere. These millions of droplets of suspended water form clouds in the sky or fog at ground level. Water condenses into droplets only when there are small dust particles present around which the droplet can form.[7]

3. Precipitation

When the temperature and atmospheric pressure are right, the small droplets of water in clouds form larger droplets and precipitation occurs. The raindrops fall to earth. As a result of evaporation, condensation and precipitation, water travels from the surface of the earth goes into the atmosphere, and returns to earth again.

4. Surface Runoff

Much of the water that returns to earth as precipitation runs off the surface of the land, and flows downhill into streams, rivers, ponds and lakes. Small streams flow into larger streams, then into rivers, and eventually the water flows into the ocean. Surface runoff is an important part of the water cycle because, through surface runoff, much of the water returns again to the oceans, where a great deal of evaporation occurs.[8]

5. Infiltration

Infiltration is an important process where rainwater soaks into the ground, through the soil and underlying rock layers. Some of this water ultimately returns to the surface at springs or in low spots downhill. Some of the water remains underground and is called groundwater. As the water infiltrates through the soil and rock layers, many of the impurities in the water are filtered out. This filtering process helps clean the water.

6. Transpiration

One final process is important in the water cycle. As plants absorb water from the soil, the water moves from the roots through the stems to the leaves. Once the water reaches the leaves, some of it evaporates from the leaves, adding to the amount of water vapor in the air. This process of evaporation through plant leaves is called transpiration. In large forests, an enormous amount of water will transpire through leaves.

However, the complete hydrologic cycle, is much more complex. The atmosphere gains water vapor by evaporation not only from the oceans but also from lakes, rivers, and other water bodies, and from moist ground surfaces. Water vapor is also gained by sublimation from snowfields and by transpiration from vegetation and trees.

Part 5 Application of Water Cycle

Because water is absolutely necessary for sustaining life and is of great importance in industry, men have tried in many ways to control the hydrologic cycle to their own advantage. An obvious example is the storage of water behind dams in reservoirs, in climates where there are excesses and deficits of precipitation (with respect to water needs) at different times in the year. Another method is the attempt to increase or decrease natural precipitation by injecting particles of dry ice or silver iodide into clouds. This kind of weather modification has had limited success thus far, but many meteorologists believe that a significant control of precipitation can be achieved in the future.[9]

Other attempts to influence the hydrologic cycle include the contour plowing of sloping farmlands to slow down runoff and permit more water to percolate into the ground, the construction of dikes to prevent floods and so on. [10]The reuse of water before it returns to the sea is another common practice. Various water supply systems that obtain their water from rivers may recycle it several times (with purification) before it finally reaches the rivers mouth. Men also attempt to predict the effects of events in the course of the hydrologic cycle. Thus, the meteorologist forecasts the amount and intensity of precipitation in a watershed, and the hydrologist forecasts the volume of runoff.

*** * * * * * * * * * * * * * * * Explanations * * * * * * * * * * * * * * * * ***

*** * * * * * * * * * * * * * * * * New Words and Expressions * * * * * * * * * * * * * * ***

1. atmosphere [ˈætməsfiə] n. 大气圈
2. hydrosphere [ˈhaidrəsfiə] n. 水圈
3. lithosphere [ˈlɪθəsfiə] n. 岩石圈
4. biosphere [ˈbaiəsfiə] n. 生物圈
5. hydrologic cycle n. 水文循环
6. evaporation [iˌvæpəˈreiʃ(ə)n] n. 蒸发，脱水
7. transpiration [ˌtrænspiˈreiʃən] n.[生]（叶面的）水汽散发；蒸腾作用；[物]流逸
8. evapotranspiration [iˈvæpəuˌtrænspiˈreiʃən] n. 土壤中水分损失总量
9. infiltration [ˌinfilˈtreiʃən] n. 渗入，渗透
10. surface water hydrology 地表水文学
11. flooding [ˈflʌdiŋ] n. 泛滥，灌溉；溢流，变色；产后出血
12. drought [draut] n. 干旱，旱灾
13. groundwater hydrology 地下水水文学
14. geohydrology [ˈdʒi:əuhaiˈdrɔlədʒi] n. 水文地质学
15. beneath [biˈni:θ] adv. 在……下面 prep. 在……之下
16. oceanography [ˌəuʃiəˈnɔgrəfi] n. 海洋学
17. meteorology [ˌmi: tjəˈrɔlədʒi] n. 气象学，气象状态
18. facet [ˈfæsit] n.（宝石等）刻面，小平面； vt. 在……上面
19. glacier [ˈglæsjə] n. 冰川
20. permeability [ˈpə: miəbiliti] n. 渗透性
21. bedrock [ˈbedˈrɔk] n. [矿]岩床，基础
22. fluid [ˈflu:id] adj. 流体的，流动的
23. contamination [kənˌtæmiˈneiʃən] n.污染
24. hydrology [haiˈdrɔlədʒi] n. 水文学，水文地理学
25. pathway [ˈpɑ:θwei] n. 路，径
26. snowfall [ˈsnəufɔ: l] n. 降雪，降雪量
27. soak [səuk] v. 浸，泡，浸透；n. 浸透
28. reservoir [ˈrezəvwɑ:] n. 水库，蓄水池
29. seepage [ˈsi: pidʒ] n. 泄漏，渗水量
30. meteorologist [ˌmi: tjəˈrɔlədʒist] n.气象学家
31. oceanographer [ˌəuʃiəˈnɔgrəfə] n. 海洋学家
32. specialist [ˈspeʃəlist] n. 专家，专门医师
33. aquifer [ˈækwəfə] n. 含水土层，蓄水层；地下含水层
34. condensation [ˌkɔndenˈseiʃən] n. 凝结，凝聚；浓缩
35. droplet [ˈdrɔplit] n. 小滴
36. atmospheric [ˌætməsˈferik] adj. 大气的
37. water vapor 水汽
38. drift [drift] n. 漂移，漂流物，观望，漂流，吹积物，趋势； v. 漂移，漂流，吹积
39. gaseous [ˈgæsiəs] adj. 气体的，含气体的
40. evaporate [iˈvæpəreit] v. 蒸发，消失
41. impurity [imˈpjuəriti] n. 杂质，不纯
42. surface [ˈsə: fis] n. 表面，平面 adj. 表面的，肤浅的
43. stream [stri:m] n. 流，水流，人潮 v. 使流出，流动
44. surface runoff 地表径流，地面径流
45. rainwater [ˈreinwɔ: tə(r)] n. 雨水
46. underlying rock 下垫岩石

47. underground ['ʌndəgraund] adj. 地下的，秘密的；
n. 地下，地铁，地道，秘密活动
48. groundwater ['graʊnd, wɔ: tə, -, wɔtə]
n. 地下水
49. filter ['filtə] v. 过滤，渗透，走漏
n. 筛选，滤波器，过滤器，滤色镜
50. enormous [i'nɔ: məs] adj. 巨大的，庞大的，[古]极恶的，凶暴的
51. transpire [træns'paiə] v. 蒸腾，散发，使…蒸发，排出；泄露，为人所知；发生
52. moist [mɔist] adj. 潮湿的，湿润的
53. snowfield ['snəufi: ld] n. 雪原，雪地
54. vegetation [ˌvedʒi'teiʃən] n. 植物，草木
55. absolutely ['æbsəlu:tli] adv. 绝对地，完全地；独立地；确实地
56. sustaining [səs'teiniŋ] adj. 维持的，支持的，持续的
57. storage ['stɔridʒ] n. 蓄水量；储藏量，库存量；蓄电；储藏，[电脑]存储，存储器
58. climate ['klaimit] n. 气候，气候区；风气，气氛
59. contour ['kɔntuə] n. 等高线，轮廓，周线，电路，概要；
vt. 画轮廓（画等高线）
60. silver iodide 碘化银微粒
61. recycle ['ri:'saikl] v. 使再循环，再制
62. purification [ˌpjuərifi'keiʃən] n. 净化
63. watershed ['wɔ:təʃed] n. 转折点，重要关头；流域，分水岭，分水线

*** * * * * * * * * * * * * * * * * Complicated Sentences * * * * * * * * * * * * * * * ***

1. Hydrology is the study of occurrence, distribution, and movement of water on, in, and above the earth.

【译文】水文学是研究地表、地下和地球上空的水的产生、分布和运动规律的学科。

【说明】句中 study 的意思是“学科”，“on, in, and above the earth”中三个介词带有相同的宾语，前面两个介词的宾语省略了。

2. Also included in hydrology is the study of motion of water and water-borne constituents-material carried either as dissolved quantities or in separate phases.

【译文】水文学还包括关于水及水中携带物质成分（不论作为溶解物，还是以分离状态携带）的运动形式的研究内容。

【说明】这是个倒装句，正常的语序为“the study of … is also included in hydrology”。其中 borne 是动词 bear（负担，具有）的过分分词，phase 在这里表示“状态”。

3. All of the physical, chemical and biological processes involving water as it travels its various path in the atmosphere, over and beneath the earth's surface and through growing plants, are of the interest to those who study the hydrologic cycle.

【译文】水循环时有许多路径：大气、土壤的表面和地下以及生长的植物。那些从事水循环研究的人们对这些与水有关的物理、化学和生物过程很感兴趣。

【说明】句子的主语为 All of the physical, chemical and biological processes，谓语为 are of the interest to those who study the hydrologic cycle。it 和 its 都指代 water; who study the hydrologic cycle 为定语从句，修饰前面的 those。

4. It may soak into the soil to be evaporated directly from the soil surface as dries or be

transpired by growing plants.

【译文】它可以浸湿土壤，再直接从土壤表面蒸发或通过生长的植物蒸腾。

【说明】句中 or 连接 soak 和 be transpired。原文为被动句式，但是按照汉语翻译的习惯，把被动句式改成主动句式，这不仅没有改变原文的意思，而且解释得更清楚、明了。

5. Precipitation is either absorbed into the ground or runs off into rivers.

【译文】降落的雨水有些可能被地面吸收，也有些可能流入河流。

【说明】either…or…是并列连词，表示“或者……或者……；要么……要么……”。并列的谓语为 is absorbed 和 runs off。

6. Water that runs off into rivers flows into ponds, lakes, or oceans where it evaporates back into the atmosphere.

【译文】流入河流的水又会流进池塘、湖泊或海洋，并会再次蒸发，回到大气中去。

【说明】句中 that runs off into rivers 是修饰 water 的定语从句。Where 从句修饰 ponds, lakes, or oceans, where 的用法相当于 in which。

7. Water condenses into droplets only when there are small dust particles present around which the droplet can form.

【译文】水凝结成小水珠的条件是大气中有微小的尘土颗粒，水汽在尘土颗粒周围形成小水珠。

【说明】本句由 when 引导的条件状语从句，从句中 which the droplet can form 修饰 small dust particles。

8. Surface runoff is an important part of the water cycle because, through surface runoff, much of the water returns again to the oceans, where a great deal of evaporation occurs.

【译文】地面径流是水循环中很重要的一部分，因为通过地表径流，大部分的水最后返回到海洋，在海洋的上空会形成更多的蒸发。

【说明】本句由 because 引导的原因状语从句。where 相当于 in which，而 which 引导的定语从句修饰 ocean。

9. This kind of weather modification has had limited success thus far, but many meteorologists believe that a significant control of precipitation can be achieved in the future.

【译文】虽然这种改造气候的方法迄今只取得了有限的成功，但许多气象学家都认为，有效地控制降水在未来是可以做到的。

【说明】句中 thus far 表示“迄今，到目前为止”。has had 中，had 为 have（有）的过去分词，具体表达“具有，得到”之意；has 为 have（表示时态的助词）的第三人称单数形式，表示句子为现在完成时。

10. Other attempts to influence the hydrologic cycle include the contour plowing of sloping farmlands to slow down runoff and permit more water to percolate into the ground, the construction of dikes to prevent floods and so on.

【译文】其他一些利用水循环的做法，如沿等高线耕作梯田，以使径流减速，让更多的水渗入地下以及修筑堤坝以防洪水等。

*** Summary of Glossary ***

1. hydrology 水文学
2. hydrologic cycle 水文循环
3. geohydrology 地下（水）水文学
4. meteorology 气象学，气象状态
5. lithosphere 岩石圈
6. hydrosphere 水圈
7. atmosphere 大气圈
8. environmental engineering 环境工程
9. precipitation 降雨
10. surface runoff 地表径流
11. evaporate 蒸发
12. condense 冷凝
13. infiltration 渗透
14. dry ice 干冰
15. silver iodide 碘化银
16. atmospheric pressure 大气压力
17. rock layer 岩石层
18. air pressure 气压

*** Exercises ***

(1) The cycle of movement of water between ________, ________, ________, and ________ is termed the hydrologic cycle.

(2) This information is essential to ________ and________ of natural and man-made channels, bridge openings and dams.

(3) Hydrology is the science that encompasses the ________, distribution, movement and ________ of the waters of the earth and their relationship with the environment within each phase of the hydrology cycle.

(4) As water goes through its cycle, it can be a solid ________, a liquid ________, or a gas ________.

(5) Water that ________ into rivers flows into ponds, lakes, or oceans where it ________ back into the atmosphere.

(6) Water vapor ________ into millions of tiny________ that form clouds.

(7) As a result of ________, ________ and ________, water travels from the surface of the Earth goes into the atmosphere, and returns to Earth again.

*** * * * * * * * * * *Word Building (4) hydr(hydro-);non- * * * * * * * * * * * ***

1. hydr，hydro- [词根]，表示：水，液，流体

hydroenergy	n.	水能
hydropower	n.	水电，水力发出的电力
hydromechanics	n.	流体力学
hydrovalve	n.	水龙头，水阀，液压开关

hydro- [词根]，表示："水力的"

hydraulic	adj.	水力学的
hydroelectric	adj.	水力发电的

2. non- [前缀] 表示：非，无，不

conducting	adj.	non-conducting	adj.	不传导的
continuous	adj.	non-continuous	adj.	不连续的，间断的
renewable	adj.	non-renewable	adj.	不可再生的，不可更新的
essential	adj.	non-essential	adj.	不重要的，非本质的
existent	adj.	non-existent	adj.	不存在的
finite	adj.	non-finite	adj.	非定形的
linear	adj.	non-linear	adj.	非线性的
metal	n.	non-metal	n.	非金属

*** * * * * * * * * * * * * * * * * Text Translation * * * * * * * * * * * * * * * * ***

第四章　水文学与水循环

第一节　水文学介绍

水文学是研究地面、地下和地球上空的水的产生、分布和运动规律的科学，因此，是一门地球科学。在大气圈、水圈、岩石圈和生物圈中水的流动循环就是水文循环。（单词个数：41）

水文学有两个大的分支。第一个分支是地表水水文学，主要研究地球表面和大气中的水。地表水水文学应用的例子有洪水和干旱。第二个分支是地下水水文学或地质水文学，主要是研究地下水的分布与流动。地下水水文学对于供水、灌溉和环境工程的应用来说很重要。对于海洋中的水的研究就是海洋学的一个分支。研究空气中的水叫气象学。（单词个数：93）

水文学也研究水及水生物质的活动——水溶解或分离阶段所携带的物质。与水文学有关的另一个方面，就是对河水与溪水流动进行预测统计。这些信息对于天然和人工渠道、桥梁和大坝的设计和评估很有必要。（单词个数：56）

水文学主要研究的是水降落到陆地后流回海洋前的循环部分，因此，气象学和海洋学与

之密切相关。水文学通过测量诸如降水的数量和强度、雪或冰川的水量、冰川的增减、溪流中的流速，以及土壤水平衡等变量来研究水循环。水文学也研究地下水流量。尽管地下水流无法看见，但水文学家可通过了解土壤和基岩的特征，包括渗透性、水如何在其他水源如河流和海洋附近的活动，以及水在地球表面流动的流体模型来推断其流动状况。水文学对于水污染，尤其是地下水和其他可饮用水的研究也很重要。水文学知识广泛用于了解垃圾埋填地、矿井径流及其他潜在污染地的水流流向、地面水和地下水的流动情况和污染程度。（单词个数：184）

第二节 什么是水文学?

水文学是研究地球上水的产生、分布、运动和特性，以及它们在每个水文循环周期内与环境的关系的科学。水文循环是一个持续过程，通过此过程，水因蒸发而净化，从地表（包括海洋）上升到大气，再回到陆地和海洋。水循环时，有许多路径，通过大气、土壤的表面和地下以及生长的植物。研究水循环的人们对这些与水有关的物理、化学和生物过程很感兴趣。降雨、降雪再回到大气，水的持续循环可以有许多路径。它能以极地冰帽的形式存在数百万年，也可以流向河流并最终流入海洋；它可以浸湿土壤再直接从土壤表面蒸发或通过生长的植物蒸发，也可以渗透土壤到地下水库（地下蓄水层）得以储存或流到井中或泉水中，或通过渗漏流回海洋。水的这种循环时间也许非常短暂，也许长达百万年。（单词个数：212）

人类开发水循环是为了自身的使用，通过泵从地下抽或从河流、湖泊抽水，从而暂时改变了水的循环周期。许多方面都要用到水，如家庭日常用水、商业用水和工业用水；通过排水沟转移废弃物；用于灌溉农田和公共场所以及水力发电。（单词个数：64）

使用后的水回到循环的另一部分：也许排放到下游，也许渗透到土壤里。用过的水通常质量差，即使是处理后也常会对下游使用者带来不利影响。（单词个数：38）

水文学家研究基本的水运动规律，以便于对水循环时（蒸发、降水、径流、渗透、地下径流及其他）的数量和质量进行描述。水文学家或水资源工程师为了对水资源进行控制、使用和管理，需要进行项目的规划、分析、设计、施工和运行管理。水资源问题也同样得到气象学家、海洋学家、地理学家、化学家、物理学家、生物学家、经济学家、政治学家、应用数学和计算机科学的专家以及很多行业工程师的关注。（单词个数：91）

第三节　水循环简介

地球上的水处于不断变化当中。不断重复的变化形成水的循环。处于循环变化中的水可能是固体（冰）、液体（水）、气体（水蒸气）。冰能转化成水或者水蒸气；水能转化成冰或者水蒸气；水蒸气也能变成冰或者水。（单词个数：59）

这些变化是如何发生的呢？增加或者降低热量就能形成整个水循环。如果给冰加热，它就会融化成水；如果给水继续加热，就会蒸发。蒸发能让液态水变成气态，即水蒸气。如果将水蒸气的热量不断降低，它就会凝结。凝结能让水蒸气变成液体。如果将液态水的热量继续降低，水就会开始凝固变成冰。（单词个数：69）

水的三态变化一般称为水文循环。在水文循环中，来自于海洋、湖泊、沼泽、河流、植物，甚至我们身体中的水都变成了水蒸气。水蒸气凝结成无数的微小颗粒，从而形成了云。最后，云中的水以雨水或雪花的形式降落下来，就称为降雨（降雪）。降雨有可能被地面吸收，也可能流入河流。被地面吸收的水分又被植物吸收，植物中水分又会从它的表面蒸发，回到大气中去。流入河流的水又会流进池塘、湖泊或海洋，并再次蒸发，回到大气中去。这样这个循环就一直进行下去。（单词个数：104）

第四节 水循环过程

在自然界中，水总是不断地从一种状态变成另一种状态。太阳辐射的热量使陆地和水面上的水变成水蒸气。这些水蒸气（一种气体）由于比空气轻，会上升直至达到高空冷气层，并在那里凝结成云。云层随风飘荡，直至遇到更冷的大气层为止。此时水便进一步冷凝，并以雨、雹或雪的形式落到地面。这样便完成了水的循环。（单词个数：83）

水文循环包括水因蒸发从海洋进入大气，通过地表径流和地下径流渗透进入土壤，然后再流回海洋。在水的循环中，有6个非常重要的组成环节，各环节如下：（单词个数：53）

1. 蒸发

在蒸发过程中，液体，这里指水，将会从它的液体状态变成气体状态。液态水变成了水蒸气。尽管较低的空气压力能促进蒸发，温度才是最根本的因素。例如，桌面上水壶里所有的水最终都将蒸发掉。这一过程可能需要几个星期。但是，如果将同样一壶水放到炉子上，或加热到沸腾的温度，水的蒸发将会快很多。在整个水循环过程中，海洋或一些水体，如湖泊、河流中的部分水，受阳光照射加热而蒸发。在蒸发过程中，水的杂质被留了下来，这样蒸发到大气中的水将会比地球上的更清洁。（单词个数：137）

2. 凝结

凝结与蒸发的过程刚好相反，气体变成液体时就发生凝结。当水汽的温度不断下降时，凝结就会产生。凝结产生的水珠微粒非常的小，它们会漂浮在大气中。无数的水珠微粒漂浮在大气中，就会形成天空中的云朵，或地面附近的一层雾。水凝结成小水珠的条件是大气中有微小的尘土颗粒，水汽在尘土颗粒周围形成小水珠。（单词个数：76）

3. 降雨

当温度和大气压处于一定状态下时，云中水珠微粒可形成更大的颗粒，就会产生降雨。雨点落向地面。在蒸发、凝结和降雨的过程中，水从地球的表面蒸发到大气中，然后又回到地面。（单词个数：51）

4. 地面径流

在降雨过程中，大部分落到地面的水会在泥土的表面汇成流，从山丘流向小溪、小河、池塘或者湖泊。小溪径流不断变大，然后汇成河，最终流向海洋。地面径流是水循环中很重要的一部分，因为通过地面径流，大部分的水最后返回到海洋，进而在海洋的上空，会形成更多的蒸发。（单词个数：74）

5. 下渗

下渗是一个很重要的过程。在这个过程中，雨水将渗透到土壤，以及更深的岩石层，来浸润地面。其中，部分水最终以喷泉，或者在低地势的地方返回到地面。有些继续保留在地表以下，形成地下水。随着水下渗穿过土壤和岩石层时，水中的许多杂质也被过滤掉。这一

浸润过程有利于水的清洁。(单词个数：71)

6. 散发（蒸腾）

散发是水循环过程中的重要一环。当植物从土壤中吸取水分，水分从植物的根和茎部到达叶子。一旦水分到达叶子，就产生蒸发，植物叶子蒸发出的水汽增加了大气中的水汽量。通过植物叶子来蒸发的过程叫做散发。在大面积的森林地区，大量的水汽通过植物的叶子散发到大气中。(单词个数：73)

然而，完整的水循环要复杂得多。由于蒸发作用，大气不仅从海洋，而且从湖泊、河流和其他水体，以及从潮湿的地表面获得水蒸气。也可从雪地中雪的升华和从植物与树木的蒸腾获得水蒸气。(单词个数：53)

第五节 水循环利用

因为水对于维持生命来说绝对必要，在工业上也很重要，所以人们为了自身的利益，试图以各种方式来控制水循环。一个明显的例子，就是在一年中不同的时间根据当地降水的多寡（按对水的需要来说）将水储存在水库中。另一种方法是，试图将干冰或碘化银微粒射入云层，来增多或减少天然降雨量。虽然这种改造气候（人工影响天气）的方法，迄今只取得了有限的成功，但许多气象学家都认为，有效地控制降水在未来是可以做到的。(单词个数：112)

其他一些影响水循环的努力包括：沿等高线耕作梯田，以使径流减速，让更多的水渗入地下；建筑堤坝以防洪水等。在水回归大海之前将它重复使用，也是一种常用的方法。自河道取水的各种供水系统可将水在最终到达河口之前，经过净化，可重复使用多次。人们还试图预测水循环过程中一些事件的结果。例如，气象学家预报一个流域的降雨量和降雨强度，水文学家预报径流量等。(单词个数：112)

Chapter 5 Principle of Hydrology—Unit Hydrographs

Part 1 Introduction

Ways to predict flood peak discharges and discharge hydrographs from rainfall events have been studied intensively since the early 1930s. One approach receiving considerable use is called the Unit Hydrograph (UH) method. It derives from a method of unit graphs employed by Sherman, in 1932. The unit hydrograph is defined as follows: if a given *X*-hour rainfall produces a 10mm depth of runoff over the given drainage area, the hydrograph showing the rates at which the runoff occurred can be considered a unit for that watershed.[1]

It is incorrect to describe a unit hydrograph without specifying the duration, *X* of the storm that produced it. An *X*-hour unit hydrograph is defined as a direct runoff hydrograph having a 10mm volume and resulting from an *X*-hour storm having a steady intensity of 10/ *X* mm/hour. A 2-hour unit hydrograph would be that produced by a 2-hour storm during which 10 mm of excess runoff was uniformly generated over the basin. A 1-day unit hydrograph would be produced by a storm having 10 mm of excess rain uniformly produced during a 24-hour period. The value *X* is often a fraction of 1 hour.

Application an *X*-hour unit graph to design rainfall excess amounts other than 10 mm is accomplished simply by multiplying the rainfall excess amount by the unit graph ordinates, since the runoff ordinates for a given duration are assumed to be directly proportional to rainfall excess.[2] A 3-hour storm producing 20 mm of net rain would have runoff rates 2 times the values of the 3-hour unit hydrograph. 5 cm in 3-hour would produce flows half the magnitude of the 3-hour unit hydrograph. This assumption of proportional flows applies only to equal duration storms.

If the duration of another storm is an integer multiple of *X*, the storm is treated as a series of end to end *X*-hour storms. First, the hydrographs from each *X* increment of rain are determined from the *X*-hour unit hydrograph. The ordinates are then added at corresponding times to determine the total hydrograph.

Part 2 Construction of Unit Hydrographs

Implicit in deriving the unit hydrograph is the assumption that rainfall is distributed in the same temporal and spatial pattern for all storms. This is generally not true; consequently, variations in ordinates for different storms of equal duration can be expected.

The construction of unit hydrographs for other than integer multiples of the derived duration is facilitated by a method known as the S-hydrograph. The procedure employs a unit hydrograph to form an S-hydrograph resulting from a continuous applied rainfall. The unit hydrograph theory can be applied to ungauged watersheds by relating unit hydrograph features to watershed characteristics. As a result of the attempted synthesis of data, these approaches are referred to as synthetic unit hydrograph methods. The need to alter duration of a unit hydrograph encouraged studies to define the shortest possible storm duration, that is, an instantaneous unit rainfall. The concept of instantaneous unit hydrograph (IUH) can be used in construction unit hydrographs for other than the derived duration.[3]

Methods of deriving unit hydrographs vary and are subject to engineering judgment. The level of sophistication employed to unravel the problem depends largely on the kind of issue in question.[4] Several methods useful in the determination of unit hydrographs will be discussed. They are subdivided into staring with unit hydrographs obtained from field data and manipulating them by S-hydrograph methods and constructing synthetic unit hydrograph.

Data collection preparatory to deriving a unit hydrograph for a gauged watershed can be extremely time consuming. To develop a unit hydrograph, it is desirable to acquire as many rainfall records as possible within the study area to ensure that the amount and distribution of rainfall over the watershed is accurately known. Preliminary selection of storms to use in deriving a unit hydrograph for a watershed should be restricted to the following:

1) Storms occurring individually, that is, simple storm structure.

2) Storms having uniform distribution of rainfall throughout the period of rainfall excess.

3) Storms having uniform spatial distribution over the entire watershed.

These restrictions place both upper and lower limits on size of the watershed to be employed. An upper limit of watershed size of approximately 2000 km^2 is overcautious, although general storms over such areas are not unrealistic and some studies if areas up to 3000 km^2 have used the unit hydrograph technique.[5] The lower limit of watershed extent depends on numerous other factors and cannot be precisely defined. A general rule of thumb is to assume about 10 km^2. Fortunately, other hydrologic techniques help resolve unit hydrographs for watersheds outside this range.

The preliminary screening of suitable storms for unit hydrograph formation must meet more restrictive criteria before further analysis.

1) Duration of rainfall event should be approximately 10%-30% of the drainage area lag time.

2) Direct runoff for the selected storm should be greater than 5 mm.

3) A suitable number of storms should be analyzed to obtain an average of the ordinates for a selected unit hydrograph duration. Modifications may be made to adjust unit hydrograph durations by means of S-hydrographs or IUH procedures.

4) Direct runoff ordinates for each storm should be reduced so that each event represents 10 mm of direct runoff.

5) The final unit hydrograph of a specific duration for the watershed is obtained by averaging ordinates of selected events and adjusting the result to obtain 10 mm of direct runoff.

Construction the unit hydrograph in this way produces the integrated effect of runoff resulting from a representative set of equal duration storms. Extreme rainfall intensity is not reflected in the determination. If intense storms are needed, a study of records should be made to ascertain their influence upon the discharge hydrograph by comparing peaks obtained utilizing the derived unit hydrograph and actual hydrographs form intense storm.

Essential steps in developing a unit hydrograph for an isolated storm follow:

1) Analyze the stream flow hydrograph to permit separation of surface runoff from groundwater flow.

2) Measure the total volume of surface runoff (direct runoff) from the storm producing the original hydrograph equal to the area under the hydrograph after groundwater base flow has been removed.

3) Divide the ordinates of direct runoff hydrograph by total direct runoff volume in inches and plot these results versus time as unit graph for the basin.

4) Finally, the effective duration of the runoff-producing rain for this unit graph must be found from the hyetograph (time history of rainfall intensity) of the storm used.

Procedures other than those listed are required for complex storms or in developing synthetic unit graphs when few data are available. Unit hydrographs can also be transposed from one basin to another under certain circumstances.

* * * * * * * * * * * * * * * * Explanations * * * * * * * * * * * * * * * *

* * * * * * * * * * * * * * * New Words and Expressions * * * * * * * * * * * * * * *

1. predict [priˈdikt] v. 预知，预言，预告
2. peak discharges 最大下泄流量，最大泄水量；最大排出量
3. hydrograph [ˈhəidrəgrɑ:fs] n. 过程线，水位线；水利图表
4. intensive [inˈtensiv] adj. 加强的，密集的；精工细作的，集约的
5. derive [diˈraiv] vt. 取得；追溯起源
   vi. 起源，衍生
6. occur [əˈkə:] vi. 发生，出现，存在；被想起（到）
7. incorrect [ˌinkəˈrekt] adj. 不正确的
8. specify [ˈspesifai] vt. 明确说明，具体指定
9. duration [djuəˈreiʃən] n. 持续，持续期间
10. storm [stɔ:m] n. 暴雨（雪）；
    vt. 猛攻
    vi. 怒气冲冲地走
11. volume [ˈvɔlju:m] n. 体积，容量；卷，册，书卷；音量，响度
12. excess [ikˈses] n. 过量，过度；v. 超越
13. uniformly [ˈju: nifɔ:mli] adv. 一致地
14. fraction [ˈfrækʃən] n. 小部分，片断；分数
15. amount [əˈmaunt] n. 数（量），总额
    vi. 合计；接近
16. accomplish [əˈkʌmpliʃ] vt. 达到（目的），完成（任务），实现（计划）
17. multiply [ˈmʌltiplai] v. （使）相乘，（使）增加
18. magnitude [ˈmægnitju: d] n. 重要性，重大；巨大，广大
19. correspond [ˈkɔriˈspɔnd] vi. 相符合；相类似；通信
20. total [ˈtəutl] n. 总数
    adj. 全部的，完全的
    v. 合计

21. implicit [im'plisit] adj. 无疑问的；含蓄的；内含的
22. temporal and spatial 时间的和空间的
23. variation [ˌveəri'eiʃən] n. 变化，变动；变体，变种；变奏（曲）
24. construction [kən'strʌkʃən] n. 建造，建设；建造物，建筑物；结构
25. facilitate [fə'siliteit] vt. 使变得（更）容易，使便利
26. S-hydrograph S 曲线
27. continuous [kən'tinjuəs] adj. 连续不断的，不断延伸的
28. unravel [ʌn'rævəl] vt. 解开
29. alter ['ɔ: ltə] v. 改变，改动，变更
30. instantaneous [ˌinstən'teiniəs] adj. 瞬间的，即刻的
31. restriction [ri'strikʃən] n. 限制，限定，约束
32. upper and lower 上、下限
33. overcautious [əuvə'kɔ: ʃəs] adj. 过分小心，保守
34. numerous ['nju:mərəs] adj. 众多的，许多的
35. precisely [pri'saisli] adv. 精确地；刻板地
36. modification [ˌmɔdifi'keiʃən] n. 修改，修饰
37. integrate ['intigreit] v.（使）成为一体，（使）合并
38. ascertain [ˌæsə'tein] vt. 查明，弄清
39. plot [plɔt] v. 绘图；密谋 n. 故事情节；密谋；小块土地
40. hyetograph ['haiitəugrɑ:f] n. 时间雨量曲线，雨量记录表，雨量分布图

*** * * * * * * * * * * * * * * * * Complicated Sentences * * * * * * * * * * * * * * * * ***

1. The unit hydrograph is defined as follows: if a given X hour rainfall produces a 10mm depth of runoff over the given drainage area, the hydrograph showing the rates at which the runoff occurred can be considered a unit for that watershed.

【译文】单位线定义如下：如果在给定的 *X* 小时内，流域上均匀地产生了 10mm 深的径流，则在该流域出口断面形成的地面径流过程线即为单位线。

【说明】文中 if 引导的 if 从句，即条件状语从句，表假设；which 引导的定语从句，其中 which 指代前面的 area。

2. Application an *X*-hour unit graph to design rainfall excess amounts other than 10 mm is accomplished simply by multiplying the rainfall excess amount by the unit graph ordinates, since the runoff ordinates for a given duration are assumed to be directly proportional to rainfall excess.

【译文】采用 *X* 小时的单位线计算并非等于 10mm 的雨量过程时，可简单采用净雨深乘以单位线的纵坐标，因为对一个给定的时段来说，单位线假定径流与净雨直接成比例。

【说明】在 since 引导的从句中，the runoff ordinates for a given duration are assumed to be directly proportional to rainfall excess 为原因，而前面的 by multiplying the rainfall excess amount by the unit graph ordinates 为结果，表示该原因下可以怎么做。

3. The need to alter duration of a unit hydrograph encouraged studies to define the shortest possible storm duration, that is, an instantaneous unit rainfall. The concept of instantaneous unit hydrograph (IUH) can be used in construction unit hydrographs for other

than the derived duration.

【译文】如果需要转化单位过程线的历时，就需要研究可能最短降雨历时——瞬时单位降雨过程，也就是利用瞬时单位过程线（简称瞬时单位线）来推导任一时段单位线。

【说明】原文中，that 指代前面的 the shortest possible storm duration。而后一句为被语句，但是按照汉语翻译习惯译为主动句，使句子更加清楚明了。

4. The level of sophistication employed to unravel the problem depends largely on the kind of issue in question.

【译文】用于解决问题的技巧水平在很大程度上取决于问题结果的类型。

【说明】句中 sophistication 为"技巧，完善"，unravel 表示"拆开，解决"，depend on 表示"依赖，取决于"，issue 表示"a solution to a problem"。

5. An upper limit of watershed size of approximately 2000 km^2 is overcautious, although general storms over such areas are not unrealistic and some studies if areas up to 3000 km^2 have used the unit hydrograph technique.

【译文】对上限面积，比较保守地选 2000km^2，尽管实际降雨面积没有这么大。并且一些研究也表明：如果上限面积加大到 3000km^2，就要使用相关的单位线技术了。

【说明】原文为并列句式，其中"although"为连词，英语和汉语的使用习惯不同，汉语一句话中可以使用许多连词，而英语中，一句话只能有一个连词；因此，在翻译中要特别注意这个问题。

******************** **Summary of Glossary** ******************

1. hydrograph	水位曲线，水位图
2. rainfall	降雨，降雨量
3. hyetograph	雨量图，雨量分布图
4. upper and lower limits	上下限
5. *X*-hour unit hydrograph	*X* 小时单位线
6. S-hydrograph	S 曲线法

******************** **Abbreviations (Abbr.)** ******************

1. UH	Unit Hydrograph	单位线
2. IUH	Instantaneous Unit Hydrograph	瞬时单位线

********************** **Exercises** ********************

(1) Ways to predict flood ________ and discharge hydrographs from ________ events have been studied ________ since the early 1930s.

(2) The ordinates are then added at ________ times to determine the ________ hydrograph.

(3) This is generally not true; ________, ________ in ordinates for different storms of equal duration can be expected.

*** * * * * * * * * * * * Word Building (5) –ment; -mit * * * * * * * * * * * ***

1. -ment [名词后缀]，表示：状态，行为或其结果；有关的设备、器具

achieve	vt.	achievement	n.	完成；成就，功绩
develop	v.	development	n.	发展
equip	vt.	equipment	n.	装备，设备，器材，装置
measure	v.	measurement	n.	测量，(量得的)尺寸，测量过程
move	v.	movement	n.	运动，动作
supply	vt.	supplement	n.	补遗，补充，附录
embank	v.	embankment	n.	堤防，筑坝
place	n./v.	placement	n.	放置，布置
judge	n./v.	judgement	n.	判决，评价
settle	n./v.	settlement	n.	沉降，下陷
align	v.	alignment	n.	轴线

2. -mit [词根]，表示：送，发，委

e-(向外)	emit	vt.	发出
inter	intermit	v.	间断
com-(共同)	commit	vt.	委派
sub-(下面)	submit	vt.	递交
per-(加强)	permit	v.	许可

*** * * * * * * * * * * * * * * * * Text Translation * * * * * * * * * * * * * * * * ***

第五章　水文学原理——单位线

第一节　概　　述

自 20 世纪 30 年代初以来，预测洪峰流量和流量过程线的方法已经研究得很透彻。其中一种比较公认的方法是单位过程线法，它起源于 1932 年谢尔曼（Sherman）提出的单位线，单位线定义如下：在给定的 *X* 小时内，流域上均匀地产生了 10mm 深的径流，则在该流域出口断面形成的地面径流过程线即为单位线。（单词个数：86）

若没有给定历时，即产生单位过程线的暴雨时间为 *X*，是不能描述单位线过程的。每个 *X* 小时单位线都是 10mm 深的径流过程线。径流深（10mm）除以 *X* 小时可得暴雨强度，即雨强为 10/*X*(mm/h)。2h 单位过程线是根据流域上每 2h 净雨达到 10mm 的降雨绘制成的。1 天单位线是根据 1 天内流域内净雨达到 10mm 的降雨绘制成的。*X* 的值通常是 1 小时的几分之一。（单词个数：106）

采用 X 小时的单位线计算并非等于 10mm 的雨量过程时，可简单采用净雨深乘以单位线的坐标，因为对一个给定的时段来说，单位线假定径流与净雨直接成比例。产生 20mm 净雨深的 3 小时暴雨产生 2 倍于 3 小时单位线数值的径流。3 小时内的 5mm 净雨深会产生 3 小时单位线一半的流量。这种成比例假设只适用于相同历时的降雨。（单词个数：94）

如果另一场降雨的历时是 X 的整数倍，这场降雨将被处理为一系列首尾相连的 X 小时的暴雨系列。首先，每 X 历时降雨增量的单位过程线将由 X 小时单位过程线确定。然后在相应时间叠加到纵坐标，以确定总的过程线。（单词个数：52）

第二节 S 曲 线 方 法

推导单位过程线时，假设所有降雨具有相同的时间和空间分布，这通常是不正确的。因此，需要预测相同历时的不同降雨的纵坐标的变化。（单词个数：41）

利用 S 单位过程线法可方便地推导非整数倍数的单位过程线，一般是通过作图来实现，该方法利用单位线来绘制出一个由连续降雨产生的 S 形过程线。因此，用于无资料地区时，可以利用单位线理论及流域特征与单位线特征的相关性，来推导其产流过程。这是数据合成的过程，有时也称为合成单位线方法。若要转化单位过程线的历时，就需要研究可能最短降雨历时——瞬时单位降雨过程，也就是利用瞬时单位过程线（简称瞬时单位线）来推导任一时段的单位线。（单词个数：121）

推导单位线方法很多，具体应用时主要取决于应用的实际工程情况。能否进一步解决问题，在很大程度上还依赖于所要解决的具体问题的种类和复杂程度。这里将讨论几种确定单位线行之有效的方法，通常可将这些方法划分为：由现场数据得到单位线、利用 S 单位线方法得到单位线和绘制综合单位线。（单词个数：64）

对有观测资料的流域来说，推导单位线时所需要的数据收集准备工作是很花费时间的。为了尽可能完善单位线，需要获取研究区域内尽可能多的降雨记录资料，以便确保全面掌握流域降雨总量及其分布情况。选择用来推导单位线的降雨过程时，主要考虑下列原则：（单词个数：72）

（1）发生的降雨是独立的，即简单的降雨组成。（降雨独立性要求）（单词个数：8）

（2）整个降雨过程的分布一致。（时间一致性要求）（单词个数：12）

（3）全流域降雨要有统一的空间分布。（空间一致性要求）（单词个数：9）

满足上述要求，就需要对流域面积的大小有上、下限的要求。对上限面积，比较保守的选 $2000km^2$，尽管现实中通常的降雨没有如此大的降雨面积，并且一些研究也表明：如果上限面积加大到 $3000km^2$ 就要使用相关的单位线技术了；下限面积受许多其他因素影响，无法准确界定，通常假设为 $10km^2$。所幸，流域面积超出这个范围的单位线，可利用一些其他的单位线技术解决。（单词个数：93）

在进一步分析绘制单位线之前，最初选择的合适降雨审查必须满足以下限制性准则：

（1）降雨持续时间应该占整个流域汇流时间的 10%～30%。（单词个数：15）

（2）所选降雨的直接径流深应大于 5mm。（单词个数：13）

（3）要选择适当数目的降雨过程，来得到所选单位过程线历时的平均纵坐标，一般需利用 S 曲线或瞬时单位线做些来修正所推求的单位线的历时。（单词个数：38）

（4）为了使每次降雨代表 10mm 的直接径流，应该将直接径流纵坐标减小。（单词个数：20）

（5）流域内某一特定时期的最终单位线，可将所选降雨纵坐标平均化得到，然后校核得到 10mm 的直接径流深的单位线。（单词个数：31）

用这种方法绘制的单位线体现了等历时降雨的综合效果，不能反映流域的极限降雨强度。如果需要考虑强降雨，则需要通过比较利用单位线获得的峰值和实际强降雨峰值做一个对比研究，来探讨它们对单位线的影响。（单词个数：66）

利用独立降雨过程推求单位线的具体步骤如下：（单词个数：12）

（1）分析径流过程线，将地表径流和地下径流分离开。（单词个数：15）

（2）计算降雨地表径流总量，即降雨总量等于分离地下径流后的降雨产生的径流过程线以下的面积。（单词个数：31）

（3）通过径流总量来分离径流过程线，时间为横坐标，径流量（一般以英尺为单位）为纵坐标绘制流域单位线。（单词个数：27）

（4）最后，这个单位线的有效产流时间一定要从所用降雨的雨量图（雨强的历时记录）中找到。（单词个数：28）

当没有水文观测数据可用，降雨情况又复杂，或者需要推导综合单位线时，除了上面已列出的，还需要一些更复杂的其他程序。在某种情况（如径流相似区）下，可将某一流域的单位线移植到另一流域。（单词个数：35）

Chapter 6 Sediments

Part 1 Origin and Formation of Sediment

Sediment is any particulate matter that can be transported by fluid flow and which eventually is deposited as a layer of solid particles on the bed or bottom of a body of water or other liquid. Sedimentation is the deposition by settling of a suspended material.

Sediments are also transported by wind (eolian) and glaciers. Desert sand dunes and loess are examples of aeolian transport and deposition. Glacial moraine deposits and till are ice transported sediments. Simple gravitational collapse also creates sediments such as talus and mountain side deposits as well as karst collapse features.

Seas, oceans, and lakes accumulate sediment over time. The material can be terrigenous (originating on the land) or marine (originating in the ocean). Deposited sediments are the source of sedimentary rocks, which can contain fossils of the inhabitants of the body of water that were, upon death, covered by accumulating sediment.[1] Lake bed sediments that have not solidified into rock can be used to determine past climatic conditions.

Organic agents of weathering are primarily burrowing animals and also roots and trunks of trees which wedge the rocks apart. After the parent rocks are disintegrated, the material is transported from one place to another and deposited by streams, wind, or glaciers. The material is called alluvium if transported and deposited by streams, loess if transported and deposited by wind and glacial drift if it is transported and deposited by glacier.[2]

1. Stream erosion and deposition

The sediment load carried by streams comes from various sources. In the hilly areas the streams pick up coarse material from the talus (Talus is a geological term describing a heap or a sheet of coarse rock that has accumulated at the foot of the hill or on a steep slope). Landslides also contribute to the load carried by streams. However, the major portion of sediment load carried by streams comes from the erosion of material in the drainage basin; a certain amount also originated as a result of weather of rocks from the bed and banks of the stream. The size of the sediment transported in any reach is dependent on the geology of the basin as well as the distance of the reach from the source.[3] The amount of the sediment load carried depends on the size of material, discharge, slope, and channel and catchment characteristics.

When there is reduction either in the discharge or in the slope of an equilibrium stream, the stream cannot transport the material supplied to it and excess material is deposited. These deposits give rise to various formations depending on the mode of deposition and they are called flood plains, alluvial fans, deltas, etc.

2. Wind erosion and deposition

Arid and semi-arid regions are characterized by relatively low and infrequent rainfall. As such, stream flows in these regions are small, unless, of course, the area is small and is traversed by streams flowing into the area from an area of appreciable rainfall. However, in general, stream erosion in arid and semi-arid regions is relatively small.

On the other hand, wind erosion becomes a predominant factor. High velocity winds, carrying fine sand with them, are effective agents of wind erosion. When wind blows over deserts and ploughed fields, fine sand and dust particles are carried away while the coarser material is left behind. This process is called deflation. The dust that is carried by the wind is transported to great distances. When the wind velocity is reduced, this material is deposited as loess. Large deposits of loess are found in China, the Mississippi valley, Europe, and some other parts of the world.

Part 2 The Size and Shape of Sediment

1. Size

Of all the properties of sedimentary particles, sediment size is one of the most important and a commonly used property. Therefore, it needs detailed discussion. If all the particles were spheres, then specifying the diameter would suffice. However, the sediment particles composing stream beds are of numerous shapes from round to flat and to needlelike. This extreme irregularity in their shape defies ordinary classification. Therefore, diameter as an index of size loses its usual significance. For these reasons, the particle size is usually defined by its volume, fall velocity, size of the sieve mesh, or by its intercepts. The method of measurement of the size by its volume and fall velocity are based on the premise that measurements made can be expressed as the diameter of an equivalent sphere. The following definitions of sizes are used in practice.

1) Nominal diameter, d_n, of a particle is defined as the diameter of a sphere which is having the same volume as the particle. The nominal diameter gives an idea of the physical size of the particle. However, in several problems one is interested in the size as related to the mobility of the particle in the fluid; this gives rise to the definitions of fall diameter and sedimentation diameter.

2) Fall diameter of a particle is defined as the diameter of a sphere of relative density 2.65 and having the same standard fall velocity as that of the particle. The standard fall velocity of a particle is the terminal fall velocity of the particle in quiescent distilled water of infinite extent and at 24℃.

3) Sedimentation diameter is the diameter of the sphere of the same relative density and the same terminal settling velocity as the given particle in the same sedimentation fluid under the same condition.

4) Sieve diameter. In most cases a series of sieves is used for separation of sediment into various size grades provided particles are large than 0.0625 mm in diameter. Sieve openings are square in shape. Since a long particle with a very small cross-sectional area can pass through a sieve, sieves classify particles on the basis of their least cross-sectional area and thus this classification is not based purely on size.[4] Sieve diameter d of a particle is the size of the sieve opening through which a given particle will just pass. Thus, the nominal and sieve diameters of a sphere are always equal. For

naturally-worn particles sieve diameter is found to be slightly smaller than the nominal diameter.

5) Triaxial size. Several geologists and hydraulic engineers have defined particle size based on measurement along three mutually perpendicular axes of a particle if *a* is the longest or the major axis, *b* the intermediate axis and *c* the shortest or the minor axis, then one can take the intermediate axis *b* as the index. In such a case correlation can be found between the shape of the particle, nominal diameter and intermediate axis *b*. The relative magnitudes of *a*, *b* and *c* represent the shape of the particle, and as will be discussed later, they are used as an index of shape. It may be mentioned that the description of particle size in this manner is not precise and differences in interpretation and variations in measuring techniques are common. For example, since the axes are perpendicular, values of *a*, *b*, and *c* for given particle depend on which axis is chosen first.

6) Size terms for particles. It is essential to standardize the terms related to particle size so that they could be used to convey definite ideas regarding the size. Udden's geometrical scale of sediment sizes and Wentworth's modification of Udden's scale are considered to be the basic works on this subject. The subcommittee on Sediment Terminology of the AGU has accepted the Wentworth scale with exception that the term "granules" has been deleted. The classification accepted by the Subcommittee, which is quite widely used by hydraulic engineers, is given in Table 6.1.[5] The sizes, given in terms of sieve diameters for coarse material, for the finer material sedimentation diameter, is recommended in Table 6.1.

Table 6.1 Grade scale for size terms

Range in mm	Class name		Range in mm	Class name	
4096-2048	Very Large	Boulder	1-1/2	Coarse	Sand
2048-1024	Large		1/2-1/4	Medium	
1024-512	Medium		1/4-1/8	Fine	
512-256	Small		1/8-1/16	Very Fine	
256-128	Large	Cobbles	1/16-1/32	Coarse	Silt
128-64	Small		1/32-1/64	Medium	
64-32	Very Coarse	Gravel	1/64-1/128	Fine	
32-16	Coarse		1/128-1/256	Very Fine	
16-8	Medium		1/256-1/512	Coarse	Clay
8-4	Fine		1/512-1/1024	Medium	
4-2	Very Fine		1/024-1/2048	Fine	
2-1	Very Coarse	Sand	1/2048-1/4096	Very Fine	

2. Shape

Hydraulic engineers, geologists and soil engineers are interested in defining the shape of coarse sediment particles for various reasons. The shape of the particle influences the mean velocity of the flow at which the particle on the bed moves, the fall velocity, the stability of beaches, and bed load transport.[6] Geologists are interested in the shape of coarse particles because its study throws light on the method of transport and on the deposit to which they belong. Specialists of soil mechanics have found that the shape of the particles is a significant variable in determining the porosity, permeability and cohesion of soil.

The shape of the particle depends initially on the source rock and the weathering process. Subsequently, however, the shape is modified by abrasion, corrosion and breakage. The shape of the particle as it finally reaches a sedimentary deposit is governed by such shape-sorting mechanisms as crushing, splitting, chipping, grinding and chemical weathering.

Since naturally-worn sediment particles assume many shapes, such geometrical shapes as cube, sphere, cylinder, cone, ellipsoid become inadequate to describe the shape of the particles. Furthermore, such description becomes more or less qualitative. On the other hand, what is required is a single parameter which will be amenable to quantitative analysis. Many such parameters have been introduced by various investigators and Durand has classified these parameters into the following categories:

1) Coefficients based on the volume;

2) Coefficients based on the projected area;

3) Coefficients based on major, intermediate and minor axes.

* * * * * * * * * * * * * * * * * Explanations * * * * * * * * * * * * * * * * *

* * * * * * * * * * * * * * * * * New Words and Expressions * * * * * * * * * * * * * * *

1. dune [dju: n] n. 沙丘
2. moraine [mɔ'rein] n. 冰碛
3. gravitational collapse 引力坍缩（物）
4. talus ['teiləs] n. 碎石堆
5. mountain slide 山体滑坡
6. terrigenous [te'ridʒinəs] adj. 陆源的
7. organic [ɔ:'gænik] adj. 有机(体)的，
8. burrow ['bʌrəu] n. 地洞
   vt. 挖掘（洞穴）；钻进
   vi. 挖洞；翻寻
9. wedge [wedʒ] vt. 把…楔牢，塞入
   n. 楔（子）
10. alluvium [ə'lu: viəl] n. 冲积层；冲积土，冲积物
11. glacial drift 浮冰
12. landslide ['lændslaid] n. [地] 崩塌，山崩，地滑，塌方
13. drainage basin 排水区
14. equilibrium [ˌi:kwi'libriəm] n.平衡
15. alluvial fans 冲积扇
16. arid and semi-arid 干旱和半干旱
17. traverse ['trævə: s] vt. 横渡，横越
18. plough [plau] v. 犁，耕
   n. 犁，犁形工具
19. deflation [di'fleiʃən] n. 风力侵蚀
20. sphere [sfiə] n. 球(体)；范围，领域
21. diameter [dai'æmitə] n. 直径
22. suffice [sə'fais] vi. 足够
23. needlelike ['ni: dllaik] n. 针状(物)
24. sieve mesh 筛分网，筛分孔
25. equivalent [i'kwivələnt] n. 等价物，意义相同的词
   adj. 相等的，相当的
26. nominal diameter 等容粒径
27. mobility [mou'biliti] n. 流动性
28. sedimentation diameter 沉积直径，泥沙粒径
29. quiescent distilled water 静水，蒸馏

过的纯水

30. sieve diameter 筛孔直径
31. triaxial [traiˈæksiəl] adj. 三轴的
32. perpendicular axes 垂直轴线
33. granule [ˈgrænju: l] n. 小粒，微粒
34. porosity [ˈpɔ: rəs] n. 多孔，空隙
35. permeable [ˈpə:miəbəl] adj. 可渗透的
36. cohesion [kəuˈhi: ʒən] n. 内聚力
37. abrasion [əˈbreiʒən] n. 表面磨损
38. cylinder [ˈsilində] n. 圆柱体，圆筒；汽缸，泵（或筒）体
39. ellipsoid [iˈlipsɔid] n. 椭球

******************* Complicated Sentences *******************

1. Deposited sediments are the source of sedimentary rocks, which can contain fossils of the inhabitants of the body of water that were, upon death, covered by accumulating sediment.

【译文】沉淀物是沉积岩的来源，其中可能包含死后被沉积的沉淀物所覆盖的风化了的人类尸骨。

【说明】句中 which 指代前面所提到的 deposited sediments，而不是 sedimentary rocks。

2. The material is called alluvium if transported and deposited by streams, loess if transported and deposited by wind and glacial drift if it is transported and deposited by glacier.

【译文】如果是通过溪流输移和沉积的泥沙，则形成冲积层；如果是通过风输移和沉积的泥沙，则形成黄土；如果是通过冰川漂移和沉积的泥沙，则形成冰碛。

【说明】原文使用了省略，在 loess if transported…和 glacial drift if it is transported…前面省略了 the material is called，在翻译成汉语时，要把略省的部分补充完整。

3. The size of the sediment transported in any reach is dependent on the geology of the basin as well as the distance of the reach from the source.

【译文】河流携带泥沙的规模取决于流域的地理、地质条件以及从泥沙源到河流的距离。

【说明】原文为并列句式，其中并列句的连词是 as well as。

4. Since a long particle with a very small cross-sectional area can pass through a sieve, sieves classify particles on the basis of their least cross sectional area and thus this classification is not based purely on size.

【译文】当一些细长（横截面小）的颗粒可以穿过筛子时，筛分粒径只是建立在这些细长颗粒的最小横截面的基础上，因此，这种分类不是纯粹地基于颗粒大小的。

【说明】原文为 since 引导的时间状语从句，翻译时，since 译为“当……时”。

5. The classification accepted by the Subcommittee, which is quite widely used by hydraulic engineers, is given in Table 6.1.

【译文】分委员会所接受的分类方法，也是水利水电工程师广泛使用的方法，已列在表 6.1 中。

【说明】句中 which 指代前面的 the classification。

6. The shape of the particle influences the mean velocity of the flow at which the particle on the bed moves, the fall velocity, the stability of beaches, and bed load transport.

【译文】颗粒的形状影响着水流在河床中的平均流速、颗粒的沉落速度，河岸的稳定及河床的淤积。

【说明】原文中 which 引导定语从句，其中 which 指代前面的 velocity of the flow。"which" 后面的句子结构为"中心词+修饰语"结构，其中 the particle on the bed moves, the fall velocity, the stability of beaches, and bed load transport 为修饰词，修饰前面的 velocity of the flow。

*** * * * * * * * * * * * * * * * * * Summary of Glossary * * * * * * * * * * * * * * * * * ***

1. sediment	沉积物
2. flood plain	冲积平原
3. alluvial fan	冲积扇
4. alluvium	冲积层，淤积层
5. delta	三角洲
6. loess	黄土
7. glacial drift	冰碛
8. dune	沙丘
9. desert	沙漠
10. landslide	山体滑坡
11. terminal settling	最终沉降

*** * * * * * * * * * * * * * * * * * * Abbreviations (Abbr.) * * * * * * * * * * * * * * * * * * ***

AGU	American Geophysical Union	美国地球物理协会

*** Exercises ***

(1) Desert sand ______ and ______ are examples of aeolian transport and deposition.

(2) ______ regions are characterized by relatively low and infrequent rainfall.

(3) The method of ______ of the size by its volume and fall velocity are based on the premise that measurements made can be expressed as the diameter of an ______ sphere. The following definitions of sizes are used ______.

(4) Fall diameter of a particle is defined as the diameter of a sphere of relative ______ 2.65 and having the same standard fall velocity as that of the particle.

*** * * * * * * * * * * * * * Word Building (6) geo-; rot- * * * * * * * * * * * * * ***

1. geo- [词根]，表示地球、大地

geohydrology	n.	水文地质学，地下水水文学
geotechnique	n.	土工学，岩土学
geochemistry	n.	地球化学
geodynamics	n.	地球动力学

2. rot- [词根]，表示旋转

rotary	adv.	旋转地
autorotate	vi.	自转
rotoplug	n.	旋塞

******************* Text Translation *******************

第六章　泥　　沙

第一节　泥沙的起源与形成

泥沙是指那些经水流携带，并最终被作为一层固体颗粒在河床上或河底部沉积下来的各种细微颗粒。泥沙淤积是沉积了的悬浮物质。（单词个数：46）

泥沙也可随风（风成的）或冰川的移动而移动。沙漠中的沙丘和黄土的运输及沉积就是风作用的例子。冰川冰碛沉积和冰渍都是冰川作用的结果。简单的引力坍缩也造成泥沙塌陷，形成由岩石碎块组成的一个倾斜的堆岩屑堆或山体滑坡以及岩溶塌陷而形成不规则的石灰岩地区的喀斯特特征。（单词个数：48）

随着时间的推移，大海、海洋和湖泊中的泥沙也在不断累积。泥沙的来源可以是陆源（原产于土地），也可能是海运的（原产于海洋）。沉淀泥沙源是沉积岩，其中可能包含被积累沉淀所覆盖的风化了的人死亡后的尸骨。湖底尚未凝固成岩石的泥沙，可以用来分析确定过去的气候条件。（单词个数：68）

风化过程最先是从动物藏身的洞穴，或被树木的树干和根系的楔入而四分五裂的岩石开始的。在母岩的解体之后，碎块就（随风或水）从一个地方转移到另一个地方，并储存于溪流、风或冰川中。如果是通过溪流运输和沉积的泥沙，则形成冲积层；如果是通过风运输和沉积的泥沙，则形成黄土；如果是通过冰川漂移和沉积的泥沙，则形成冰碛。（单词个数：72）

1. 河流冲刷与积淤

河流中的泥沙来自各种各样的来源。丘陵地区的河流从碎石堆（碎石堆是地质术语，描述积累在山脚下或在陡峭的斜坡处的石堆或粗糙的岩石层）中获得粗料。山体滑坡也会使泥沙进入到河流中。然而，河流中的大部分泥沙来自整个流域的河床侵蚀；也有一部分是由于天气的变化，来源于河流的岩床和河岸（风化或冲刷）。河流携带泥沙的规模取决于流域的地理、地质条件以及从泥沙源到河流的距离。河流携带泥沙的大小则受泥沙颗粒的大小、流量（的大小）、低坡坡度和河道（的大小）以及集水特点的影响。（单词个数：153）

当河流流量减少或者河流处于稳定的缓流时，河流不能将水中的泥沙运送到下游，其中的一部分就会沉积下来。不同的沉积模式形成各种各样的沉积形式，如冲积平原、冲积扇、三角洲等。（单词个数：53）

2. 风力侵蚀和沉积

干旱和半干旱地区的特点是地势相对较低，降雨稀少。因此，这些地区的河流流量是很小的。当然，如果该地区很小，并且又有着丰富的降雨量地区的河流从中穿过的话，则另当别论。不过，一般而言，河流的侵蚀在干旱和半干旱地区相对较小。（单词个数：61）

另一方面，干旱和半干旱地区的风力侵蚀是主导因素。携带着细沙的高速风，很有可能造成风力侵蚀。当风吹在沙漠和田地上时，带走了细砂和灰尘颗粒，地面就只剩下粗沙，这一过程称为风力侵蚀。风能把细沙运到很远的距离。当风速下降时，这种细沙就沉积下来，也就是黄土。在中国、密西西比河流域、欧洲和世界上其他一些地区都发现了大面积的黄土沉积。（单词个数：96）

第二节 泥沙的大小和形状

1. 泥沙颗粒大小

在泥沙颗粒的所有属性中，大小是最重要和最常用的属性。因此，有必要详细讨论。如果所有的颗粒都是球形的，只要具体给出它们的的直径就可以了。然而，河床中的泥沙颗粒是由不同的形状组成的，它们有圆的、有扁平的，还有锥形的。这种不同形状的泥沙颗粒就不能用常规（直径）的分类方法。因此，直径为大小的指标失去了传统的意义。由于这些原因，粒径通常是由它们的体积、下降的速度、规模、筛格筛分或其截面来定义的。该方法测量它的体积大小和下降速度是基于这样一个前提下的，即它们的体积大小可以用一定大小的直径的球体来表示。下面就介绍实际中应用的几种颗粒尺寸的定义方法：（单词个数：137）

1）等容粒径，用符号 d_n 来表示，它定义为一个与颗粒有着相同体积的球体的直径来表示。等容粒径给出了一个物理尺寸的粒子大小的想法。然而，在考虑问题时，最让人感兴趣的是颗粒的大小会影响水体的流动性，这就引起了沉落直径和沉淀直径的定义（的探讨）。（单词个数：70）

2）沉落直径，其定义是一个相对密度为 2.65 的球体直径，并具有相同的下降速度的颗粒。该粒子标准下降速度是指温度为24℃时，在无限延伸的水中由静止自由下落的最终速度。（单词个数：55）

3）沉淀直径，是指在相同的流体条件下，与沉淀物有相同的相对密度和相同的最终下降速度的球体的直径。（单词个数：33）

4）筛分直径。在大多数情况下，用于直径筛分法的设备中提供了颗粒直径大于 0.0625mm 的筛分设备，用于筛分不同粒径的沉淀物。筛子做成正方形的开口形状。一些细长（横截面小）的颗粒可以穿过筛子，而筛子区分这些颗粒是在所有这些颗粒能够穿过的筛子横截面的基础之上，因此这种分类不是纯粹的基于颗粒的大小的。颗粒的筛分直径 d 是指筛子打开时通过筛子的颗粒的尺寸大小。因此，在等容粒径和筛分直径都是相等的。对于自然磨损颗粒来说，筛分直径要稍低于等容直径。（单词个数：122）

5）三轴粒径。一些地质学家和水利水电工程师用三轴粒径来确定测量的粒径大小，即粒径可以用沿三个互相垂直轴线的粒子直径来描述，如果是最长或长轴用 a 表示，中间轴用 b 表示，c 表示最短或短轴，那么可以采取中间轴 b 作为指标。在这种情况下，可以通过等容粒径和中间轴 b 之间的相关性，确定粒径的大小。a、b 和 c 的大小代表粒子相对的形状，它们被用作形状指数，稍后再进一步讨论。值得一提的是，粒径以这种方式描述并不准确，不同的测量技术和条件下，误差是不可避免的。例如，当各个轴都相互垂直时，给定颗粒的 a、b 和 c 的值的大小取决于首先选定哪个轴。（单词个数：153）

6）颗粒大小规范。关键是要规范相关的粒径，使它们可以被用来传达明确的颗粒大小。最具代表性的是乌登的几何大小模型和温特沃斯修改乌登的模型。美国地球物理学会泥沙组委员会已接受了温特沃斯模型，另外“颗粒（模型）”已被骈除。美国地球物理协会的泥沙分委员会所接受的分类方法，也是水利水电工程师广泛使用的方法，已列在表 6.1 中。这是关于粗颗粒的筛分直径，但同时也为精细材料沉积直径提供依据。（单词个数：111）

表 6.1　　泥 沙 粒 径 分 类

粒径变化幅度（mm）	岩石种类		粒径变化幅度（mm）	岩石种类	
4096-2048	非常大	卵石	1-1/2	粗	砂
2048-1024	大		1/2-1/4	中	
1024-512	中		1/4-1/8	细	
512-256	小		1/8-1/16	非常细	
256-128	大	鹅卵石	1/16-1/32	粗	淤泥
128-64	小		1/32-1/64	中	
64-32	非常粗	砾	1/64-1/128	细	
32-16	粗		1/128-1/256	非常细	
16-8	中		1/256-1/512	粗	黏土
8-4	小		1/512-1/1024	中	
4-2	非常小		1/1024-1/2048	细	
2-1	大	砂	1/2048-1/4096	非常细	

2. 泥沙颗粒形状

水利水电工程师、地质学家和土壤工程师都对确定粗泥沙颗粒形状感兴趣有多方面的原因。粒子的形状影响水流在河床中的平均流速、该粒子的沉落速度、河岸的稳定及河床的淤积。地质学家最关注粗颗粒的形状，因为对粗颗粒的研究重点是其输移和沉积。土力学专家研究发现，在确定土壤的孔隙度、渗透率和凝聚力时，粒子的形状是一个重要的变量。（单词个数：106）

粒颗的形状取决于最初的源岩和风化过程，但后来形状逐渐被磨损、腐蚀和断裂。颗粒的最终形状，即最终沉积形状，是由粉碎、分裂、切、磨和化学风化等形状分类机制所决定的。（单词个数：52）

由于自然磨损，许多泥沙颗粒有许多形状，立方体、球、圆柱、圆锥、椭球等几何形状不足以描述颗粒的形状。此外，这种描述或多或少地成了一种定性描述。另一方面，现在值得一提的是，需要确定一个简单的可进行定量分析的参数。许多研究者提出各种各样的参数，杜兰德已将这些参数划分为以下类别：（单词个数：70）

1）基于体积的参数；（单词个数：5）

2）基于投影面积的参数；（单词个数：6）

3）基于主轴、中轴和次轴的参数。（单词个数：8）

Chapter 7 River

Part 1 Introduction

A river is a natural watercourse, usually freshwater, flowing toward an ocean, a lake, a sea or another river. In a few cases, a river simply flows into the ground or dries up completely before reaching another body of water. Small rivers may also be called by several other names, including stream, creek, brook, rivulet, and rill; there is no general rule that defines what can be called a river. Many names for small rivers are specific to geographic location; one example is Burn in Scotland and North-east England. Sometimes a river is said to be larger than a creek, but this is not always the case, due to vagueness in the language.

A river is part of the hydrological cycle. Water within a river is generally collected from precipitation through surface runoff, groundwater recharge, springs, and the release of stored water in natural ice and snowpacks (i.e. from glaciers).[1]

The water in a river is usually confined to a channel, made up of a stream bed between banks. In larger rivers there is also a wider floodplain shaped by flood-waters over-topping the channel. Flood plains may be very wide in relation to the size of the river channel. This distinction between river channel and floodplain can be blurred especially in urban areas where the floodplain of a river channel can become greatly developed by housing and industry.[2]

The term upriver is referred to the direction leading to the source of the river, which is against the direction of flow. Likewise, the term downriver describes the direction towards the mouth of the river, in which the current flows.

The river channel typically contains a single stream of water, but some rivers flow as several interconnecting streams of water, producing a braided river. Extensive braided rivers are now found in only a few regions worldwide, such as in south most Alabama and the South Island of New Zealand. They also occur on peneplains and some of the larger river deltas. Anastomosing rivers are similar to braided rivers and are also quite rare. They have multiple sinuous channels carrying large volumes of sediment.

A river flowing in its channel is a source of energy which acts on the river channel to change its shape and form. According to Brahm's law (sometimes called Airy's law), the mass of objects that may be flown away by a river is proportional to the sixth power of the river flow speed. Thus, when the speed of flow increases two times, it can transport 64 times larger (i.e. more massive) objects. In mountainous torrential zones this can be seen as erosion channels through hard rocks and the creation of sands and gravels from the destruction of larger rocks. [3] In U shaped glaciated valleys, the subsequent river valley can often easily be identified by the V shaped channel that it has carved. In the middle reaches where the river may flow over flatter land, meanders may form through erosion of the river banks and deposition on the inside of bends. Sometimes the river will cut off a loop, shortening the channel and forming an oxbow lake or billabong. Rivers that carry large amounts of sediment may develop conspicuous deltas at their mouths, if conditions permit. Rivers whose mouths are in saline tidal waters may form estuaries.

Throughout the course of the river, the total volume of water transported downstream will often be a combination of the free water flow together with a substantial contribution flowing through sub-surface rocks and gravels that underlie the river and its flood plain. For many rivers in large valleys, this unseen component of flow may greatly exceed the visible flow.

Part 2 Classification

Although the following classes are a useful way to visualize rivers, there are many other factors at work. Gradient is controlled largely by tectonics, but discharge is controlled largely by climate, and sediment load is controlled by various factors including climate, geology in the headwaters, and the stream gradient.

1. Youthful river

A river with a steep gradient that has very few tributaries and flows quickly. Its channels erode deeper rather than wider. (Examples: Brazos River, Trinity River, Ebro River)

2. Mature river

A river with a gradient that is less steep than those of youthful rivers and flows more slowly. A mature river is fed by many tributaries and has more discharge than a youthful river. Its channels erode wider rather than deeper. (Examples: Mississippi River, St. Lawrence River, Danube River, Ohio River, River Thames)

3. Old river

A river with a low gradient and low erosive energy. Old rivers are characterized by flood plains. (Examples: Huang He River, Ganges River, Tigris, Euphrates River, Indus River, Nile River)

4. Rejuvenated river

A river with a gradient that is raised by tectonic uplift. The straight-line distance from the beginning to the end of most rivers is about one third their actual length. The way in which a river's characteristics vary between the upper course and lower course of a river is summarized by the Bradshaw model. [4] Most rivers flow on the surface; however subterranean rivers flow underground in caves or caverns. Such rivers are frequently found in regions with limestone geologic formations.

5. Intermittent River (or ephemeral river)

An intermittent river (or ephemeral river) only flows occasionally and can be dry for several years at a time. These rivers are found in regions with limited or highly variable rainfall, or can occur due to geologic conditions such as having a highly permeable river bed. Some ephemeral rivers flow during the summer months but not in the winter. Such rivers are typically fed from chalk aquifers which recharge from winter rainfall. In the UK these rivers are called Bournes and give their name to place such as Bournemouth and Eastbourne.

Part 3 Uses

Rivers have been used as a source of water, for food, for transport, as a defensive barrier, as a source of hydropower to drive machinery, and as a means of disposing of waste.

Rivers have been used for navigation for thousands of years. The earliest evidence of navigation is found in the Indus Valley Civilization, which existed in northwestern Pakistan around 3300 BC.[5] Riverine navigation provides the cheapest means of transport, and is still used extensively on major rivers of the world like the Amazon, the Ganges, the Nile, the Mississippi, and the Indus.

In some heavily-forested regions such as Scandinavia and Canada, lumberjacks use the river to float felled trees downstream to lumber camps for further processing, saving much effort and cost by transporting the huge heavy logs by natural means.[6]

Rivers have been a source of food since pre-history. They can provide a rich source of fish and other edible aquatic life, and are a major source of fresh water, which can be used for drinking and irrigation. It is therefore no surprise to find most of the major cities of the world situated on the banks of rivers. Rivers help to determine the urban form of cities and neighbourhoods and their corridors often present opportunities for urban renewal through the development of foreshore ways such as riverwalks. Rivers also provide an easy means of disposing of waste-water and, in much of the less developed world, other wastes.

Fast flowing rivers and waterfalls are widely used as sources of energy, via watermills and hydroelectric plants. Evidence of watermills shows them in use for many hundreds of years such as in the Orkneys at Dounby click mill. Prior to the invention of steam power, water-mills for grinding cereals and for processing wool and other textiles were common across Europe. In the 1890s the first machines to generate power from river water were established at places such as Cragside in Northumberland and in recent decades there has been a significant increase in the development of large scale power generation from water, especially in wet mountainous regions such as Norway.[7]

The rocks and gravel generated and moved by rivers are extensively used in construction. In parts of the world this can generate extensive new lake habitats as gravel pit re-fill with water. In other circumstances it can destabilise the river bed and the course of the river and cause severe damage to spawning fish populations which rely on stable gravel formations for egg laying.[8]

The beauty of rivers and their surroundings contributes to tourist income in many parts of the world from Shakespear's Avon to the wilds of Alaska's glacier streams.In upland rivers, rapids with whitewater or even waterfalls occur. Rapids are often used for recreation, such as whitewater kayaking.

Rivers have been important in determining political boundaries and defending countries. For example, the Danube was a long-standing border of the Roman Empire, and today it forms most of the border between Bulgaria and Romania. The Mississippi in North America and the Rhine in Europe are major east-west boundaries in those continents. The Orange and Limpopo Rivers in southern Africa form the boundaries between provinces and countries along their routes.

* * * * * * * * * * * * * * * * * * Explanations * * * * * * * * * * * * * * * * * *

* * * * * * * * * * * * * * * * * New Words and Expressions * * * * * * * * * * * * * * * *

1. watercourse [ˈwɔ: təkɔ: s] n. 水道，河道，航道
2. brook [bruk] n. 小溪
   vt. 容忍 [常用于否定句或疑问句]
3. rill [ril] n. 小河，小溪；
   vi. 小河般地流
4. geographic [dʒi: əˈgræfik] adj. 地理(学)的
5. vagueness [ˈveignis] n. 模糊，含糊，暧昧，茫然
6. snowpack [ˈsnəupæk] n. 积雪
7. braid[breid] vt. 混合，交错，编织，编结
8. peneplain [ˈpi:neplein] n. 准平原
9. anastomosing [əˈnæstəməuzing] n. 网状河流
10. sinuous [ˈsinjuəs] adj. 蜿蜒的
11. oxbow[ˈɑksˌboʊ] n.由河流中的马蹄形弯曲形成的环；牛轭的U形项圈
12. billabong [ˈbɪləˌbɔŋ] n.〈澳〉死河(指支流)，死水潭
13. torrential [təˈrenʃəl] adj. 汹涌的，奔流的，猛烈的
14. estuary [ˈɛstjʊ(ə)ri] n. 河口，潮汐河口，一条大河的潮汐河口
15. meander [miˈændə] vi. 蜿蜒而行
16. tectonics [tekˈtɔniks] n. 地质学
17. tributary [ˈtribjutəri] adj. 支流的，进贡的
18. recharge [ˈri: tʃɑ:dʒ] n.（水）补给
19. upriver [ˈʌpˈrivə] n. 向上游
20. downriver [daʊnˈrɪvər] n. 向下游
21. headwaters [ˈhedwɔ:tə] n. 源头，河源
22. geologic [dʒiəˈlɔdʒik] adj. 地质的，地质学的
23. ephemeral [iˈfemərəl] adj. 生命短暂的，朝生暮死的
24. lumberjack [ˈlʌmbədʒæk] n. 伐木工
25. edible aquatic 可食用的水生生物
26. foreshore way [fɔ:ˈʃædəu] n. 林荫路
27. grinding cereal 谷物研磨，切削麦片
28. watermill [ˈwɔ: təmil] n. 水磨坊
29. destabilise [di:ˈsteibiˌlaiz] vt. 使动摇
30. spawn [spɔ: n] vi. 产卵
31. whitewater kayaking 激流皮艇

* * * * * * * * * * * * * * * * * Complicated Sentences * * * * * * * * * * * * * * * * *

1. Water within a river is generally collected from precipitation through surface runoff, groundwater recharge, springs, and the release of stored water in natural ice and snowpacks (i.e. from glaciers).

【译文】流域内的水一般是汇集来的，包括降水、地表径流、地下水补给、泉水以及自然冰和积雪（即冰川）融化所形成的水。

【说明】"surface runoff, … snowpacks(i.e. from glaciers)." 都是介词 through 的宾语。

2. This distinction between river channel and floodplain can be blurred especially in urban areas where the floodplain of a river channel can become greatly developed by housing and industry.

【译文】漫滩和河槽的区别很模糊，特别是有些河道的漫滩可能成为人们定居和工业扩大

的城镇。

【说明】句中 where 引导定语从句，修饰前面的“Urban areas”。

3. In mountainous torrential zones this can be seen as erosion channels through hard rocks and the creation of sands and gravels from the destruction of larger rocks.

【译文】在山丘暴雨区，经常可以见到侵蚀的河道，就是由于水流穿过坚硬的岩石或冲击大块的岩石所造成的大量砂和砾石。

【说明】句中的“and”等同于“therefore”，且 the creation 前面省略了“there is”。

4. The way in which a river's characteristics vary between the upper course and lower course of a river is summarized by the Bradshaw model.

【译文】河流特征从上游到下游的变化方式可以用 Bradshaw 模型概括。

【说明】原文为 which 引导的从句，指代前面的 way (方式)。upper course 和 lower course 分别为“上游”和“下游”。

5. The earliest evidence of navigation is found in the Indus Valley Civilization, which existed in northwestern Pakistan around 3300 BC.

【译文】有关航运的最早记录是在印度河文明文献中发现的，它出现于公元前 3300 年左右的巴基斯坦西北边境。

【说明】which 引导的定语从句，修饰前面的 Indus Valley Civilization。

6. In some heavily-forested regions such as Scandinavia and Canada, lumberjacks use the river to float felled trees downstream to lumber camps for further processing, saving much effort and cost by transporting the huge heavy logs by natural means.

【译文】在一些森林茂密的地区，如斯堪的纳维亚半岛和加拿大，伐木工通过河流运输砍伐的树木到下游木材营地，再作进一步处理。这种通过自然途径运输巨大的重型原木，节省了很大的精力和成本。

【说明】原文 saving much effort and cost by transporting the huge heavy logs by natural means 为被动语态，但是按照汉语翻译习惯译为主动语态。

7. In the 1890s the first machines to generate power from river water were established at places such as Cragside in Northumberland and in recent decades there has been a significant increase in the development of large scale power generation from water, especially in wet mountainous regions such as Norway.

【译文】自从 1890 年在诺森伯兰的克拉格塞德地区首次制造了利用河水发电的机器设备，近几十年来出现了大规模水力发电，且在迅速增长，尤其是像挪威这样的山区水能资源丰富的国家。

【说明】原文中的长句是并列句式，用 and 连接，可以独立翻译，阅读时可分别理解。

8. In other circumstances it can destabilise the river bed and the course of the river and cause severe damage to spawning fish populations which rely on stable gravel formations for egg laying.

【译文】其他情况下，则会扰动河床和河道，并对那些依靠稳定的砾石层产卵的鱼类会造成很大危害。

【说明】原文中有两个 and，第一个 and 是连接两个并列的两个短语 the river bed 和 the

course of the river，第二个 and 是连接前后两个并列分句的；后一分句由 which 引导，修饰前面的 fish populations。

*** * * * * * * * * * * * * * * * * * Summary of Glossary * * * * * * * * * * * * * * * * * ***

1. floodplain 泛滥平原，涝原，漫滩
2. river channel 河槽
3. river bed 河床
4. upper course 上游
5. lower course 下游
6. source of the river 河源
7. mouth of the river 河口
8. navigation 航运
9. valley （山）谷，流域
10. groundwater recharge 地下水补给
11. saline tidal 潮水域

*** * * * * * * * * * * * * * * * * * Abbreviations (Abbr.) * * * * * * * * * * * * * * * * * ***

i.e. 拉丁文 id est，相当于 that is，in other words，即，也就是

*** * * * * * * * * * * * * * * * * * Exercises * * * * * * * * * * * * * * * * * ***

(1) A river is a natural __________, usually freshwater, flowing toward an ocean, a lake, a sea or another river.

(2) Small rivers may also be called by several other names, including __________, creek, __________, __________, and rill; there is no general rule that defines what can be called a river.

(3) The water in a river is usually confined to a __________, made up of a stream bed between banks.

*** * * * * * * * * * * * Word Building (7) alter-; circum- * * * * * * * * * * * ***

1. alter- [前缀]，表示：改变

alterable	adj.	可变更的
alternative	adj.	可替换的
alternating	adj.	交变的
alternation	n.	变化
alternator	n.	交流发电机

2. circum- [前缀]，表示：周围，环境

circumstance	n.	环境
circumferetial	adj.	环向的
circumsolar	adj.	围绕太阳的
circumfluence	n.	环向水流

******************* Text Translation *******************

第七章 河 流

第一节 概 述

河流是一个自然水道，通常是淡水，流向大洋、湖泊、大海或其他河流。少数情况下，河流在流入另一个水体之前，就全部下渗或完全枯竭。中小型河流会有数个名称，如溪流、小湾、小河、溪和细沟；河流的定义没有一般通用的规则。许多小河流因其地理位置而得名，举个例子，在苏格兰和英格兰东北部波恩附近的河流，叫波恩河。一般河流要比小溪大，但因为语言的含糊性，情况并非总是如此。（单词个数：113）

河流是水文循环的一部分。流域内的水一般是汇集降水，包括地表径流、地下水补给、泉水以及自然冰和积雪（即冰川）融化的水而形成的。（单词个数：38）

河里的水通常在一个两岸间河床形成的河槽内流动。较大的河流还会有由洪水漫过河槽所形成的更宽阔的漫滩。漫滩因河道的大小不同而不同，也可能会非常广阔。漫滩和河槽的区别很模糊，特别是有些河道的漫滩可能成为人们定居和工业扩大的城镇。（单词个数：79）

向上游是指河流的源头方向，与水流方向相对。同理，向下游是指河口方向，与水流方向相同。（单词个数：40）

河道里通常只有一支主流，但一些河流间是相通的，形成蜿蜒前进的辫状河。全世界范围内，大多数的辫状河流只在少数几个地区发现，例如在阿拉巴马州南部的大部分地区和新西兰南岛。在侵蚀平原和一些较大的河流三角洲地区也可以发现它们。网状河流和辫状河流比较相似，也十分罕见，它们蜿蜒曲折的河道通常携带有大量的泥沙。（单词个数：83）

作用于河道并且改变其形状和形式的河水流动是一种能量来源。据布拉姆定律（有时也称为艾里定律），被河水冲走物体的质量与河水流速的6次方成比例关系。因此，当流速增加2倍时，它的运输能力则增大64倍（即更大的规模）。在山区暴雨区，经常可以见到侵蚀的河道，就是由于水流穿过坚硬的岩石或冲击大块的岩石造成了大量砂和砾石。在U形冰川峡谷，往往可以很容易地通过侵蚀的V形河道来形成连续的河谷。在河流流过平原的中游，可能通过侵蚀河岸或河道内的大量沉积形成弯道。有时河流将截断水流循环，缩短河道（裁弯取直），形成了牛轭U形的弯管或死河。携带大量泥沙的河流在河口可能会有明显的三角洲。条件允许的情况下，河流会在有潮水域形成河口湾。（单词个数：197）

在河流的整个运动过程中，水流向下游携带的往往是由看到的自由水流和看不到的水流中的岩石沙砾所组成。对于许多大峡谷中的河流，这看不见的部分的流量可能大大超过有形流动。（单词个数：58）

第二节 分 类

虽然下面的分类对于河流的形象化是一种有效的方式，但还有许多其他因素在起作用。梯度主要受构造地质学控制，而泄流量主要由气候控制，泥沙量由气候、源头地质情况、河流梯度等多种因素控制。（单词个数：49）

1. 幼年河流

河流坡度大、支流少，且一般流速大。它的河道向下冲刷而不是向两边。[例如：布拉索斯河、圣三一河（也叫三合河）、埃布罗河]（单词个数：28）

2. 成年河流

比幼年河流的坡度稍缓，流速慢一些。成年河流一般有许多支流，且比年幼河流有更大的流量。河道向两边冲刷而不是向下。（例如：密西西比河、圣劳伦斯河、多瑙河流域、俄亥俄河、泰晤士河）（单词个数：53）

3. 老年河流

老年的河是指低坡度并且冲刷能量小的河。老年河流的特点是形成冲积平原。（例如：黄河、恒河、底格里斯河、幼发拉底河、印度河、尼罗河）（单词个数：30）

4. 再生河

河流的坡度由构造抬升形成。大多数河流从河源到河口的直线距离是实际长度的三分之一。河流特征从早期到晚期的变化方式，可以用Bradshaw模型概括。多数河流在地表上流动，而地下河流在洞穴或地下溶洞中流动。在石灰岩地质构造地区，可以很容易发现这种河流。（单词个数：81）

5. 间歇性河流（或季节性河流）

间歇性河流（或季节性河流）只偶尔流动并且一次可以干涸好几年。这些河流经常出现在降雨有限的地区，或降雨极不均匀地区，或可能因某种地质条件出现，如高渗透的河床。一些季节性河流只在夏季而不在冬季流动。冬季，这种河流通常由降水补给的白垩纪含水层来供水。在英国，这些河流称为Bournes，它们是由博内茅斯（Bournemouth）和伊斯特本（Eastbourne）这些地名而定名的。（单词个数：91）

第三节 用 途

河流已被用来作为水源、饮用水、运输载体、作为防御屏障、作为水电能源来运转机器，并且成为水解废物的一种手段。（单词个数：33）

河流用于航运已有数千年之久。有关航运最早的证据是在印度河发现的，它出现于公元前3300年左右的巴基斯坦西北边境。河流航运是最便宜的运输手段，至今仍广泛应用于世界

上的主要河流，像亚马逊河、恒河、尼罗河、密西西比河和印度河。（单词个数：61）

在一些森林茂密的地区，如斯堪的纳维亚半岛和加拿大，伐木工使用河流运输砍伐的树木到下游木材营地，再作进一步处理。通过使用这种自然途径运输巨大的重型原木，从而节省了很大的精力和成本。[6]（单词个数：39）

河流自古以来一直是重要的食物来源。它们可以提供丰富的鱼类和其他可食用的水生生物，也可作为饮用和灌溉淡水的主要来源。因此，世界大部分主要城市坐落于河流的两岸就不足为奇了。河流有助于确定城市和街区的形态，并且通过利用浅滩（如河边小路）来给城市的发展提供机会。在许多欠发达地区，河流还是一种简单处理废水和其他废物的载体。（单词个数：107）

快速水流和瀑布通过水磨坊和水力机组，而被广泛用作动力和水电能源。已有证据显示：水磨坊的使用已有几百年的历史了，如在 Orkneys，Dounby 的单机轧机。它早于蒸汽动力的发明，用水磨机研磨谷物和加工羊毛及其他纺织品，就更是遍及整个欧洲。自 1890 年代起，在诺森伯兰和克拉格塞德地区，就制造了利用河水发电的机器，近几十年来出现了大规模水力发电，而且还在迅速增加，尤其是像挪威这样的多水国家。[7]（单词个数：109）

河流产生并携带的岩石和砾石可以广泛应用于建筑。世界上一些地区，砂石坑重新填满水就形成了新的湖泊栖息地。其他情况下，扰动河床和河道对依靠稳定的砾石层产卵的鱼类也会造成很大危害。[8]（单词个数：64）

世界许多地区，从莎士比亚的埃文河畔到阿拉斯加的冰川河流，美丽的河流和周围的环境都为旅游收入作出了贡献。山地河流、急流甚至瀑布激流经常出现，因此，急流常常用于娱乐，如激流皮艇。（单词个数：46）

河流对于确定重要的政治边界和捍卫国家也有很大作用。例如，多瑙河是罗马帝国的一个长期的边界，如今它构成保加利亚和罗马尼亚的大部分边境。北美的密西西比河和欧洲的莱茵河是各自大陆的主要东西方的边界。非洲南部的奥兰治河和林波河构成沿河各省和各国的边界。（单词个数：70）

Chapter 8 Flood

Part 1 Introduction

A flood is an overflow of water, an example of water submerging land, a deluge.

In many arid regions of the world, the soil has very poor water retention characteristics, or the amount of rainfall exceeds the ground's ability to absorb water. When a rainfall does occur, it can sometimes result in a sudden flood of water filling dry streambeds known as a "flash flood".

Many rivers that flow over relatively flat land border on broad flood plains. When heavy rainfall or melting snow causes the river's depth to increase and the river to overflow its banks, a vast expanse of shallow water can rapidly cover the adjacent flood plain.

Floods are part of the natural cycle of things. The benefits of natural floods almost certainly outweigh the negative aspects.

In areas largely inhabited by people, there are both positive and negative environmental effects of flooding. Floods can distribute large amounts of water and suspended river sediment over vast areas. In many areas, this sediment helps replenish valuable topsoil components to agricultural lands and can keep the elevation of a land mass above sea level.[1] An example of the latter case is the Mississippi delta. Before the Mississippi and associated rivers were controlled in levees in southern Louisiana, the river would frequently spill their banks. These processes made the lands of the Mississippi delta. This area is slowly subsiding with time and without the continued replenishement of sediment from river floods, much of it has dropped to elevations below natural sea level.[2] Thus, one could say that not allowing floods is negative for this area. Our society has chosen instead to create a vast and complex system to keep Mississippi waters from reaching these lands. The lands remain dry but each year they subside more, making it ultimately more and more difficult to keep that way.

On the negative side, floods make an enormous impact on the environment and society. Floods destroy drainage system in cities, causing raw sewage to spill out into bodies of water. Also, in cases of severe floods, buildings can be significantly damaged and even destroyed. This can lead to catastrophic effects on the environment as many toxic materials such as paint, pesticide and gasoline can be released into the rivers, lakes, bays, and ocean, killing maritime life. Floods may also cause millions of dollars worth of damage to a city, both evicting people from their homes and ruining businesses. Floods cause significant amounts of erosion to coasts, leading to more frequent flooding if not repaired.

Part 2 Types of Flooding

1. Coastal flooding

Hurricanes and tropical storms can produce heavy rains, or drive ocean water onto land. Beaches and coastal houses can be swept away by the water. Coastal flooding can also be produced by sea waves called tsunamis, giant tidal waves that are created by volcanoes or earthquakes in the ocean.[3]

2. River flooding

Flooding along river is a natural event. Some floods occur seasonally when winter snows melt and combine with spring rains. Water fills river basins too quickly, and the river will overflow its banks. Often the land around a river will be covered by water for miles around.

3. Urban flooding

As land is converted from fields or woodlands to roads and parking lots, it loses its ability to absorb rainfall. Urbanization increases runoff 2 to 6 times over what would occur on natural terrain. During periods of urban flooding, streets can become swift moving rivers, while basements can become death traps as they fill with water.[4]

4. Flash floods and arroyos

An arroyo is a water-carved gully or a normally dry creek found in arid or desert regions. When storms appear in these areas, the rain water cuts into the dry, dusty soil creating a small, fast-moving river. Flash flooding in an arroyo can occur in less than a minute, with enough power to wash away sections of pavement.

Part 3 Cause of Flood

Too much rain in one area causes the water to rise to a dangerous level. The causes of flooding include full reservoirs; ice-covered rivers, frozen ground, saturated soil, and so on.

1. Deep snow melt

Deep snow melts into large amount of water. Deep snow rarely causes flooding by itself. So, look for heavy rain and rapid heating, combined with rapidly melting snow.

2. Rain or frozen or melt Soil

Frozen soil cannot absorb as much as unfrozen soil. Rain or rapid snow melt on top of frozen soil can cause more flooding than if the soil were not frozen. Saturated soil can't absorb rain and water from melting snow. The excess water becomes runoff and rapidly flows into rivers and streams. Unsaturated soil acts like a sponge, absorbing some of the water from rain or snow melt. Usually, heavy rain or rapid snow melt combined with saturated soil causes flooding most frequently.

3. Full reservoirs

Reservoirs can alleviate river flooding by absorbing the water from the river. This would reduce the height in which the water rises downstream of the reservoir. If the reservoir is already full, then it cannot absorb any water from dense rivers and thus causes flooding.

4. Ice dams

As rain or melting snow fills the river, ice at the surface cracks and breaks up into chunks that float downstream. These chunks of ice can form a dam as they run into barriers (bridges) along the rivers. The ice dams cause water to rise rapidly behind them. If the dam suddenly breaks, water can also flood downstream.

5. Widespread heavy rain

This is perhaps the most important and influential factor of them all. Long period of heavy rain can cause flooding even if all other factors are unfavorable for flooding. Often, heavy rain is a cause of some of the factors listed above such as wet soil, high stream levels and full reservoirs. The Midwest flooding during the summer of 1993 and the Southeast flooding caused by the remnants of Tropical Strom Alberto in 1994 are a couple of examples of flooding caused by heavy rain.

6. High river and stream levels

Streams or rivers that are already at bankfull can be a precursor to major flooding. Heavy rain or rapid snow that flows into an already full river will cause the river to overflow its banks and flood nearby locations. High river levels, such as those in the Ohio Valley in the spring of 1997, make forecasters very nervous. A prolonged dry spell, however, can alleviate flooding concerns.

7. Monsoon rainfalls

Monsoon rainfalls can cause disastrous flooding in some equatorial countries, such as Bangladesh, due to their extended periods of rainfall.

8. Hurricanes

Hurricanes have a number of different features which, together, can cause devastating flooding. One is waves of up to 8 meters high, caused by the leading edge of the hurricane when it moves from sea to land. Another is the large amounts of precipitation associated with hurricanes. The eye of a hurricane has extremely low pressure, so sea level may rise a few meters in the eye of the storm.

9. Melting snow

Under some rare conditions associated with heat waves, flash floods from quickly melting mountain snow has caused loss of property and life.

10. Undersea earthquakes

Undersea earthquakes, eruptions of island volcanoes may all engender a tidal wave that causes destruction to coast areas.

Part 4 Variability of Flooding

The flow of water that streams must carry varies day to day, season to season and year to year.[5] Sometimes there are long spells of little or no rain in an area and flows slow to a trickle. At other times, the same area might have a wet period in which one storm follows another. Floods vary in size depending on such things as the intensity of rain, the area over which the rain falls, the rate at which snow melts, and/or other factors.[6]

The amount of previous rain on a watershed or the current storage in a reservoir may also play a major role in the potential for flooding. Heavy rain on a dry watershed may not result in any flooding while just a small amount of rain on an already saturated ground may cause a flood. Sometimes the several factors combine to cause only a minor rise in the areas' streams and, at other times, combine to cause destructive floods. The occurrence of factors that determine the size of floods is largely a random one. Large floods can occur at any time and may often be impossible to predict more than a short time ahead of the event.

1. Frequency of flooding

There's no way to predict when the next flood will come or how big it will be. However, past flooding gives some clues as to what to expect.

Engineers studying past floods use statistics to estimate the chance that floods of various sizes will occur. For example: A relatively common small flood of a certain size might be expected from experience to occur 33 times over a 100-year period. It would be expected to happen on an average of once every three years or, said another way, have a 1 in 3 or 33% chance of happening in any particular year. It would be called the 3-year flood or the 33 percent chance flood.

A large (more unusual, less frequent) flood found to occur on the average of 10 times in 100 years would be called the 10 percent chance flood or the 10-year flood.

The flood so large and unusual that it only occurs on the average of once every hundred years would have a one percent chance of occurring in any particular year and be called the 100-year flood or 1 percent chance flood.

This doesn't mean of course that a 10 percent flood occurs exactly once every 10 years. A rainy year might have several 10 percent floods and then might not be another for many years. Similarly, two or more large floods, like the 100-year flood or even the 500-year flood could occur back to back. The percentage chance of a flood occurring is based on the average of what is expected over a long period of time.

The chance of a flood of a certain size occurring and then the same or bigger flood happening right away is like flipping a coin.[7] Just because heads come up doesn't mean that the next try has to be tails. Each time the coin is flipping there is a 50-50 chance for either heads or tails. In the same way, when one flood has passed, the chances are reset. A 1 percent flood has a 1percent chance of occurring in any one year. And, as soon as it does happen, the chances are still 1 percent that it will occur again sometime during the following 365 days.

2. Flood plains

Flood plains are low areas subject to flooding from time to time. Most flood plains are adjacent to streams, lakes or oceans although almost any area can flood under the right circumstance. The amount of land inundated by a flood depends on the flood's magnitude. Approximately seven percent of the nation's land area almost as big as Texas is subject to severe flooding.

The action of water on the land and interaction of water with vegetation produce flood plains which differ appreciably from one another and from uplands in their soils, drainage systems and vegetation. Beaches and small river valleys are usually easily recognizable as flood plains to people with a trained eye. Less obvious flood plains occur in dry washes and on alluvial fans in arid parts of the western United States, around prairie potholes, in areas subject to high groundwater levels, and in low lying areas where water may accumulate.

Flood plains are designated by the size of the flood that will cover them. For example, the 10-year flood plain is the land covered by the 10-year flood and the 100-year flood plain is the land covered by the 100-year flood. The likelihood of property being flooded varies depending on how high it is above the stream. Buildings on the 10-year flood plain can be expected to flood on the average of once every 10 years while buildings on higher ground will be flooded less often.

* * * * * * * * * * * * * * * * * Explanations * * * * * * * * * * * * * * * * *

* * * * * * * * * * * * * * * New Words and Expressions * * * * * * * * * * * * * *

1. overflow [ˈəuvəfləu] n. 泛滥，溢出；溢流口
   v. 淹没，泛滥；充满；溢出
2. deluge [ˈdelju: dʒ] n. 洪水，暴雨
   v. 使泛滥，淹，浸，压倒
3. arid region 干旱地区
4. flood plain 洪泛平原，洪泛区，洪积平原
5. shallow water 浅水
6. adjacent [əˈdʒeisnt] adj. 邻近的，接近的
7. inhabit [inˈhæbit] vt. 居住于，存在于，占据，栖息
8. elevation [ˌeliˈveiʃən] n. 海拔，上升，高地，正面图，提高，仰角
9. levee [ˈlevi] n. 防洪堤，码头，大堤
   v. 筑防洪堤于
10. replenishment [riˈpleniʃmənt] n. 补给，补充
11. subside [səbˈsaid] v. 下沉，沉淀，平息，减退，衰减
12. drainage [ˈdreinidʒ] n. 排水，排泄，排水装置，排水区域，排出物，消耗
13. sewage [ˈsju: idʒ] n. 脏水，污水
14. bay [bei] n. 海湾；狗吠声；绝路
15. maritime life 海生物
16. coastal [ˈkəustl] adj. 海洋的，海岸的，沿海的，沿岸的
17. hurricane [ˈhʌrikən] n. 飓风
18. sea wave 海浪
19. tsunami [tsju: ˈnɑ:mi] n. 海啸
20. volcano [vɔlˈkeinəu] n. 火山
21. earthquake [ˈə: θkweik] n. 地震，[喻]在震荡，在变动
22. swift [swift] adv. 迅速地，敏捷地
    n.[鸟] 雨燕，（梳棉机等的）大滚筒
    adj. 迅速的，快的，敏捷的，立刻的
23. saturated soil 饱和土壤，饱水土壤
24. sponge [spʌndʒ] n. 海绵，海绵状物，（医）棉球，纱布
    v. 用海绵等洗涤、擦拭或用海绵吸收（液体），依赖某人生活
25. alleviate [əˈli: vieit] vt. 减轻，使（痛苦等）易于忍受
26. float [fləut] n. 漂流物，浮舟，漂浮，
    vi. 漂流，浮动，漂浮
    vt. 使漂浮，容纳，淹没
27. tropical [ˈtrɔpikl] adj. 热带的
28. bankfull [bæŋkfull] n. 满水时期
    adj. 水位齐岸的

29. forecaster [fɔ:ˈkɑ:stə] n. 预报员
30. prolonged [prəˈlɔŋd] adj. 持续很久的，长时期的，长时间的
31. equatorial [ˌekwəˈtɔ: riəl] n. 赤道仪
adj. 赤道的，近赤道的
32. sea level 海拔，海平面
33. eruption [iˈrʌpʃən] n. 爆发，火山灰；[医] 出疹
34. tidal wave 潮汐波，浪潮
35. snows melt 融雪
36. destructive [diˈstrʌktiv] adj. 破坏（性）的
37. statistics [stəˈtistiks] n. 统计数据，统计学，统计资料，统计数字
38. particular [pəˈtikjulə] n. 细节，详细
adj. 特定的，特殊的，特别的，详细的，精确的，挑剔的
39. percentage [pəˈsentidʒ] n. 百分率
40. inundate [ˈinəndeit] v. 淹没，浸水，泛滥
41. approximately [əˈprɔksimətli] adv. 近似地
42. interaction [ˌintərˈækʃən] n. 相互作用，相互影响
43. upland [ˈʌplənd] n. 高地，丘陵地，丘阜
adj. 高地的，山地的
44. valley[ˈvæli] n. 山谷，流域，溪谷
45. recognizable [ˈrekəgnaizəbl] adj. 可辨认可认知的，可承认的
46. accumulate [əˈkju: mjuleit] vi. 积聚，累积，堆积
47. likelihood [ˈlaiklihud] n. 可能，可能性

****************** Complicated Sentences ******************

1. In many areas, this sediment helps replenish valuable topsoil components to agricultural lands and can keep the elevation of a land mass above sea level.

【译文】在许多农田地区，这些沉淀物有助于补充有益的表层土成分，并使大面积土地保持在海平面以上。

【说明】and 连接前后两个动词 helps 和 can keep。

2. This area is slowly subsiding with time and without the continued replenishment of sediment from river floods, much of it has dropped to elevations below natural sea level.

【译文】随着时间的推移，该地区没有了来自河流洪水的沉淀物的持续补充，就慢慢下陷，大部分地方甚至下陷到海平面以下。

【说明】句子中的 it 指代 this area。

3. Coastal flooding can also be produced by sea waves called tsunamis, giant tidal waves that are created by volcanoes or earthquakes in the ocean.

【译文】海岸洪水也会因为由海底火山或地震造成巨大的潮浪（海啸）而发生。

【说明】called 引导后面分词短语作定语，修饰 sea waves；that 也是定语从句修饰 tidal waves。

4. During periods of urban flooding, streets can become swift moving rivers, while basements can become death traps as they fill with water.

【译文】当城市发生洪灾时，街道里充满了快速流动的水流，而建筑物底部充满水时就会成为死亡陷阱。

【说明】此句中的 while 表达了转折的含义，后面的 as 引导的条件状语从句。

5. The flow of water that streams must carry varies day to day, season to season and year to year.

【译文】河水流量日复一日、季复一季、年复一年地发生变化。

【说明】注意句中 day to day 的构词方式。

6. Floods vary in size depending on such things as the intensity of rain, the area over which the rain falls, the rate at which snow melts, and/or other factors.

【译文】洪灾规模不同，取决于降雨强度、降雨面积、融雪快慢和（或）其他因素。

【说明】句中 over which the rain falls 为定语从句，修饰 the area; at which snow melts 为定语从句，修饰 the rate。

7. The chance of a flood of a certain size occurring and then the same or bigger flood happening right away is like flipping a coin.

【译文】一定规模的洪灾发生后，紧接着有同样或更大洪灾发生的概率就像是掷硬币。

【说明】句中 the same 指代 flood。

********************* Summary of Glossary *********************

1. flood	洪水
2. storm	暴风雨
3. hurricane	飓风
4. tsunami	海啸
5. volcano	火山
6. earthquake	地震
7. vegetation	植被
8. sediment	沉淀物, 沉积
9. downstream	下游的，向下游
10. severe flood	大洪水
11. the Mississippi delta	密西西比三角洲
12. toxic material	有毒物质
13. coastal flooding	海潮，风暴潮，沿海洪水
14. flash flooding	暴洪，山洪暴发
15. tropical storm	热带风暴
16. water-carved gully	水流冲成的沟渠
17. river basin	流域
18. devastating flooding	灾难性洪水，洪灾
19. heavy rain	暴雨，大雨
20. melting snow	融雪
21. undersea earthquake	海底地震
22. chance flood	洪水频率

23. arid part 干旱地区
24. current storage （水库）当前蓄水量
25. size of flood 洪水规模

* Exercises *

(1) When a _________ does occur, it can sometimes result in a sudden _________ of water filling dry streambeds known as a _________.

(2) When heavy _________ or _________ causes the river's depth to increase and the river to _________ its banks, a vast _________ of shallow water can rapidly cover the adjacent flood plain.

(3) Floods may also cause millions of dollars' worth of _________ to a city, both evicting people from their _________ and ruining _________.

* * * * * * * * * * * * Word Building (8) -age; magni- * * * * * * * * * * * *

1. -age [名词词尾]，接在数量单位名词后表示：数量的总和，集合

mile	n.	mileage	n.	英里数，英里里程
ton	n.	tonnage	n.	吨位；总吨数
drain	v.	drainage	n.	排水量

-age [名词后缀]，表示：状态，行为或其结果

leak	v.	leakage	n.	漏，泄漏，渗漏
link	v.	linkage	n.	连接
pack	v.	package	n.	包裹，包
pass	v.	passage	n.	通过，经过，通道
short	adj.	shortage	n.	短缺，匮乏；缺乏之量
use	vt.	usage	n.	使用，用法
gate	n.	gatage	n.	闸门开度
sew	v.	sewage	n.	污水，下水道
sull	v.	sullage	n.	污泥，淤泥

2. magni- [前缀]，表示：大，长

-fication	magnification	n.	扩大，放大倍率
-ficent	magnificent	adj.	华丽的，高尚的，宏伟的
-fy	magnify	vt.	放大，扩大，赞美，夸大，夸张
-tude	magnitude	n.	大小，幅度；数量，巨大，量级

***************** Text Translation *****************

第八章 洪 水

第一节 概 述

洪水指水的泛滥，如淹没陆地的浩瀚水面，或大面积的洪水。（单词个数：15）

世界上的许多干旱地区，土壤保持水的能力很差，或者降雨量超出地面吸水能力。一旦发生降雨，有时会导致流入干涸河川的水突然泛滥，称为“暴涨的洪水”。（单词个数：50）

许多在相对平坦陆地上流动的河流一般会与广阔的泛滥平原毗连。大雨或融雪使河水水位上升，河水泛滥溢出河堤时，浩瀚的水面会迅速漫过邻近的洪泛平原。（单词个数：45）

洪水是事物自然循环的一部分，天然洪水的好处会超过其坏处。（单词个数：20）

在居住人口很多的地区，洪水既有正面影响，也有负面的环境影响。洪水可将大量的水和河流悬浮的沉淀物大面积地重新分配。在许多农田地区，这种沉淀物有助于补充有益的表层土成分，能够使大面积土地保持在海平面以上。密西西比河三角洲就是一个例子。在南路易斯安那州的防洪堤之中，控制了密西西比河及相关河流，这些河流经常溢出堤岸，最终形成了密西西比河三角洲。但随着时间的推移，该地区没有了来自河流洪水的沉淀物的持续补充，地面慢慢下陷，部分地方甚至下陷至海平面以下。因此，人们会说，没有洪水对该地区不利。而社会做出的通常选择是，要创建一个大而复杂的系统来阻止密西西比河的水流向这些地方，从而使这些地方仍然干旱，但每年下陷更多了，最后，使之保持原状也越来越难。（单词个数：175）

不利的一面表现在：洪水对社会和环境有巨大的影响。洪水破坏城市的排水系统，造成未经处理的污水流入水体。同样，如果洪水严重，建筑物会受到严重破坏甚至摧毁。这可能导致对环境灾难性的影响，因为许多像油漆、杀虫剂和汽油这样有毒的物质会释放到河流、湖泊、海湾和海洋，伤害海洋生物。洪水也会对城市造成数以百万美元的破坏，使人们无家可归，破坏商业。洪水造成海洋大量的侵蚀，如不进行修复，还会导致更多更频繁的洪水。（单词个数：114）

第二节 洪 水 种 类

1. 海洋洪水

飓风和热带风暴会造成暴雨或使海洋的水流到陆地。海滩和海边房屋会被水冲走。海洋洪水也会因为由海底火山或地震造成巨大的海浪（海啸）而引起。（单词个数：49）

2. 河流洪水

河流洪水比较普遍。当冬天的雪融化与春天的雨水共同作用时，就会发生一些季节性洪

水。水很快填入河槽，河水会漫过河堤。通常，河流周围几千米地方的土地都被水淹没。（单词个数：47）

3. 城市洪水

当土壤从农田或林区转变成道路和停车场时，其吸收雨水的能力就消失了。城市化使得径流与天然地面相比增大了 2～6 倍。当城市发生洪灾时，街道里充满了迅速流动的水流，而建筑物地下室充满水时，就会成为死亡陷阱。（单词个数：56）

4. 暴涨的洪水和小溪

小溪是指干旱或沙漠地区才可见到的水冲刷出来的集水沟，或一般干涸的小河。当这些地方发生暴雨时，雨水钻进干燥的泥土飞扬的土壤，形成小的快速流动的小河。在小溪上，暴涨的洪水不到 1 分钟就可以发生，也有足够的力量冲走人行道的某些地方。（单词个数：58）

第三节 洪 水 的 起 因

某个地区过多的雨水会使水上升到危险水位。发生洪水的原因包括水库爆满、河流冰盖、地面冻结和土壤饱和等。（单词个数：30）

1. 深雪融化

深雪融化成大量的水。深雪本身一般不会形成洪水。只有在下大雨时，温度迅速上升，再加上雪迅速融化时才会发生。（单词个数：28）

2. 雨水或冻结或潮湿的土壤

冻结的土壤比不冻结的土壤吸收的水分少。和不冻结土壤相比，雨水或冻结土壤中迅速融化的雪更容易造成洪水。饱和的土壤不能吸收雨水和雪融化的水，多余的水就变成径流并很快流入河流和小溪。未饱和的土壤就像海绵，吸收一些雨水或雪融化的水。通常，大雨或雪迅速融化加上土壤饱和会频繁地造成洪水。（单词个数：83）

3. 水库漫坝

水库可暂时容纳河流的水，以减轻河流的洪灾。水库下泄流量加大时则可降低其水位。如果水库已经爆满，它就不能容纳河流中的任何水，因而产生洪灾。（单词个数：45）

4. 冰坝

雨水或融化的雪流进河流时，表面的冰破裂变成碎块随河水漂流。当这些碎块冰顺着河流流到障碍物（桥梁）时，就形成冰坝。冰坝使水位迅速上升。如果冰坝突然劈裂，也会造成洪灾。（单词个数：59）

5. 大面积暴雨

这也许是最重要、最有影响的洪灾因素。即使其他因素不造成洪水，长时间的大雨也会造成洪水。大雨通常也是发生上述列出洪灾因素的原因，如湿土、高水位和水库漫坝等。1993 年夏天中西部的洪水和 1994 年热带风暴艾伯托的残余风暴造成的东南部的洪水，是大雨造成洪灾的两个例子。（单词个数：88）

6. 河流和溪流高水位

堤坝已满的溪流或河流可能是大部分洪水的先兆。流入已满河流的大雨或快速融雪会使河水溢出堤坝淹没周围地区。如 1997 年春天俄亥俄州流域的高水位预报，使得人非常紧张。较长时间的持续干旱，可减轻对洪水的担忧。（单词个数：72）

7. 台风降雨

由于降雨持续时间长，台风降雨会造成一些赤道附近的国家，如孟加拉国灾难性的洪灾。（单词个数：23）

8. 飓风

飓风有许多不同特点，这些特征同时发生，则会造成破坏性的洪灾。一个是飓风从海洋移向陆地时，其漩涡中心可引起的高达 8m 的浪；另一个是与飓风有关的大量的降雨。飓风中心气压相当低，因此，处于飓风眼的海平面可能会升高几米。（单词个数：73）

9. 雪融

当一些罕见的与热浪有关的情况发生时，山上的雪迅速融化而造成的山洪暴发，可能导致财产和人员的伤亡损失。（单词个数：26）

10. 海啸

海底地震、岛屿火山喷发都会产生海啸，造成对沿海地区的破坏。（单词个数：22）

第四节 洪水泛滥的多变性

河水流量日复一日、季复一季、年复一年地发生变化。有时候某个地区长时间少雨或无雨，水流量很少。而其他时间，同样还是这个地区或许出现汛期，暴雨接二连三。洪灾规模也不相同，取决于雨量多少、降雨面积、融雪快慢或其他因素。（单词个数：85）

流域前期降雨量或水库的储水量也是引起洪水泛滥的原因。干旱流域的大雨也许不会引起任何洪灾，但地下水已饱和的流域，也许很少的雨量却会造成洪灾。有时候，几种因素共同作用，却只会使河水上涨一点点；而其他时间、共同作用却会造成破坏性的洪灾。决定着洪灾大小的各种因素很不确定。严重的洪灾可能随时发生，且常常在发生前不可预测。（单词个数：116）

1. 洪灾泛滥频率

虽然我们无法预测下次洪灾何时会发生，或规模有多大。但是，可根据过去发生的历史洪灾总结一些经验。（单词个数：28）

对过去的历史洪水进行研究的工程师利用统计数据，计算各种规模大小的洪灾发生的概率。例如：凭经验，某一规模相对较小的洪灾或许一百年会发生 33 次。换句话说，平均每三年发生一次，或平均 3 年发生 1 次或每年有 33%的概率发生。可叫作 3 年一遇的洪水或洪灾的概率为 33%。（单词个数：86）

严重的（很不平常，频率低）洪灾平均每一百年发生 10 次，叫作洪灾概率为 10%或平均 10 年发生一次的洪灾。某些特定的年份里，也许平均一百年只发生 1 次的稀遇的、频率低的洪灾，其概率为 1%，叫作百年一遇洪水，或洪灾概率为 1%。（单词个数：72）

当然，这并不是说 10%概率的洪灾正好每 10 年发生一次。雨水多的年份也许这种洪灾会发生好几次，也有可能多年也不发生一次。同样，两个或更多严重的洪灾，如 100 年一遇或甚至 500 年一遇的洪灾也许会接连发生。洪灾发生的百分比概率取决于较长一段时间里的平

均值。(单词个数：77)

某一规模的洪灾发生后，紧接着有同样或更大洪灾发生的概率，就像是掷硬币。出现正面，并不意味着下次一定是反面。每掷一次硬币，正面与反面出现的概率都为 50%。同样，某个洪灾发生了，其发生的概率也是如此。百年一遇的洪灾任何一年都有 1%发生的概率。一旦真的发生，下个一年的 365 天里仍然有 1%的发生概率。(单词个数：108)

2. 洪泛平原

泛滥平原是指那些地势较低、并随时会出现水泛滥的地区。虽然几乎任何一个地方都有可能出现洪水泛滥，但多数洪泛平原毗邻河流、湖泊或海洋。受洪水淹没的陆地面积大小取决于洪水的大小。美国全国约有 7%的陆地，相当于一个得克萨斯州，有较大的发生洪灾的概率。(单词个数：61)

水对陆地的作用和水与植物的相互作用造成了洪泛平原。这些洪泛平原各自大不相同，高地上的土壤、排水系统和植物都不一样。通常，人们凭经验很容易识别出海边和小河上的洪泛平原。一些不太明显的洪泛平原，如在美国西部干旱地区的干洼地和冲积扇、大草原周围的河床、地下水位较高的地方、积水的低地上等，都可能看到。(单词个数：86)

洪泛平原的范围由淹没它们的洪水的大小决定。例如，10 年一遇的洪泛平原是指可能平均 10 年淹没一次的陆地，而 100 年一遇的洪泛泛平原是可能平均 100 年淹没一次的陆地。受洪水淹没的财产取决于其位于河面之上有多高。10 年一遇洪泛平原上的建筑物可能平均 10 年会受到一次洪水淹没，而位于较高地面上的建筑物就较少受到淹没。(单词个数：82)

Chapter 9 Water Pollution

Part 1 Global Water Pollution

Estimates suggest that nearly 1.5 billion people lack safe drinking water and that at least 5 million deaths per year can be attributed to waterborne diseases. With over 70 percent of the planet covered by oceans, people have long acted as if these very bodies of water could serve as a limitless dumping ground for wastes. Raw sewage, garbage, and oil spill have begun to overwhelm the diluting capabilities of the oceans, and most coastal waters are now polluted. Beaches around the world are closed regularly, often because of high amounts of bacteria from sewage disposal, and marine wildlife is beginning to suffer.

Perhaps the biggest reason for developing a worldwide effort to monitor and restrict global pollution is the fact that most forms of pollution do not respect national boundaries. The first major international conference on environmental issues was held in Stockholm, Sweden, in 1972 and was sponsored by the United Nations (UN). This meeting, at which the United States took a leading role, was controversial because many developing countries were fearful that a focus on environmental protection was a mean for the developed world to keep the undeveloped world in an economically subservient position.[1] The most important outcome of the conference was the creation of the United Nations Environmental Program (UNEP).

Clearly, the problems associated with water pollution have the capabilities to disrupt life on our planet to a great extent. Congress has passed laws to try to combat water pollution thus acknowledging the fact that water pollution is, indeed, a serious issue. But the government alone cannot solve the entire problem. It is ultimately up to us, to be informed, responsible and involved when it comes to the problems we face with our water. We must become familiar with our local water resources and learn about ways for disposing harmful household wastes so they don't end up in sewage treatment plants that can't handle them or landfills not designed to receive hazardous materials.[2] In our yards, we must determine whether additional nutrients are needed before fertilizers are applied, and look for alternatives where fertilizers might run off into surface waters. We have to preserve existing trees and plant new trees and shrubs to help prevent soil erosion and promote infiltration of water into the soil. Around our houses, we must keep litter, pet waste, leaves, and grass clipping out of gutters and storm drains. These are just a few of many ways in which we, as human, have the ability to combat water pollution. If these measures are not taken and water pollution continues, life on earth will suffer severely.

Part 2 Cause of Pollution

Many causes of pollution including sewage and fertilizers contain nutrients, such as nitrates and phosphates. In excess levels, nutrients over stimulate the growth of aquatic plants and algae. Excessive growth of these types of organisms consequently clogs our waterways, use up dissolved oxygen as they decompose, and block light to deeper waters. This, in turn, proves very harmful to aquatic organisms as it affects the respiration ability of fish and other invertebrates that reside in water.

Pollution is also caused when silt and other suspended solid, such as soil, wash off plowed fields, construction and logging sites, urban areas, and eroded river banks when it rains.[3] Under natural conditions, lakes, rivers, and other water bodies undergo eutrophication, an aging process that slowly fills in the water body with sediment and organic matter. When these sediments enter various bodies of water, fish respiration becomes impaired, plant productivity and water depth become reduced, and aquatic organisms and their environments become suffocated. Pollution in the form of organic enters waterways in many different forms as sewage, as leaves and grass clippings, or as runoff from livestock feedlots and pastures. When natural bacteria and protozoan in the water break down this material, they begin to use up the oxygen dissolved in the water. Many types of fish and bottom-dwelling animals cannot survive when levels of dissolved oxygen drop below two to five parts per million. When this occurs, it kills aquatic organisms in large numbers which leads to disruptions in the food chain.

Part 3 Classifying Water Pollution

The major sources of water pollution can be classified as municipal, industrial, agricultural and additional forms of water pollution.

1. Municipal water pollution

Municipal water pollution consists of waste water from homes and commercial establishments. For many years, the main goal of treating municipal wastewater was simply to reduce its content of suspended solids, oxygen-demanding materials, dissolved inorganic compounds, and harmful bacteria. In recent years, however, more stress has been placed on improving means of disposal of the solid residues from the municipal treatment processes. The basic methods of treating municipal wastewater fall into three stages: primary treatment, including grit removal, screening, grinding, and sedimentation; secondary treatment, which entails oxidation of dissolved organic matter by means of using biologically active sludge, which is then filtered off; and tertiary treatment, in which advanced biological methods of nitrogen removal and chemical and physical methods such as granular and activated carbon absorption are employed. The handling and disposal of solid residues can account for 25 to 50 percent of the capital and operational costs of a treatment plant.

2. Industrial water pollution

The characteristics of industrial waste waters can differ considerably both within and among industries. The impact of industrial discharges depends not only on their collective characteristics, such as biochemical oxygen demand and the amount of suspended solids, but also on their content of specific inorganic and organic substances. Three options are available in controlling industrial wastewater. Control can take place at the point of generation in the plant; wastewater can be pretreated for discharge to municipal treatment sources; or wastewater can be treated completely at the plant and either reused or discharged directly into receiving waters. Wastewater treatment raw sewage includes waste from sinks, toilets, and industrial processes. Treatment of the sewage is required before it can be safely buried, used, or released back into local water systems. In a treatment plant, the waste is passed through a series of screens, chambers, and chemical processes to reduce its bulk and toxicity. The three general phases of treatment are primary, secondary, and tertiary. During primary treatment, a large percentage of the suspended solids and inorganic material is removed from the sewage. The focus of secondary treatment is reducing organic material by accelerating natural biological processes. Tertiary treatment is necessary when the water will be reused; 99 percent of solids are removed and various chemical processes are used to ensure the water is as free from impurity as possible.[4]

3. Agricultural water pollution

Agriculture, including commercial livestock and poultry farming, is the source of many organic and inorganic pollutants in surface waters and groundwater. These contaminants include both sediment from erosion cropland and compounds of phosphorus and nitrogen that partly originate in animal wastes and commercial fertilizes.[5] Animal wastes are high in oxygen-demanding material, nitrogen and phosphorus, and they often harbor pathogenic organisms. Wastes from commercial feeders are contained and disposed of on land; their main threat to natural waters, therefore, is from runoff and leaching. Control may involve settling basins for liquids, limited biological treatment in aerobic or anaerobic lagoons, and a variety of other methods.

4. Additional forms of water pollution

These are three additional forms of water pollution existing in the forms of petroleum, radioactive substances, and heat. Petroleum often pollutes water bodies in the form of oil, resulting from oil spills. These large-scale accidental discharges of petroleum are an important cause of pollution along shore lines. Besides the supertankers, off-shore drilling operations contribute a large share of pollution. One estimate is that one ton of oil is spilled for every million tons of oil transported. This is equal to about 0.0001 percent. Radioactive substances are produced in the form of waste from nuclear power plants, and from the industrial, medical, and scientific use of radioactive materials. Specific forms of waste are uranium and thorium mining and refining. The last form of water pollution is heat. Heat is a pollutant because increased temperatures result in the deaths of many aquatic organisms. These decreases in temperatures are caused when a discharge of cooling water by factories and power plants occurs.

Part 4 Effects and Danger of Water Pollution

The effects of water pollution are not only devastating to people but also to animals, fish, and birds. Polluted water is unsuitable for drinking, recreation, agriculture, and industry. It diminishes the aesthetic quality of lakes and rivers. More seriously, contaminated water destroys aquatic life and reduces its reproductive ability.[6] Eventually, it is a hazard to human health. Nobody can escape the effects of water pollution.

Virtually all water pollutants are hazardous to humans as well as lesser species; sodium is implicated in cardiovascular disease, nitrates in blood disorders. Mercury and lead can cause nervous disorders. Some contaminants are carcinogens. DDT is toxic to humans and can alter chromosomes. PCBs cause liver and nerve damage, skin eruptions, vomiting, fever, diarrhea, and fetal abnormalities. Along many shores, shellfish can no longer be taken because of contamination by DDT, sewage, or industrial wastes.

Dysentery, salmonellosis, cryptosporidium, and hepatitis are among the maladies transmitted by sewage in drinking and bathing water. In the United States, beaches along both coasts, riverbanks, and lake shores have been ruined for bathers by industrial wastes, municipal sewage, and medical waste. Water pollution is an even greater problem in the Third World, where millions of people obtain water for drinking and sanitation from unprotected streams and ponds that are contaminated with human waste.[7] This type of contamination has been estimated to cause more than 3 million deaths annually from diarrhea in Third world countries, most of them children.[8]

* * * * * * * * * * * * * * * * * Explanations * * * * * * * * * * * * * * * * *

* * * * * * * * * * * * * * * * New Words and Expressions * * * * * * * * * * * * * * * *

1. waterborne [ˈwɔ: təbɔ: n] adj. 水传播的，水上的，水运的
2. dumping [ˈdʌmpiŋ] n. 倾倒，倾销
3. overwhelm [ˈəuvəˈwelm] vt. 淹没，覆没，打击，制服，压倒
4. dispose [disˈpəuz] v. 处理，处置，销毁
5. marine [məˈri: n] adj. 海的，海产的，航海的，船舶的，海运的
   n. 舰队，水兵，海运业
6. issue [ˈisju:] n. 出版，发行，（报刊等）期，论点，问题，结果，（水、血等）流出
   vi. 发行，造成…结果，进行辩护
   vt. 使流出，放出，发行（钞票等），发布（命令），出版（书等），发给
7. sponsor [ˈspɔnsə] vt. 发起，主办，赞助
   n. 主办人，发起人，保证人
8. controversial [ˌkɔntrəˈvə: ʃəl]. adj. 争议的，争论的
9. fearful [ˈfiəful] adj. 担心的，可怕的
10. subservient [sʌbˈsə: viənt] adj. 屈从的，有帮助的，有用的，奉承的
11. disrupt [disˈrʌpt] v. 使中断，使分裂，使瓦解，使陷于混乱，破坏
12. handle [ˈhændl] v. 处理，买卖，操作
    n. 柄，把手
13. nutrient [ˈnju: triənt] n. 养分，滋养物
    adj. 营养的，滋养的
14. fertilizer [ˈfə: tilaizə] n. 肥料
15. alternative [ɔ: lˈtə: nətiv] n. 可供选择的办法，替代物
16. shrub [ʃrʌb] n. 灌木，灌木丛
17. gutter [ˈgʌtə] n. 水槽，檐槽，排水沟；贫民区；装订线
18 algae [ˈældʒi:] n. 藻类，海藻

19. clog [klɔg] v. 障碍，阻塞
20. invertebrate [in'və: tibrit] n. 无脊椎动物，无骨气的人
    adj. 无脊椎的，无骨气的
21. suffocate ['sʌfəkeit] vt. 使窒息，噎住，闷熄
    vi. 被闷死，窒息，受阻
22. feedlot ['fi: dlɔt] n. 饲育场
23. protozoan [ˌprəutəu'zəuən] n. [动] 原生动物
24. food chain 食物链
25. compound ['kɔmpaund] v. 混合，调合，妥协
    n. 化合物，混合物，复合词
26. residue ['rezidju:] n. 残余，渣滓，滤渣，残数，剩余物
27. grinding ['graindiŋ] n. 碾碎
    adj. 磨的，摩擦的，碾的
28. oxidation [ɔksi'deiʃən] n. [化] 氧化
29. sludge [slʌdʒ] n. 软泥，淤泥，矿泥，煤泥
30. granular ['grænjulə] adj. 粒状的，小粒的
31. release [ri'li: s] n. 释放，发行
    v. 释放，让与，准予发表，发射
32. bulk [bʌlk] n. 体积，大部分，大多数，大块，大批，容积
    vi. 越来越大
33. toxicity [tɔk'sisiti] n. 毒性
34. contaminant [kən'tæminənt] n. 污染物
35. aerobic [ˌeiə'rəubik] adj. 需氧的，有氧的，增氧健身法的
36. anaerobic lagoon 厌氧生物礁湖
37. nuclear power plant 核电站
38. uranium [juə'reiniəm] n. 铀
39. reproductive ['ri: prə'dʌktiv] adj. 再生的，复制的，生殖的
40. sodium ['səudiəm] n. [化] 钠
41. cardiovascular [ˌkɑ:diəu'væskjulə] adj. 心脏血管
42. chromosome ['krəuməsəum] n. [生物] 染色体
43. liver ['livə] n. 肝脏
44. shellfish ['ʃelfiʃ] n. 水生贝壳类动物贝，甲壳类动物
45. cryptosporidium [ˌkriptəuspɔ:'ridiəm] n. 隐孢子虫
46. diarrhea [ˌdaiə'riə] n. 痢疾，腹泻

* * * * * * * * * * * * * * * * * Complicated Sentences * * * * * * * * * * * * * * * * *

1. This meeting, at which the United States took a leading role, was controversial because many developing countries were fearful that a focus on environmental protection was a means for the developed world to keep the undeveloped world in an economically subservient position.

【译文】美国在此次会议上扮演了主要角色。该会议产生很多争议，因为许多发展中国家担心，强调环境保护会成为发达国家置不发达国家于经济屈从地位的手段。

【说明】句中主语 This meeting 由 at which 引导的定语从句修饰。because 引导的原因状语从句中又含有 that 引导的宾语从句。

2. We must become familiar with our local water resources and learn about ways for disposing harmful household wasters so they don't end up in sewage treatment plants that can't handle them or landfills not designed to receive hazardous materials.

【译文】我们必须熟悉本地水资源，了解家庭有害废弃物的处理方法，这样，它们就不会存在于无法处理它们的污水处理厂，或存在于并不接受危险物的埋填地。

【说明】句中 that can't handle them 为定语从句，修饰前面的 sewage treatment plants; not designed to receive hazardous materials 是 landfills 的后置定语；or 连接 sewage treatment plants 和 landfills; they 和 them 都指代 harmful household wastes。familiar with 表示"对…熟悉"。

3. Pollution is also caused when silt and other suspended solid, such as soil, wash off plowed fields, construction and logging sites, urban areas, and eroded river banks when it rains.

【译文】下雨时，淤泥和土壤一类的固体冲刷耕过的田地、施工现场、城区和侵蚀过的河堤，也会造成污染。

【说明】本句 when it rains 为 when 引导的时间状语从句，从句内还含有一 when 引导的时间状语从句。

4. Tertiary treatment is necessary when the water will be reused; 99 percent of solids are removed and various chemical processes are used to ensure the water is as free from impurity as possible.

【译文】第三阶段很有必要，可以去除 99%的固体物，通过各种各样的化学处理以确保尽可能干净的水。

【说明】前句中 when 引导原因状语从句,此时相当于 since/now that 引导的原因状语从句。后句 ensure 后省略了 that，即 that the water is as free from impurity as possible 是 that 引导的宾语从句。as free as possible 是 as...as...结构"尽……所能"。

5. These contaminations include both sediment from erosion cropland and compounds of phosphorus and nitrogen that partly originate in animal wasters and commercial fertilizers.

【译文】这些污染物一部分来自被冲蚀的农田，另一部分产生于动物粪便和商业肥料，如一些氮磷化合物。

【说明】句中第一个 and 与前面的 both 配合，即 both...and...；第二个 and 用来连接 phosphorus 和 nitrogen; that partly originate in animal wasters and commercial fertilizers 为定语从句，修饰 compounds of phosphorus and nitrogen。

6. More seriously, contaminated water destroys aquatic life and reduces its reproductive ability.

【译文】更为严重的是，污染过的破坏水中的动植物，降低它们的再生能力。

【说明】该句是由 and 连接的两个简单句，后句省略了主语 contaminated water。Its 代指 aquatic life。

7. Water pollution is an even greater problem in the Third World, where millions of people obtain water for drinking and sanitation from unprotected streams and ponds that are contaminated with human waste.

【译文】水污染问题在第三世界更为严重，那里数以百万计的人们从那些受到人类废弃物污染的未得到保护的溪流和池塘中获取饮用水和卫生用水。

【说明】该句中 where 引导的地点状语从句修饰 the Third World。从句中 that 引导的定语从句修饰 unprotected streams and ponds。

8. This type of contamination has been estimated to cause more than 3 million deaths annually from diarrhea in Third World countries, most of them children.

【译文】在第三世界国家，据估计，这种污染造成因腹泻而死亡的人数，每年超过 300 万，其中大部分为儿童。

【说明】句中 children 前面省略了 are。

******************* **Summary of Glossary** *******************

1. pollution	污染
2. pollutant	污染物
3. contamination	污染，污染物
4. toxicity	毒性
5. bacteria	细菌
6. aquatic organism	水生有机物
7. protozoan	原生动物
8. suspended solid	淤泥，悬浮固体
9. dissolved oxygen	溶解氧
10. silt	淤泥
11. sewage disposal	污水处理
12. waterborne disease	水传播疾病
13. nutrient	营养品，滋养物
14. fertilizer	肥料（尤指化学肥料）
15. plowed field	耕地
16. soil erosion	土壤侵蚀
17. surface water	地表水
18. to a great extent	很大程度上
19. safe drinking water	安全饮用水
20. water pollution	水污染

******************* **Abbreviations (Abbr.)** *******************

1. UNEP	the United Nations Environmental Program	联合国环境计划
2. DDT	dichloro-diphenyl-trichloroethane	滴滴涕，二氯二苯三氯乙烷
3. PCBs	polychlorinated biphenyls	多氯联苯

*************** **Exercises** ***************

(1) Converting _______ to freshwater is generally too _______ to be used for _______, agricultural or _______ purpose.

(2) Raw sewage, _______, and _______ have begun to _______ the diluting capabilities of the oceans.

(3) Congress has passed laws to try to _______ water pollution thus the fact that water pollution

is, indeed, a serious issue.

(4) We have to ________ existing trees and plant new trees and shrubs to help prevent soil erosion and promote ________ of water into the soil.

(5) This, in turn, proves very ________ to aquatic organisms as it affects the ability of fish and other that reside in water.

(6) Under natural conditions, lakes, rivers, and other water bodies undergo ________, an aging ________ that slowly fills in the water body with ________ and organic matter.

************ Word Building (9) ab-; a- ************

1. ab- [前缀]，表示：偏离，脱离或离开，可译为：不，无，反，异

normal	n./adj	abnormal	adj.	反常的，变态的
react	vi.	abreact	vt.	使消散，发泄

2. a- [前缀]，表示：偏离，脱离或离开，不，非

seasonal	adj.	aseasonal	adj.	无季节性的
seismic	adj.	aseismic	adj.	无地震的，耐震的
septic	n./adj.	aseptic	adj.	防腐的，无菌的
sexual	adj.	asexual	adj.	无性的，无性生殖的
symmetrical	adj.	asymmetrical	adj.	不均匀的，不对称的
synchronously	adv.	asynchronously	adj.	异步地，不同时的
pathy	n.	apathy	n.	缺乏感情，冷漠

****************** Text Translation ******************

第九章　水　污　染

第一节　水污染概述

据估计，大约有近 15 亿人的安全饮用水量不足，每年至少 500 万人死于水污染引起的疾病。地球表面 70%以上为海洋所覆盖，长期以来，人们一直认为这些水体好像是可以无限制地倾倒废弃物的场所。未经处理的污水、垃圾和溢出的石油已开始摧毁海洋的稀释能力，现在大部分的海水都受到了污染。因为污水处理产生的大量细菌，全世界的海滩常常定期关闭。海洋野生物也开始受灾。（单词个数：103）

大多数污染的形式并不受国界的影响，也许是基于这个事实全球努力监控并限制全球性污染。第一次关于环境问题的国际会议，由联合国主持，于 1972 年在瑞典的斯德哥尔摩举行。美国在此次会议上扮演了主要角色，该会议有很多争议，因为许多发展中国家担心，强调环境保护，会成为发达国家置不发达国家于经济屈从地位的手段。此会的最重要的成果就是制订了联合国环境规划（UNEP）。（单词个数：110）

很明显，水污染会在很大程度上破坏我们这个星球上的生命。（美国）国会已通过法律试图同水污染作斗争，承认水污染是一个必须面对的事实。但光靠政府并不能解决所有问题。最终，当水问题来临时，应该由我们来认识、负责并参与。我们必须熟悉本地水资源，了解家庭有害废弃物的处理方法，这样，它们就不会存在于无法处理它们的污水处理厂，或存在于并不接受危险物的埋填池。在自家院子里，在施肥前，我们应该确定植物是否需要其他养分供应，并做好防止肥料流入地下水的措施。我们必须保护现有的树木，种植新树和灌木以防土壤侵蚀，提倡滴灌。在家周围，我们必须让废物、宠物粪便、树叶和草片远离沟槽，不让它们随暴雨到处流淌。这些只是我们人类战胜水污染的众多方法中的一部分。如果不采取这些措施，水污染将继续，地球上的生命将严重遭殃。（单词个数：220）

第二节 水污染的原因

许多污染源，包括污水和肥料，都是含有像硝酸盐和磷酸盐这类滋养物质。数量过多时，滋养物质会过多地刺激水生植物和水藻的生长。这类有机体的过多生长最终会堵塞水路、耗尽其分解时产生的氧气、阻止光线照射深水。这反过来说明了它对水生生物非常有害，因为它影响了鱼类及其他生活在水中的无脊椎动物的呼吸。（单词个数：75）

下雨时，雨中淤泥和土壤一类的固体冲刷耕过的田地、施工现场、城区和侵蚀过的河堤，也会造成污染。自然条件下，湖泊、河流和其他水体都要经历沉淀物和有机物慢慢填充水体这一养分富有的衰老过程。这些沉淀物进入各种水体，会使鱼类呼吸受到影响、植物生产量减少、水深度变浅、水中的有机体窒息而死，其环境也受到影响。有机体的污染以许多不同的形式进入水体，如污水、叶子和草片或牲畜饲养和喂养的流出物等。水中的自然细菌和原生动物破坏这种有机体时，他们开始渐渐消耗水中的溶解氧。当分解的氧气减少到 2%～5%时，许多种类的鱼类和底部的微生物将无法生存。当这类情况发生时，它杀死了大量的水生生物从而破坏了食物链。（单词个数：174）

第三节 水污染的分类

水污染主要可分为城市水污染、工业水污染、农业水污染和其他形式的水污染。（单词个数：19）

1. 城市水污染

城市水污染包括家庭和商业用的废水。多年来，城市废水处理的主要目的就是简单地减少悬浮固体的含量、耗氧物、分解无机化合物和有害细菌。但近些年，改进城市处理过程中的固体残留物的处理方法越来越受到重视。处理城市废水的基本方法分三个阶段：第一阶段初步处理，包括去除沙粒、筛分、碾碎及沉淀；第二阶段，利用微生物的氧化作用来消耗沉积物使之溶解，之后过滤掉；第三阶段，使用先进的生物除氮法及过滤粒状物和碳吸收等物理和化学方法。固体残留物的处理要占到污水处理厂费用的 25%～50%。（单词个数：152）

2. 工业水污染

工业废水因行业不同而差异很大。工业排出物的影响不仅取决于如生化氧气需求和悬浮固体的含量等特点，而且取决于特定的有机体和无机体的含量。控制工业废水有三种方法：从工厂源头开始；预先处理废水；在工厂彻底处理，即要么再次使用废水，要么直接把废水排到能够承受的水域中。原始污水包括阴沟、厕所和工业生产过程中的废弃水。在安全掩埋、使用和排回当地水系之前，需要对废水进行处理。在污水处理厂，对废水进行一系列的筛分、过滤和化学处理以减少体积和毒性。处理基本分三个阶段：第一阶段，将大部分的悬浮固体和无机体从污水中去除；第二阶段，加速自然生物对有机体的分解作用以减少有机体数量；如果水需要再被利用时，第三阶段是很有必要；可以通过各种各样的化学处理去除 99%的固体物，以确保水尽可能干净。（单词个数：227）

3. 农业水污染

商用牲畜和家禽的饲养等农业生产，是造成地面水和地下水的许多有机和无机污染的源头。这些污染物既包括腐蚀的农作物的沉积物，也包括部分动物粪便和商业肥料中的氮和磷的化合物。动物粪便含氮和磷等高耗氧物质，并且常常导致有机体生病。而商业肥料的废弃物通常是存放在地面上，并且也是在地面上处理的，因而径流和沥滤对地面的冲刷，就可能将这些废弃物质带到水体，而对天然水造成威胁。这些污染可以通过液体沉淀池、有限的大气生物或厌氧生物礁湖及许多其他方法来处理。（单词个数：105）

4. 其他形式的水污染

还有石油、放射性物质和热三种形式的水污染。石油由于溢出常常污染水体。石油大量的意外泄漏，是海洋沿岸污染的一个重要原因。除了超大型油轮意外泄漏外，沿海钻油也是造成污染的主要原因。据估计，每运输 100 万 t 的石油就有 1t 溢出。这个比例达到了 0.0001%。放射性物质以废弃物形式产生于核电站和工用、医用及科学用的放射性材料。废弃物的具体形式有采矿与冶炼的铀、钍。最后一种水污染是热污染。热污染是一种污染物质，因为温度升高导致许多水中有机体的死亡。如果工厂与电厂排放冷水，温度又会降低。（单词个数：158）

第四节 水污染的影响与危害

水污染的影响不仅包括对人类的破坏，也包括对动物、鱼类和鸟类的破坏。污染过的水不能饮用、不能用于景观娱乐、不能用于农业和工业。污染水会降低湖泊与河流的水质。更为严重的是，污染过的水会破坏水中的动植物，降低它们的再生能力。最终，对人类的健康造成伤害。没有人能逃避水污染的影响。（单词个数：65）

事实上，所有水污染物质对人类及较小的物种都有害。钠会导致心血管病，以及由于硝酸盐而引起血液紊乱。汞和铅等物质可造成神经紊乱。有些污染物质还是致癌物质。滴滴涕（DDT）对人体有毒并会改变水性染色体。多氯联苯（PCB）损害肝脏和神经、引起皮肝症、呕吐、发热、腹泻和胎儿畸形。由于受到 DDT、污水或工业废弃物的污染，许多海滩再也看不到水生贝壳动物了。（单词个数：75）

痢疾、沙门氏菌病、似隐孢菌和肝炎，是污水传播到饮用水和浴水中而导致的疾病。对游泳者来说，由于工业废弃物、城市污水和医疗废弃物的影响，美国的海边、河边和湖边，都已经遭到破坏。水污染问题在第三世界更为严重，在那里，数以百万计的人们，从那些已经被人类废弃物污染的、并且未得到处理和保护的溪流和池塘中，获取饮用水和卫生用水。在第三世界国家，据估计，这种污染造成每年超过300万人因腹泻而死亡，其中大部分是儿童。(单词个数：98)

Chapter 10 Dams

Part 1 Introduction

A dam is a structure built across a stream, river or estuary to retain water. Some dams are tall and thin, while others are short and thick. Dams are made from a variety of materials such as rock, steel and wood. Dams affect society in many positive ways. Dams gather drinking water for people. Dams help farmers bring water to their farms. Dams help create power and electricity from water. Dams keep areas from flooding. Dams create lakes for people to swim in and sail on. Dams also aid in navigation by making waters deeper and calmer.

The first dam for which there are reliable records was built on the Nile River sometime before 4000 B.C. It is used to divert the Nile and provide a site for the ancient city of Memphis. The oldest dam still in use is the Almanza Dam in Spain, which was constructed in the sixteenth century. With the passage of time, materials and methods of construction have been improved, making possible the erection of such large dams as the Nurek Dam which was constructed in the Former Soviet Union on the Vaksh River near the border of Afghanistan.[1] This dam will be 1,017 ft. (333m) high, of earth and rock fill.

The failure of a dam may cause serious loss of life and property; consequently, the design and maintenance of dams are commonly under government surveillance. In the United States over 30 000 dams are under the control of state authorities. The 1972 Federal Dam Safety Act (PL 92-267) requires periodic inspections of dams by qualified experts. The failure of the Teton Dam in Idaho in June 1976 added to the concern for dam safety in the United States.

Part 2 Types of Dams

Dams are classified on the basis of the type and materials of construction, as arch dams, buttress dams, gravity dams, and embankment dams.

1. Arch dam

An arch dam is a curved dam which is dependent upon arch action for its strength. Arch dams are thinner and therefore require less material than any other type of dam. Arch dams are good for sites that are narrow and have strong abutments.

These are the main forces on an arch dam: force of the reservoir water, weight of concrete and both forces together. There are other forces that may act on an arch dam: internal water pressure, temperature variations in the concrete, earthquake loads, and settlement of the foundation or abutments.

The first multiple arch dam of reinforced concrete had been completed in 1908. It impounded the Hume Lake fluming reservoir on the Ten Mile creek in the California Sierra Nevada Mountains. The dam was designed and supervised during construction by John S. Eastwood (1857-1924). Eastwood's designs were very economic: they required less concrete than an equivalent gravity dam and cost less, too. Overall, a dozen multi-arch dams were built in the following decade and a half according to Eastwood's design.

2. Buttress dam

Buttress dams are dam in which the face is held up by a series of supports. Buttress dams can take many forms the face may be flat or curved.

Usually, buttress dams are made of concrete and may be reinforced with steel bars. The buttress may be hollow or solid. The buttress of Bartlett Dam is hollow and the highest concrete buttress dams in the U.S., at 287 feet.[2]

3. Gravity dams

Gravity dams are dams which resist the horizontal thrust of the water entirely by their own weight. Concrete gravity dams are typically used to block streams through narrow gorges.

Because it is the weight holding the water back, concrete gravity dams tend to use a large amount of concrete. This can be expensive. But many prefer its solid strength to arch or buttress dams. Gravity dams can be made of earth or rock fill or concrete. These dams can be very expensive because of how much material they use.

Generally, the base of a concrete gravity dam is equal to approximately 0.7 times the height of the dam; base=0.7×height; crest=0.2×height. The shape of the gravity dam resembles a triangle. This is because of the triangular distribution of the water pressure. The deeper the water, the more horizontal pressure it exerts on the dam. [3] So at the surface of the reservoir, the water is exerting no pressure and at the bottom of the reservoir, the water is exerting maximum pressure.

A relatively new development in the construction of gravity dams is incorporation of post-tensioned steel into the structure. This helped reduce the cross section of Allt Na Lairige Dam in Scotland to only 60 percent of that of a conventional gravity dam of the same height. A series of vertical steel rods near the upstream water face, stressed by jacks and securely anchored into the rock foundation, resists the overturning tendency of this slenderer section.[4] This system has also been used to raise existing gravity dams to a higher crest level, economically increasing the storage capacity of a reservoir.

4. Embankment dam

Embankment dams are massive dams made of earth or rock. They rely on their weight to resist the flow of water.

Embankment dams usually have some sort of water proof insides (called the core), which is covered with earth or rock fill. Grass may even be grown on the earth fill. Water will seep in through the earth or rock fill, but should not seep into the core.

The main force on an embankment dam is the force of the water. The weight of the dam is also a force, but each material has a different weight. The uplift force is also acting on the embankment dam, but some of the water seeps into the dam so the force is not the same as on a concrete dam.

The embankment dam is the only dam type that is not made of concrete. Embankment dams may be made of earth or rock, both of which are pervious to water—that is, water can get into it.[5] The water will seep into the core material and should stop at the seepage line. The core material is usually more watertight than the rock or earth that is on the outside of the dam, but the core material is still not totally impervious to water. Concrete is not only truly impervious either, but is does not allow as much seepage as these materials do.

The selection of the best type of dam for a given site is a problem in both engineering feasibility and cost. Feasibility is governed by topography, geology and climate. For example, because concrete spalls when subjected to alternate freezing and thawing, arch and buttress dams with thin concrete sections are sometimes avoided in areas subject to extreme cold. The relative cost of the various types of dams depends mainly on the availability of construction materials near the site and the accessibility of transportation facilities. Dams are sometimes built in stages with the second or later stages constructed a decade or longer after the first stage.

The height of a dam is defined as the difference in elevation between the roadway, or spillway crest, and the lowest part of the excavated foundation. However, figures quoted for heights of dams are often determined in other ways. Frequently the height is taken as the net height above the old river bed.

Part 3 Forces on Dams

A dam must be relatively impervious to water and capable of resisting the forces acting on it. The most important of these forces are gravity (weight of dam), hydrostatic pressure, uplift, ice pressure, and earthquake forces. These forces are transmitted to the foundation and abutments of the dam, which react against the dam with an equal and opposite forces, the foundation reaction.[6] The effect of hydrostatic pressure caused by sediment deposits in the reservoir and of dynamic forces caused by water flowing over the dam may require consideration in special cases.[7]

The weight of a dam is the product of its volume and the specific weight of the material. The line of action of this force passes through the center of mass of the cross section. Hydrostatic forces may act on both the upstream and downstream faces of the dam. The horizontal component H_h of the hydrostatic force is the force on a vertical projection of the face of the dam, and for unit width of dam it is

$$H_h = \gamma h^2 / 2 \tag{10-1}$$

Where: γ is the specific weight of water; h is the depth of water. The line of action of this force is $h/3$ above the base of the dam. The vertical component of the hydrostatic force is equal to the weight of water vertically above the face of the dam and passes through the center of gravity of this volume of water.[8]

Water under pressure inevitably finds its way between the dam and its foundation and creates uplift pressures. The magnitude of the uplift force depends on the character of the foundation and the construction methods. It is often assumed that the uplift pressure varies linearly from full hydrostatic pressure at the upstream face (heal) to full tail-water pressure at the downstream face (toe). For this assumption the uplift force U (Fig.10.1) is

$$U=\gamma(h_1+h_2)t/2 \tag{10-2}$$

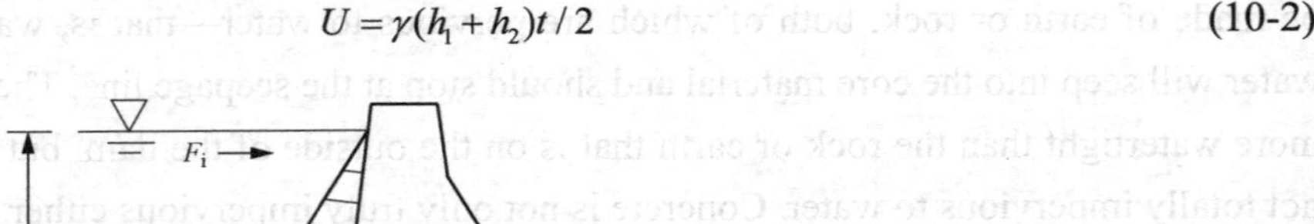

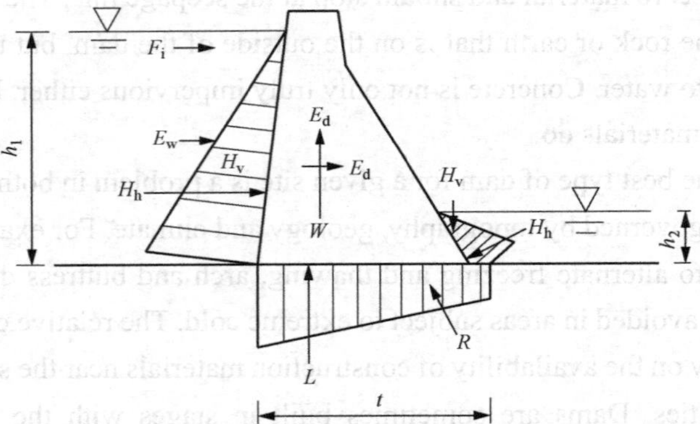

Fig.10.1 Free-body diagram of the section of a gravity

where: t is the base thickness of the dam; $h_1$and h_2 are the water depths at the heel and toe of the dam, respectively. The uplift force will act through the center of area of the pressure trapezoid.

Actual measurements on dams indicate that the uplift force is much less than that given by Eq. (10-2). Various assumptions have been made regarding the distribution of uplift pressures. The U.S. Bureau of Reclamation sometimes assumes that the uplift pressure on gravity dams varies linearly from two-thirds of full uplift at the heel to zero at the toe.[9] Drains are usually provided near the heel of the dam to permit the escape of seepage water and relieve uplift.

Part 4 Several Disadvantages to Dams

There are a number of disadvantages to existing dams:

1. Dams detract from natural settings, ruin nature's work.
2. Dams have inundated the spawning grounds of fish.
3. Dams have inhibited the seasonal migration of fish.
4. Dams have endangered some species of fish.
5. Dams may have inundated the potential for archaeological findings.
6. Reservoirs can foster diseases if not properly maintained.
7. Reservoirs water can evaporate significantly.
8. Some researchers believe that reservoirs can cause earthquakes.

There are also disadvantages to building new dams:

1. The reservoir created by the dam may inundate land, crops, cities and villages.
2. People may be displaced by the reservoir and have to find new homes.

3. The reservoir may cause instability of the hillsides.

Each dam may <u>stir up</u> its own controversy because of some impact it has had or will have on its surroundings.

* * * * * * * * * * * * * * * * * Explanations * * * * * * * * * * * * * * * * *

* * * * * * * * * * * * * * * * * New Words and Expressions * * * * * * * * * * * * * * * * *

1. Memphis ['memfis] 孟菲斯（埃及尼罗河畔城市，美国城市）
2. Almanza Dam 阿尔曼扎坝
3. Nurek Dam 努列克坝
4. Vaksh River 瓦赫什河
5. Afghanistan 阿富汗
6. arch [ɑ:tʃ] n. 拱门，弓形
   v. 成拱形弯曲，成拱形
7. arch dam 拱坝
8. buttress ['bʌtrəs] v. 支持
   n. 扶墙，拱壁，支墩
9. buttress dam 支墩坝
10. embankment dam 堤坝
11. abutment [ə'bʌtmənt] n. 拱座，桥台（支撑桥梁底部的结构）
12. multiple ['mʌltipl] n. 倍数
    adj. 多样的，多重的
13. multiple arch dam 连拱坝
14. impound [im'paund] vt. 关在栏中，拘留，扣押，没收
15. hollow ['hɔləu] n. 洞，窟窿，山谷
    adj. 空的，虚伪的，空腹的，凹的
    adv. 彻底
    vi. 形成空洞
    vt. 挖空，弄凹
16. horizontal [ˌhɔri'zɔntl] adj. 水平的，平的
17. thrust [θrʌst] n. 推力，插，戳，刺，猛推
    vt. 冲，插入，挤进，刺，戳
    vi. 插入，刺，戳，延伸，强行推进
18. gorge [gɔ: dʒ] n. 峡谷，咽喉，胃，暴食
    v. 狼吞虎咽，塞饱
19. crest [krest] n. 顶部，顶峰，浪头
    vi. 到达绝顶
    vt. 加以顶饰
20. post-tension 后张力
21. vertical ['və: tikəl] adj. 垂直的；
    n. 垂直线
22. jack [dʒæk] n. 起重器，千斤顶，（J-）杰克（男子名，也指各种男性工人），插孔，插座
    vt. 用起重器抬起，提醒，增加，放弃，提高
23. anchor ['æŋkə] n. 铁锚
24. seep [si: p] v. 渗出，渗流，漏
25. topography [tə'pɔgrəfi] n. 地形学，地形
26. thawing ['θɔ: iŋ] n. 熔化，融化
27. spillway crest 溢洪道顶，溢流堰顶
28. hydrostatic pressure 静水压力
29. inevitably [in'evitəbli] adv. 不可避免地
30. pressure trapezoid 梯形（扬）压力

* * * * * * * * * * * * * * * * * Complicated Sentences * * * * * * * * * * * * * * * * *

1. With the passage of time, materials and methods of construction have been improved,

making possible the erection of such large dams as the Nurek Dam which was constructed in the Former Soviet Union on the Vaksh River near the border of Afghanistan.

【译文】随着岁月的流逝，建筑材料和施工方法得到了改善，修建努列克这样的大坝（位于苏联瓦赫什河上，靠近阿富汗边境）才成为可能。

【说明】making 引导的分词短语做状语；后面 which 引导的定语从句修饰 Nurek Dam。

2. The buttress of Bartlett Dam is hollow and the highest concrete buttress dams in the U.S., at 287 feet.

【译文】巴特利特大坝就是空心坝，也是美国最高的混凝土支墩坝，有 287 英尺高。

【说明】形容词 hollow 与名词短语 the highest concrete buttress dams in the U.S.是并列的表语。

3. The deeper the water, the more horizontal pressure it exerts on the dam.

【译文】水越深，对水坝的水平压力就越大。

【说明】本句是“the more...the more”结构，译为“越……越……”。

4. A series of vertical steel rods near the upstream water face, stressed by jacks and securely anchored into the rock foundation, resists the overturning tendency of this slenderer section.

【译文】上游面板由墩支撑，并用千斤顶把一系列垂直钢杆锚固定在岩石基座上，以防止薄长的面板翻到。

【说明】此局为一简单句，连接 stressed 和 anchored 两个分词短语做定语修饰 rods。

5. Embankment dams may be made of earth or rock, both of which are pervious to water—that is, water can get into it.

【译文】土石坝可用土或岩石来建造，土和岩石可渗透水，即水能够通过它们。

【说明】be made of 由……组成，which 引导一定语从句修饰 earth or rock，that is 译为“即，就是，换句话说，更确切地说”。

6. These forces are transmitted to the foundation and abutments of the dam, which react against the dam with an equal and opposite forces, the foundation reaction.

【译文】这些力传给坝基和坝座，而坝基和坝座对坝体产生一个大小相等方向相反的基础反力。

【说明】句中 which 引导定语从句修饰主语 forces，具体解释了该力的大小和作用方向。

7. The effect of hydrostatic pressure caused by sediment deposits in the reservoir and of dynamic forces caused by water flowing over the dam may require consideration in special cases.

【译文】某些特殊情况下，还要考虑水库中沉积泥沙引起的静水压力的影响以及坝顶溢流所产生的动水压力作用。

【说明】该句主体结构一简单句，主语为 The effect，由 and 连接的两个短语 of hydrostatic pressure caused by sediment deposits in the reservoir 和 of dynamic forces caused by water flowing over the dam 作定语修饰。in special cases 为“某些特殊情况下”。

8. The vertical component of the hydrostatic force is equal to the weight of water vertically above the face of the dam and passes through the center of gravity of this volume of water.

【译文】静水压力的垂直分力等于坝面正上方的水重，并通过该水体的重心。

【说明】该句式是由 and 连接的两个简单句，前句的谓语是 is equal to，后句的谓语是 passes。

9. The U.S. Bureau of Reclamation sometimes assumes that the uplift pressure on gravity dams varies linearly from two-thirds of full uplift at the heel to zero at the toe.

【译文】美国垦务局认为重力坝的扬压力成直线变化，在坝踵处为全部扬压力的 2/3，到坝趾处为零。

【说明】该句 that 引导一宾语从句，varies linearly 指"呈直线变化"，from two-thirds of full uplift at the heel to zero at the toe 中介词短语 from…to…作状语。

*** * * * * * * * * * * * * * * * * * * Summary of Glossary * * * * * * * * * * * * * * * * * * ***

1. arch dam 拱坝
2. buttress dam 支墩坝
3. gravity dam 重力坝
4. embankment dam 土石坝
5. divert 改道
6. inundate 淹没
7. Nile river 尼罗河
8. drinking water 饮用水
9. core 心墙
10. multi-arch dam 连拱坝
11. uplift force 扬压力
12. ice pressure 冰压力
13. toe 坝趾
14. heel 坝踵
15. cross section 横断面
16. horizontal component 水平分力
17. periodic inspection 定期检查
18. the Former Soviet Union 苏联
19. the U.S. Bureau of Reclamation 美国垦务局
20. core material 心墙材料
21. engineering feasibility 工程可行性
22. excavated foundation 基坑

*** * * * * * * * * * * * * * * * * * * Abbreviations (Abbr.) * * * * * * * * * * * * * * * * * * ***

ASCE American Society of Civil Engineers 美国土木工程协会

*** * * * * * * * * * * * * * * * * * * Exercises * * * * * * * * * * * * * * * * * * ***

(1) A dam is a structure built across a _________, river or_________to retain water.

(2) The ________ of a dam may cause serious loss of life and ________; consequently, the design and ________ of dams are commonly under government ________.

(3) The ________ of the Teton Dam in Idaho in June 1976 added to the ________ for dam ________ in the United States.

(4) Dams are ________ on the basis of the type and materials of construction, as arch dams, ________ dams, ________ dams, and ________ dams.

* * * * * * * * * * * * Word Building (10) dis-; multi- * * * * * * * * * * * *

1. dis- [前缀]，表示：否定，相反，不

connect	v.	disconnect	v.	断开，拆开，分离
order	n./v.	disorder	vt./n	扰乱（状态），（使）失调，（使）紊乱
place	n./v.	displace	vt.	移置，转移
able	adj.	disable	vt.	使残废，使失去能力，丧失能力
advantage	n.	disadvantage	vt./n	(使)不利，缺点，劣势
agree	v.	disagree	vi.	不同意
appear	vi.	disappear	vi.	消失，不见
regarded	adj.	disregarded	adj.	无视的
courage	vt.	discourage	vt.	阻止；使气馁

2. multi- [前缀]，表示：多，重，倍

purpose	n./v.	multi-purpose project	adj.	综合利用工程
national	n./v.	multinational	adj.	跨国公司的
aspect	n./v.	multiaspect	n.	多方面
arch	n./v.	multi-arch dam	adj.	连拱坝

* * * * * * * * * * * * * * * * Text Translation * * * * * * * * * * * * * * * *

第十章 大　　坝

第一节 概　　述

大坝是跨溪流、河流或河口建造的用于挡水的建筑物。有些水坝高而薄，而另外一些水坝则矮而厚。大坝用许多材料建成，如岩石、钢材和木料。大坝对社会有许多积极的影响。大坝为人们积聚饮用水；帮助农民将水引入农场；有助于用水发电；防止洪水；提供人们游泳和航行的湖泊；也使水体更深更平静以利于航行。（单词个数：97）

据可靠记载，世界上第一座坝是公元前4000年以前在尼罗河上修建的。它使尼罗河改道，并为古老的孟菲斯城提供城址。至今仍在使用的最古老的坝是16世纪修建的西班牙阿尔曼扎坝。随着岁月的流逝，各种建筑材料和施工方法得到了改善，修建努列克这样的大坝才成为可能。该坝位于苏联境内靠近阿富汗边界的瓦赫什河上，是一座高达1017英尺（333m）的土石坝。（单词个数：110）

大坝失事可能造成生命财产的严重损失。因此，坝的设计和维修通常是在政府监督下进行的。美国有3万多座坝由各州政府控制着。1972年（美国）联邦大坝安全法（PL92-367）规定，必须由合格的专家对大坝进行定期检查。在1976年6月爱达荷州提堂大坝失事后，美国对大坝安全更为关切。（单词个数：76）

第二节 大坝的类型

按其形式和建筑材料，大坝分为拱坝、支墩坝、重力坝和土石坝。（单词个数：23）

1. 拱坝

拱坝是依靠拱作用的拱形坝。拱坝较单薄，因而比其他任何种类的坝所用的筑坝材料都少。拱坝对于狭窄而又有牢固支撑的地方很适合。（单词个数：44）

拱坝所受的力主要有水库的水作用力、混凝土的重量及这两种力的合力。此外，还有内部水压力、混凝土抗震负荷中温度的变化应力以及支座反力。（单词个数：49）

第一座混凝土加固的连拱坝于1908年建成。该坝为休姆湖蓄水，此湖用引水槽引加利福尼亚内华达山脉10英里外的支流水库的水。此坝在建造期间由约翰S. Eastwood（1857～1924）设计并监督。Eastwood的设计非常经济：它比相应的重力坝需要的混凝土少，成本也低。在此后的15年中，总的说来，又建造了12座多拱坝，都是采用Eastwood的设计。（单词个数：83）

2. 支墩坝

支墩坝是表面有一系列支撑的坝。支墩坝形式多样——表面可分为平坦的或曲线的。（单词个数：29）

支墩坝常用混凝土做成，可用钢柱加固。支墩坝可为空心或实心。巴特勒坝就是空心坝，是美国最高的混凝土支墩坝，有287英尺高。（单词个数：41）

3. 重力坝

重力坝指完全依靠自身的重量抵挡水平向水流推力的水坝。很典型地，重力坝用来阻止小型河流通过狭窄的沟壑。（单词个数：29）

因重力坝靠重量阻挡水，重力坝要用大量的混凝土，这个造价较昂贵。但同拱坝或支墩坝相比，多数人还是喜欢其实心牢固的坝型。重力坝的筑坝材料可以是土料、岩石或混凝土。这类坝因多用材料而造价较高。（单词个数：60）

通常，混凝土重力坝坝基的宽度约为坝高的0.7倍：坝基=0.7×坝高；坝顶宽=0.2×坝高。重力坝形状类似于三角形。这是因为水压的三角形分布。水越深，对水坝的水平压力就越大。因此，在水库表面，水并没有什么压力，但在水库底部，水压最大。（单词个数：79）

重力坝建设的相对较新的发展就是将后张力的钢材溶入建筑物中。对苏格兰的阿尔特娜·莱雷格坝来说，有助于将其横截面减小到同样高度的普通重力坝的60%。上游水面附近用支墩加固，并安全地支撑在岩石基座上的一系列垂直钢杆，防止纤瘦的坝体截面发生倾倒的趋势。该系统也已经用来将现有重力坝顶部提升到较高的水平，很经济地增加了水库的储水能力。（单词个数：100）

4 土石坝

土石坝是结实的土料或岩石筑成大坝。它们靠自身重力抵挡水流。（单词个数：21）

土石坝通常有某种防水的内部结构（叫作心墙），为泥土或岩石填料所覆盖。泥土填料上甚至会长出草。水会渗透泥土或岩石填料，但不能渗透到心墙。（单词个数：47）

土石坝上主要的力是水的作用力，坝重也是一种力，但每种材料重量不同。浮托力也是作用于土石坝的力，但有些水渗透到坝里，所以该力与混凝土坝不同。（单词个数：60）

土石坝是唯一的一种未用混凝土建造的水坝。筑堤坝可用土或岩石来建造。土和岩石可渗透水，即水能够经过它们。水会渗透到心墙并在渗流线处停止。心墙通常比大坝其他部位的土或岩石更防水，但心墙并非完全不渗漏。混凝土也并非完全不渗漏，但它防渗漏比这些材料好。（单词个数：102）

在既定的坝址选择最佳坝型是一个关系到工程可行性及其造价的问题。工程可行性受地形、地质及气候条件所支配。例如：由于混凝土在遭受冻融作用的交替影响时，会引起剥落，因此，在低温地区常避免采用断面单薄的混凝土拱坝或支墩坝。各类坝的造价主要取决于能否在工地附近取得建筑材料，以及各种运输工具能否进入。大坝有时分期建造，第二期或以后各期工程，往往在第一期以后 10 年或更长的时间再建造。（单词个数：105）

坝高定义为路面或溢洪道顶与基坑最低点之间的高程差。不过，引用的坝高值常常是用另外的一些方法确定的，往往取原河床以上的净高度作为坝高。（单词个数：53）

第三节 作用在坝上的力

坝必须是相对不透水的，并能经受得住作用在它上面的各种力。这些作用力中最重要的是重力（坝体重量）、静水压力、扬压力、冰压力及地震力。这些力传给坝基和坝座，而坝基和坝座则对坝体产生一个大小相等方向相反的基础反力。某些特殊情况下，还要考虑水库中沉积泥沙引起的静水压力的影响，以及坝顶溢流所产生的动水压力的作用。（单词个数：91）

坝的自重是其体积和材料比重的乘积。该力的作用线通过横剖面的形心。静水压力可同时作用在坝的上游面和下游面。静水压力的水平分力 H_h 是作用在坝面垂直投影上的力，对于单位宽度坝体而言其值为；

$$H_h = \gamma h^2 / 2 \tag{10-1}$$

式中：γ 为水的比重；h 为水深；该力的作用线在坝基以上 $h/3$ 处。静水压力的竖直分力等于坝面正上方的水重，并通过该水体的重心。（单词个数：140）

处于压力作用下的水必然要在坝和坝基之间流动，因而产生了扬压力。扬压力的大小取决于基础的特性和施工方法。经常假定扬压力从上游面（坝踵）处的全部静水压力直线变化到下游面（坝趾）处的全部尾水压力。根据这一假设，扬压力 U 为：

$$U = \gamma (h_1 + h_2) t / 2 \tag{10-2}$$

式中：t 为坝基的宽度；h_1 和 h_2 分别为坝踵和坝趾处的水深。扬压力的作用线通过压力梯形的

形心（图 10.1）。（单词个数：110）

一些坝的实测资料表明：扬压力比公式（10-2）所给出的值小得多。对扬压力的分布有各种不同的假设，美国垦务局认为重力坝的扬压力成直线变化，在坝踵处为全部扬压力的 2/3，到坝趾处为零。坝踵附近通常设有排水装置，以便排除渗流水量，减小扬压力（美国土木工程师协会）。（单词个数：90）

第四节 水坝的负面影响

现有的水坝有许多负面影响：

1. 影响自然环境，并会破坏自然。
2. 淹没了鱼类产卵的地方。
3. 限制了鱼类季节性的迁徙。
4. 危害了一些鱼类。
5. 可能淹没考古发现的潜在地。
6. 水库若不妥当管理，会造成疾病。

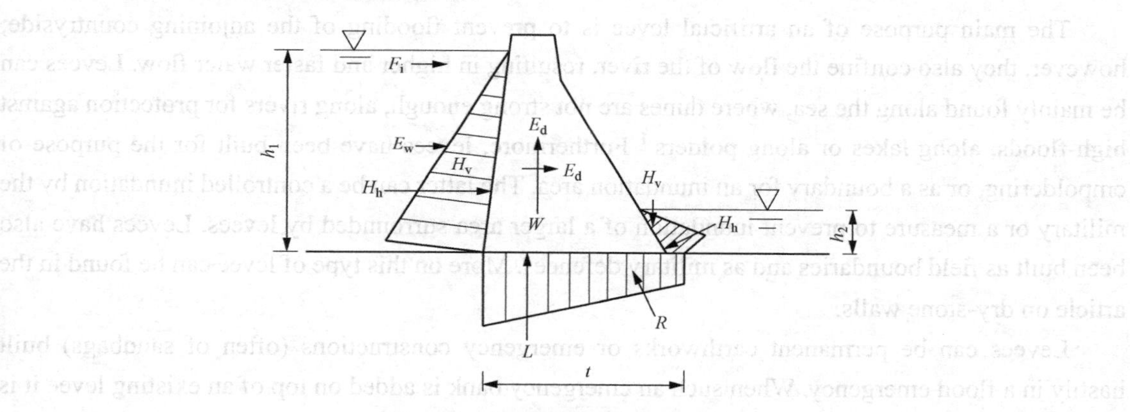

图 10.1 重力坝剖面受力示意图

7. 水库里的水会大量蒸发。
8. 有些研究者认为水库可能导致地震。（单词个数：78）

新建造的水坝也有负面影响：

1. 水坝拦水而形成的水库会淹没土地、作物、城市和乡村。
2. 建库后会淹没库区附近人们的居住地，人们被迫移民到外地。
3. 水库也可能造成山坡不稳定。

每座水坝都会因为其对周围环境已产生或将会产生的影响而引起争议。（单词个数：66）

Chapter 11 Levee

A levee, levée, dike (or dyke), embankment, floodbank or stopbank is a natural or artificial slope or wall to regulate water levels. It is usually earthen and often parallel to the course of a river or the coast.

The word levee, from the French word levée (from the feminine past participle of the French verb lever, "to raise"), is used in American English (notably in the Midwest and Deep South); it came into English use in New Orleans circa 1720.

Part 1 Artificial Levees

The main purpose of an artificial levee is to prevent flooding of the adjoining countryside; however, they also confine the flow of the river, resulting in higher and faster water flow. Levees can be mainly found along the sea, where dunes are not strong enough, along rivers for protection against high-floods, along lakes or along polders.[1] Furthermore, levees have been built for the purpose of empoldering, or as a boundary for an inundation area. The latter can be a controlled inundation by the military or a measure to prevent inundation of a larger area surrounded by levees. Levees have also been built as field boundaries and as military defences. More on this type of levee can be found in the article on dry-stone walls.

Levees can be permanent earthworks or emergency constructions (often of sandbags) built hastily in a flood emergency. When such an emergency bank is added on top of an existing levee it is known as a cradge.[2]

Levees were first constructed in the Indus Valley Civilization (in Pakistan and North India from circa 2600 B. C.) on which the agrarian life of the Harappan peoples depended. [3] Also levees were constructed over 3000 years ago in ancient Egypt, where a system of levees was built along the left bank of the River Nile for more than 600 miles (966 km), stretching from modern Aswan to the Nile Delta on the shores of the Mediterranean.[4] The Mesopotamian civilizations and ancient China also built large levee systems. Because a levee is only as strong as its weakest point, the height and standards of construction have to be consistent along its length. Some authorities have argued that this requires a strong governing authority to guide the work, and may have been a catalyst for the development of systems of governance in early civilizations. However others point to evidence of large scale water-control earthen works such as canals and/or levees dating from before King Scorpion in Predynastic Egypt during which governance was far less centralized.

Levees are usually built by piling earth on a cleared, level surface, broad at the base, they taper to a level top, where temporary embankments or sandbags can be placed.[5] Because flood discharge intensity increases in levees on both river banks, and because silt deposits raise the level of riverbeds, planning and auxiliary measures are vital. Sections are often set back from the river to form a wider channel, and flood valley basins are divided by multiple levees to prevent a single breach from flooding a large area. A levee made from stones laid in horizontal rows with a bed of thin turf between each of them is known as a spetchel.

Artificial levees require substantial engineering. Their surface must be protected from erosion, so they are planted with vegetation such as Bermuda grass in order to bind the earth together (see Fig.11.1). On the land side of high levees, a low terrace of earth known as a banquette is usually added as another anti-erosion measure. On the river side, erosion from strong waves or currents presents an even greater threat to the integrity of the levee. The effects of erosion are countered by planting with willows, weighted matting or concrete revetments. Separate ditches or drainage tiles are constructed to ensure that the foundation does not become waterlogged.

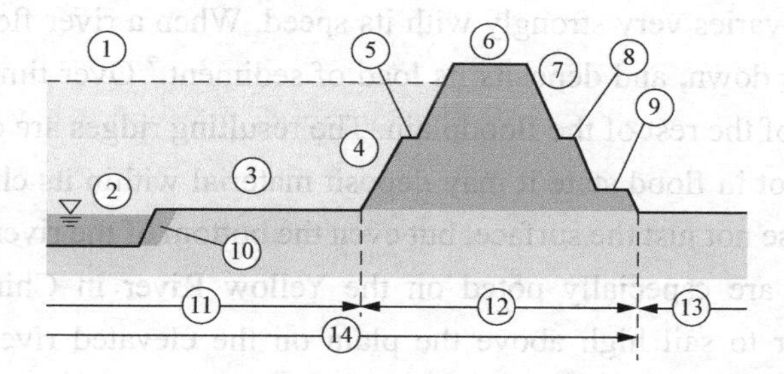

Fig.11.1 Schematic Cross Section Figure of the River Levees

①—Design High Water Level (HWL); ②—Low water channel; ③—Flood channel; ④—Riverside Slope; ⑤—Riverside Banquette; ⑥—Levee Crown; ⑦—Landside Slope; ⑧—Landside Banquette; ⑨—Berm; ⑩—Low water revetment; ⑪—Riverside land; ⑫—Levee; ⑬—Protected lowland; ⑭—River zone

1. River flood prevention

A levee keeps high water on the Mississippi River from flooding Gretna, Louisiana, in March 2005. Prominent levee systems exist along the Mississippi River and Sacramento River in the United States, and the Po, Rhine, Meuse River, Loire, Vistula, the river delta in the Netherlands and Danube in Europe.

The Mississippi levee system represents one of the largest such systems found anywhere in the world. They comprise over 3500 miles (5600 km) of levees extending some 1000 miles (1600 km) along the Mississippi, stretching from Cape Girardeau, Missouri to the Mississippi Delta. They were begun by French settlers in Louisiana in the 18th century to protect the city of New Orleans. The first Louisianan levees were about 3 feet (0.9 m) high and covered a distance of about 50 miles (80 km) along the riverside. By the mid-1980s, they had reached their present extent and averaged 24 feet (7 m) in height; some Mississippi levees are as much as 50 feet (15 m) high. The Mississippi levees also include some of the longest continuous individual levees in the world. One such levee extends southwards from Pine Bluff, Arkansas for a distance of some 380 miles (611 km).

2. Coastal flood prevention

Levees are very common on the flatlands bordering the Bay of Fundy in New Brunswick and Nova Scotia Canada. The Acadians who settled the area can be credited with construction of most of the levees in the area, created for the purpose of farming the fertile tidal flatlands. These levees are referred to as "aboiteau". In the Lower Mainland around the city of Vancouver, British Columbia, there are levees to protect low-lying land in the Fraser River delta, particularly the city of Richmond on Lulu Island. There are also levees to protect other locations which have flooded in the past, such as land adjacent to the Pitt River and other tributary rivers.[6]

Part 2 Natural Levees

Levees are commonly thought of as man-made, but they can also be natural. The ability of a river to carry sediments varies very strongly with its speed. When a river floods over its banks, the water spreads out, slows down, and deposits its load of sediment.[7] Over time, the river's banks are built up above the level of the rest of the floodplain. The resulting ridges are called natural levees.

When the river is not in flood state it may deposit material within its channel, raising its level. The combination can raise not just the surface, but even the bottom of the river above the surrounding country. Natural levees are especially noted on the Yellow River in China near the sea where oceangoing ships appear to sail high above the plain on the elevated river. Natural levees are a common feature of all meandering rivers in the world.

Part 3 Levees in Tidal Waters

The basic process occurs in tidal creeks when the incoming tide carries mineral material of all grades up to the limit imposed by the energy of the flow. As the tide overflows the sides of the creek towards high water, the flow rate at the brink slows and larger sediment is deposited, forming the levee. At the height of the tide, the water stands on the salt-marsh or flats and the finer particles slowly settle, forming clay. In the early ebb, the water level in the creek falls leaving the broad expanse of water standing on the marsh at a higher level.

The area of water on the marsh is much greater than the water surface of the creek so that in the latter, the flow rate is much greater.[8] It is this rush of water, perhaps an hour after high water, which keeps the creek channel open. The cross-sectional area of the water body in the creek is small compared with that initially over the levee which at this stage is acting as a weir. The deposited sediment (coarse on the levee and on the mud flats or salt-marsh) therefore tends to stay put so that, tide by tide, the marsh and levee grow higher until they are of such a height that few tides overflow them. In an active system, the levee is always higher than the marsh. That is how it came to be called "une rive levée", or raised shore.

Part 4 Levee Failures and Breaches

Man-made levees can fail in a number of ways. The most frequent (and dangerous) form of levee failure is a breach. A levee breach is when part of the levee actually breaks away, leaving a large opening for water to flood the land protected by the levee.[9] A breach can be a sudden or gradual failure that is caused either by surface erosion or by a subsurface failure of the levee. Levee breaches are often accompanied by levee boils, or sand boils. A sand boil occurs when the upward pressure of water flowing through soil pores under the levee (underseepage) exceeds the downward pressure from the weight of the soil above it. The underseepage resurfaces on the landside, in the form of a volcano-like cone of sand. Boils signal a condition of incipient instability which may lead to erosion of the levee toe or foundation or result in sinking of the levee into the liquefied foundation below. Complete breach of the levee may quickly follow.

Sometimes levees are said to fail when water overtops the crest of the levee. Levee overtopping can be caused when flood waters simply exceed the lowest crest of the levee system or if high winds begin to generate significant swells in the ocean or river water to bring waves crashing over the levee.[10] Overtopping can lead to significant landside erosion of the levee or even be the mechanism for complete breach. Properly built levees are armored or reinforced with rocks or concrete to prevent erosion and failure.

*** * * * * * * * * * * * * * * * * Explanations * * * * * * * * * * * * * * * * ***

*** * * * * * * * * * * * * * * * * New Words and Expressions * * * * * * * * * * * * * * * * ***

1. coast [kəust] n. 海岸，海滨
2. participle [ˈpaːtiˌsipəl] n. 分词
3. New Orleans 新奥尔良
4. adjoin [əˈdʒɔin] vt. 贴近，与……毗连
5. inundation [ˌinənˈdeiʃən] n. 淹没
6. military [ˈmilitəri] adj. 军事（用）的 n. [the~] 军队，武装力量
7. defence/defense [diˈfens] n.防御，保卫；[pl.] 防御工事；辩护
8. hastily [ˈheistili] adv. 急速地；草率地
9. emergency [iˈmə: dʒənsi] n. 紧急情况，不测事件，非常时刻
10. agrarian [əˈgreəriən] adj. 农业的
11. Mediterranean [ˌmedıtəˈreini: ən] n.地中海海岸
12. Mesopotamian Civilization 美索不达米亚文明
13. catalyst [ˈkætəlist] n. 催化剂，促使事情发展的因素，刺激因素
14. auxiliary [ɔ:gˈziljəri] adj. 辅助的，补助的；备用的，后备的
15. vital [ˈvaitl] adj. 极其重要的，生死攸关的，充满生机的
16. turf [tə: f] n. 草皮
17. terrace [ˈterəs] n. 阶梯；看台；排屋；地坪，草坪；[pl.] 梯田
18. banquette [ˈbæŋkwit] n. 护坡道，凸部，人行道，弃土堆
19. willow [ˈwiləu] n. 柳，柳树
20. mat [mæt] n. 席子，垫子
21. revetment [riˈvə:t mənt] n. 护岸
22. separate [ˈsepəreit] v. 使分离；区分；

分居

adj. 分离的，各自的

23. ditch [ditʃ] n. 沟，渠道
24. tile [tail] n. 瓦，瓷砖

vt. 铺瓦于，贴瓷砖于

25. Mississippi River 密西西比河
26. prominent [ˈprɔminənt] adj. 杰出的，突出的；凸起的
27. flatland [ˈflæt, lænd, -lənd] n. 平原，平坦地
28. spread [spred] v. 展开；散布，涂

n. 传播；幅度，范围

29. ridge [ridʒ] n. 脊，山脊，垄，埂，脊状突起
30. brink [briŋk] n. （悬崖、河流等的）边缘，边沿
31. fine[ˈfain] adj. 纤细的，尖细的，细小的
32. ebb [eb] n. 退潮，落潮

v. 退潮，落潮；减少，衰落

33. marsh [mɑːʃ] n. 沼泽，湿地
34. rush [rʌʃ] v. 冲；仓促从事；突袭

n. 冲；匆忙；繁忙时刻

35. cross-sectional adj. 过流断面的
36. tide by tide 一浪接着一浪
37. breach [briː tʃ] n. 缺口；破坏；不和

vt. 攻破；破坏

38. accompanied[əˈkʌmpənid] adj. 伴随的
39. pore [pɔː] n. 毛孔，细孔

vi. (over)仔细阅读，审视

40. sink [siŋk] n. 水槽

v. 陷入，衰退；下沉；降低；掘

41. liquefy [ˈlikwifai] v. 液化，溶解
42. overtop [ˈəuvəˈtɔp] n. 溢过顶部

* * * * * * * * * * * * * * * * * Complicated Sentences * * * * * * * * * * * * * * * * *

1. Levees can be mainly found along the sea, where dunes are not strong enough, along rivers for protection against high-floods, along lakes or along polders.

【译文】防潮堤主要用在沿海地区，那里淤积的沙丘不够稳定，防潮堤用来保护低田，或者沿河流、湖泊两岸建造堤防，防止大洪水。

【说明】词句为简单句，3 个介词 along 表达了堤防常见的位置；句中 where 引导状语从句，进一步解释沿海地区的特征。

2. When such an emergency bank is added on top of an existing levee it is known as a cradge.

【译文】当紧急筑堤是在现有堤防的顶部继续修筑时，一般称为子埝。

【说明】此句不长，但含有时间状语从句，以及一个后置的省略"that"的主语从句（that it is known as a cradge），进一步解释 emergency bank。

3. Levees were first constructed in the Indus Valley Civilization (in Pakistan and North India from circa 2600 B.C.) on which the agrarian life of the Harappan peoples depended.

【译文】最早的堤防建造于印度河文明时期(大约公元前 2600 年的巴基斯坦和印度北部)，是由哈拉帕人修筑的。

【说明】which 引导的定语从句作 depended on 的宾语，其中 on 提到了 which 前面。

4. Also levees were constructed over 3000 years ago in ancient Egypt, where a system of levees was built along the left bank of the River Nile for more than 600 miles (966 km), stretching from modern Aswan to the Nile Delta on the shores of the Mediterranean.

【译文】同时，河流两岸的堤坝也建造在距今有 3000 多年前的古埃及，曾建成了沿尼罗河左岸超过 600 英里（966km）长的堤坝，它绵延在地中海海岸地区，从阿斯旺一直到尼罗河三角洲。

【说明】句中 where 引导的定语从句修饰前面的 Egypt，后面的 streching 引导分词短语进一步解释 600 miles 的具体地点和位置。

5. Levees are usually built by piling earth on a cleared, level surface, broad at the base, they taper to a level top, where temporary embankments or sandbags can be placed.

【译文】堤防通常是在干净且平整的地面上打桩建造的，堤防的底部较宽，向上则逐渐变窄，顶部可以再加放临时土堤或沙袋。

【说明】they taper to a level top 前面省略“that”；where 引导定语从句，修饰前面的 top。

6. There are also levees to protect other locations which have flooded in the past, such as land adjacent to the Pitt River and other tributary rivers.

【译文】防潮堤还可以保护过去已被淹没的其他地区，如毗邻皮特河和其他河流支流的地区。

【说明】句中 which 指代前面的 other locations。

7. When a river floods over its banks, the water spreads out, slows down, and deposits its load of sediment.

【译文】当洪水超出堤防时，水溢出而使水的速度变慢，进而使泥沙沉积。

【说明】此句为含有时间状语从句（when）的复合句，主句中有 3 个动词（spread、slow、deposit）。

8. The area of water on the marsh is much greater than the water surface of the creek so that in the latter, the flow rate is much greater.

【译文】该湿地中的水位大大高于溪流的水位，因此，后者中水的流速更大。

【说明】原文为比较句，出现了有 than、greater 这样的比较词。

9. A levee breach is when part of the levee actually breaks away, leaving a large opening for water to flood the land protected by the levee.

【译文】当堤防决口时，部分堤防实际上已经失效，决口处大量的洪水冲向原来堤防保护的土地。

【说明】此句描述了堤防决口的具体情形，用现在分词短语（leaving）做状语表明决口形态，后面的过去分词（protected）短语为后置定语，修饰 land。

10. Levee overtopping can be caused when flood waters simply exceed the lowest crest of the levee system or if high winds begin to generate significant swells in the ocean or river water to bring waves crashing over the levee.

【译文】当洪水超出最低的堤防顶部，或者强风开始在海洋或是河流上产生海浪或波浪，冲击堤顶时，可能会发生漫顶。

【说明】两个“or”具体表达了漫顶发生的具体条件，且又用时间状语从句（when）条件状语从句（if）及分词短语（crashing）表达不同的情况。

******************** Summary of Glossary ********************

1. levee 防洪堤，码头，大堤
2. aboiteau 防潮堤，挡潮闸
3. bank 堤，岸
4. cradge 堤坝
5. empolder 海边围垦
6. polder 低洼开拓地（尤指荷兰等国围海造的低田），圩田，围垦地
7. marsh 湿地，沼泽，沼泽地

******************** Abbreviations (Abbr.) ********************

1. B.C. Before Christ 公元前
2. A.D. Anno Domini 公元 [拉丁文]

********************** Exercises ********************

(1) A levee, levée, dike (or dyke), ______ , floodbank or stopbank is a natural or artificial slope or wall to regulate water levels.

(2) Levees can be __________ earthworks or emergency constructions (often of sandbags) built ________ in a flood _________.

(3) The ability of a river to carry ________ varies very strongly with its speed. When a river floods over its banks, the water __________ out, slows down, and deposits its load of sediment.

************ Word Building (11) co-; ex- ************

1. co- [前缀]，表示：共，同，相互，联合，伴同

exist	vi.	coexist	vi.	共存
operate	n./v.	cooperate	vi.	合作，协作
efficient	adj.	coefficient	adj.	共同作用的
ordination	n.	coordination	n.	配合；协调
relation	n.	correlation	n.	相互关系，相关(性)
worker	n.	co-worker	n.	合作者，同事，帮手

2. ex- [前缀]，表示：出，外

-pand		expand	v.	膨胀
-ternal		external	adj.	外部的
-clude		exclude	vt.	除外
port	n.	export	n.	出口
-cess		excess	n.	超出
tract	n.	extract	vt.	摘出

****************** **Text Translation** ******************

第十一章　堤　　防

河堤、堤防（或堤）、挡洪水或止水墙是天然或人工的调节水位的斜坡或墙壁。它通常是土质的，而且往往平行于河流或海岸。（单词个数：38）

堤防这个词（是个外来词），起源于法语“levée”（是法语动词“抬高”的阴性过去分词，或“提高”）引入到美式英语后，主要用于中西部地区和南内部地区；大约在1720年，引入英式英语，主要在新奥尔良使用。（单词个数：42）

第一节　人工堤防

人工堤防的主要目的是防止洪水淹没毗邻的农村及田地；然而，它们也影响河流的流动，从而导致更高、更大的水流量。大部分堤防用在沿海地区，那里淤积的沙丘不够稳定，沿海岸线用堤防保护低田，或沿河流、湖泊岸边建设堤防，或防止高洪水。此外，防洪堤的建成主要为保护田地，或作为淹没区边界。近年来，堤防则是一种防止大洪水淹没的策略或措施，有时甚至是出于军事目的考虑的。因此，堤防已成为地区的边界和军事防御壁垒，这种类型的堤防一般是建在干石墙的上层。（单词个数：125）

堤防可以是永久性的土方工程，或者也可以是在发生洪水时的紧急情况下的临时建设（通常用沙包）的。当这样的紧急筑堤是在现有堤防的顶部继续修筑时，一般称为子埝或临时堤坝。（单词个数：36）

最早的堤防建造于印度河文明时期（大约公元前2600年的巴基斯坦和印度北部），是这土地上赖以生存的哈拉帕人修筑的。同时，河流两岸的堤防也建造在距今有3000多年前的古埃及，曾建成了沿尼罗河左岸超过600英里（966km）长的堤防，绵延地中海海岸地区，从阿斯旺一直到尼罗河三角洲。美索不达米亚文明和古代中国也曾建造过大型堤防系统。由于堤防自身的强弱点并存，高度和建设的标准都与它的长度相关联。因此，一些人认为，这需要一个强大的管理机构来指导堤防的修建工作，并且这可能是早期文明发展的催化剂。但是，另一些人则指出，一些大型的水控制工程实例，如运河或堤防，它们的历史可以追溯到埃及王朝统治前，早期的直布罗陀人治理时期是很疏散的，并不是很集中。（单词个数：174）

堤防通常是在干净并平整的地面上打桩建造的，堤防的基础比较宽阔，向上则逐渐变窄，顶部可以再继续临时加筑土堤或放置沙袋。由于河道两岸堤防内的泄洪流量可能逐渐增加，再者，由于泥沙沉积也会使河床升高，因此，堤防的规划和配套措施至关重要。往往是把分开的河流融合在一起，形成更宽阔的河道，洪泛区则用堤坝分成多个，以防止大面积的洪水泛滥。堤防由水平并列放置的石块及薄层草坪间隔筑成时，称为护岸。（单词个数：114）

人工堤坝是一项需要大量人力和物力的工程。其表面必须受到保护以防止侵蚀，因此，可以种植草木，如狗牙根草，用来约束住地面的泥土（不被流失）。对高堤岸边侧，一般建有比较低的平台，有时作为行人道，其实也是一种抗侵蚀的措施（见图 11.1）。从河流方面考虑，堤防内大浪或水流造成的侵蚀，是一个更大的威胁。此时可以种植柳树、加固地基或修筑混凝土护岸来防止侵蚀影响。有时也需要开挖单独的沟渠或放置排水瓦管，以确保地基不会淹水。（单词个数：104）

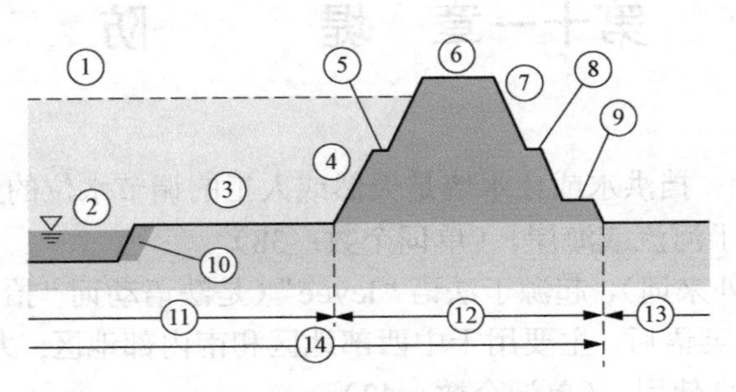

图 11.1　河堤横断面示意图

①—Design High Water Level (HWL); ②—Low water channel; ③—Flood channel; ④—Riverside Slope; ⑤—Riverside Banquette; ⑥—Levee Crown; ⑦—Landside Slope; ⑧—Landside Banquette; ⑨—Berm; ⑩—Low water revetment; ⑪—Riverside land; ⑫—Levee; ⑬—Protected lowland; ⑭—River zone

1. 河流防洪

2005 年 3 月，密西西比河的堤防有效地防御了路易斯安那州格雷特纳洪水。美国的密西西比河和萨克拉门托河、意大利北部的波河、西欧的默兹河、莱茵河、法国的卢瓦尔河、波兰的维斯瓦河、荷兰的河流三角洲以及欧洲的多瑙河等都有坚固的堤防系统。（单词个数：49）

密西西比河的堤防是世界上最大的堤防系统之一。它包括 3500 英里（5600km）的防洪堤，沿密西西比河延伸了约 1000 英里（1600km），从密苏里州的开普吉拉多一直到密西西比河三角洲。最初，该堤防是由 18 世纪在路易斯安那州的法国定居者们开始修建的，用来保护新奥尔良市。第一段路易斯安那防洪堤约 3 英尺（0.9m）高，沿河岸约 50 英里（80km）长。到 20 世纪 80 年代中期，已经达到现在的规模，即堤高平均 24 英尺（7m）；其中一些防洪堤高达 50 英尺（15m）高。密西西比河防洪堤中有一些是世界上连续最长的堤防。该堤防从阿肯色州的派恩布拉夫向南，延伸距离大约 380 英里（611km）。（单词个数：148）

2. 沿海防洪

在加拿大的与平原接壤的芬迪湾，以及新不伦瑞克省和新斯科舍省，堤防是非常普遍的。在这个地区居住的阿卡迪人的修建堤防意见均被采纳，其目的是在肥沃平原上发展农业，这些堤称为“防潮堤”。在海拔低的内地城市温哥华、不列颠哥伦比亚省都有堤防，用以保护低洼土地，如菲沙河三角洲地区，特别是里士满城市的露露岛。防潮堤还可以保护过去已被淹没的其他地区，如毗邻的皮特河和其他支流河流。（单词个数：113）

第二节　天　然　堤　防

普遍认为堤防是人造的，但它们也可以是天然的。河流通过它的快速水流运送泥沙的能力是变化多端的。当洪水超出堤防时，洪水溢出而使水流速度变慢，进而使泥沙沉积。随着时间的推移，河流两岸都形成了洪泛漫滩，由此产生的两岸山脊称为天然堤防。（单词个数：70）

当河流不在洪水地区时，可能因为在其河道中的泥沙沉积而使它的水位升高。升高的不单单是地表水位，而且甚至是周围村庄的地下水位也跟着升高。特别值得关注的天然堤防是，中国黄河的近海区出现了远洋轮船航行高于平原上的高架河现象。世界上许多游荡型的河流中，天然堤防是一种常见的现象之一。（单词个数：78）

第三节 海湾防潮堤

最常见的海湾中的潮汐流携带着不同大小的矿物，当这些矿物足以限制水流的流动能量时，就会在海湾中发生淤积现象。随着潮汐流从海湾的一端溢出直到高水位时，河流的流速减缓，而后大量的沉积物堆积起来，形成了防潮堤。在高速的潮流冲刷下，盐沼或单位细颗粒慢慢细化分解，形成黏土。初退潮时，海潮流入广阔的地域，湿地中的水位更高一些。（单词个数：102）

该湿地中的水位大大高于海湾面，因此，其流速更大。正是这种急流，大约高潮后 1h，使得海湾更开阔。海湾中这些有代表性的水体与最初漫过防潮堤的相比要少很多，此时的海堤称之为堰。因此，水中的沉积物（有粗大的块体、泥浆或盐沼）往往留在原地的，而后，一浪接着一浪，沼泽和堤防逐渐增高，直到它们达到了一个水流很难溢过的高度。潮汐活跃地区的堤防总是高于湿地，这也是为什么它们被称作“防潮堤”或“抬升的岸边”的原因。（单词个数：143）

第四节 溃堤以及决口

人造堤防可能会发生很多形式的溃决。最常见的形式是决口。当堤防决口时，部分堤防实际上已经失效，决口处大量的洪水冲向原来堤防保护的土地。决口可能突然发生或逐渐扩大而发生，它要么是由表面侵蚀引起，要么是由堤防地下渗漏引起的。堤防决口往往伴随堤防管涌，或沙涌。当穿过堤防（地下边界）土壤孔隙的水流向上的压力超过了向下的土壤重力时，出现了管涌。这种地下渗漏逐渐浮出地面时，它的形式像火山锥的沙子一样。管涌由最初的不稳定，可能导致侵蚀堤防基部或底部，或导致堤防沉没，发生地基液化。也可能迅速导致堤防完全溃决。（单词个数：166）

据说，有时候堤坝溃决是由于洪水溢过堤顶。当洪水简单地超出最低的堤防顶部，或者是强风开始在海洋或是河流上产生海浪或波浪，冲击堤顶时，可能会发生漫顶。漫顶可导致重大的滑坡侵蚀堤坝，甚至完全决口。可以适当修建岩石或混凝土等坚固的堤坝，以防止土壤侵蚀和溃堤的发生。（单词个数：87）

Chapter 12 Hydropower Plants

Part 1 Hydroelectric Complex

Hydroelectric complex includes the reservoir structures and installations that use water power to produce electricity. Cross section of a typical hydroelectric complex is shown in Fig.12.1.

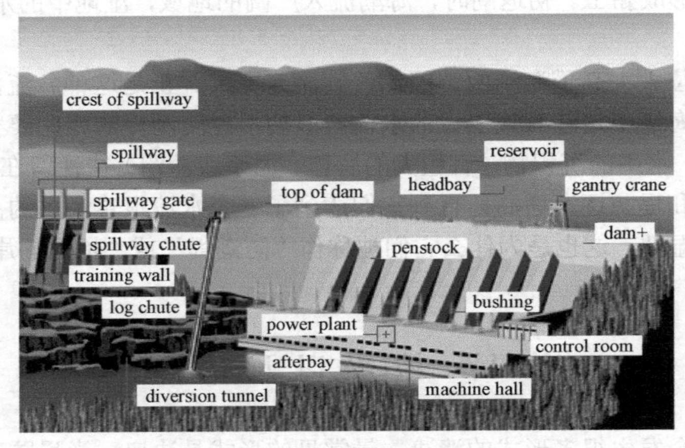

Fig.12.1 Cross Section of a Typical Hydroelectric Complex (http://www.visualdictionaryonline.com)

According to the basic functions in hydroelectric power generation, a hydroelectric complex can be separated into three major parts.

1. Dam and reservoir

Dam, the largest building in a hydroelectric complex, is the barrier built across a watercourse in order to build up a supply of water for use as an energy source. And the reservoir, basin formed by the construction of a dam, holds back a very large volume of water so that the flow rate can be controlled.[1] Top of dam, upper part of the dam, rises above the water level of the reservoir by several yards.

2. Spillway

Spillway is the channel that discharges excess water from the reservoir during flooding to avoid submerging the dam. The water discharging is controlled by spillway gate, a movable vertical panel; the reservoir's overflow is allowed to pass through when it is opened.[2] And the cement crests over which the reservoir's overflow discharge when the spillway gates are opened are called crests of spillway.

There are some inclined surfaces called spillway chutes along which discharged water flows out. The walls those separate the spillway chutes are called training wall; they are used to direct the water flow. Some underground conduits called diversion tunnel were used to divert water during construction. And there are often log chutes, structures that allow floating wood to travel from upstream to downstream of the dam.[3]

3. Power plant

Area that houses the generator units used to produce electricity is called machine hall. Area that contains the various control and monitoring devices required for the production of electricity is called control room.

Part of the reservoir immediately in front of the dam where the current originates is called headbay. Channel that carries water under pressure to the power plant's turbines is called penstock. And the area of the watercourse where water is discharged after passing through the turbines is called afterbay.

Part 2 Reservoir

The largest parts of a hydroelectric complex are often the dam and reservoir if needed. Details for the dam are introduced in Chapter 10.

A reservoir is an artificial lake used to store water. Reservoirs are often created by building a sturdy dam, usually out of concrete, earth, rock, or a mixture across a river or stream.[4] Once the dam is completed, the stream fills the reservoir. When a reservoir is mainly man-made, rather than being an adaptation of a natural basin, it may be called a cistern. A raw water reservoir does not simply hold water until it is needed. It is the first part of the water treatment process. The time the water is held for before it is released is known as the retention time. This is a design feature that allows particles and silts to settle out, as well as time for natural biological treatment using algae, bacteria and zooplankton that naturally live within the water.[5]

The more common dam across a valley relies on naturally formed features to form the water tight elements. Generally, engineers look for dam sites which are narrow with a broad area upstream; the valley sides can then act as natural walls and the broad area upstream makes a large reservoir for the height. The best place along the valley for building a dam has to be determined according to where the dam can best be tied into the valley walls and floor to form a watertight seal.

Where water is taken from a river of variable quality or quantity, it is common to construct bank-side reservoirs to store water pumped or siphoned from the river.[6] Such reservoirs are usually built partly by excavation and partly by the construction of a complete encircling bund or embankment. Both the floor of the reservoir and the bund must have an impermeable lining or core, often made of puddled clay. The water stored in such reservoirs may have a residence time of several months during which time normal biological processes are able to substantially reduce many contaminants and almost eliminate any turbidity.[7] The use of bank-side reservoirs also allows a water abstraction to be closed down for extended period at times when the river is unacceptably polluted or when flow conditions are very low due to drought.

Part 3 Other Structure for Hydroelectric Architecture

There are still some other structure designs for hydroelectric architecture.

Water intake is the structure that directs water from the headbay to the penstock. Penstocks are channels that carry water under pressure to the power plant's turbines, as shown in Fig.12.2. There is often assembly of bars, called screen, placed in front of the water intake to hold back anything that could hinder the operation of the turbine.

In order to make the turbine turn smoothly, scroll case is used to distribute water uniformly around it, which is duct shaped like a spiral staircase as shown in Fig.12.3.

Fig.12.2 Penstocks (http://www.fwee.org)

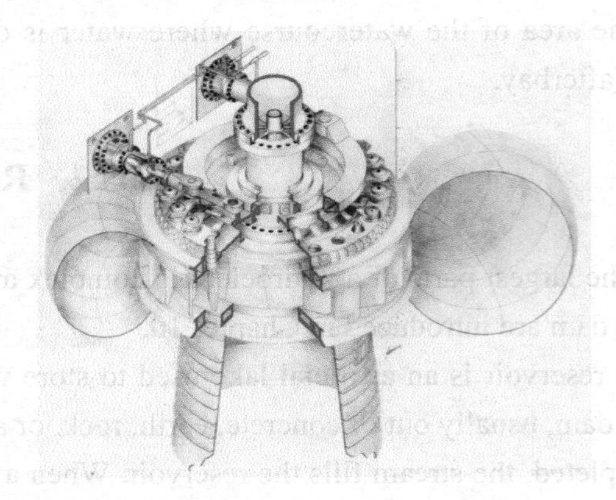

Fig.12.3 Scroll Case (http://www.itaipu.gov.br)

After the water has been used to drive the turinbe, the conduit named draft tube at the base of the turbine, as shown in Fig.12.4, is used to increase the runner's output by reducing the pressure of the water as it exits.[8]

Channel that discharges water toward the afterbay in order to return it to the watercourse is called tailrace. There are also some movable vertical panels used as gates that control the discharge of water to the tailrace, as shown in Fig.12.5.

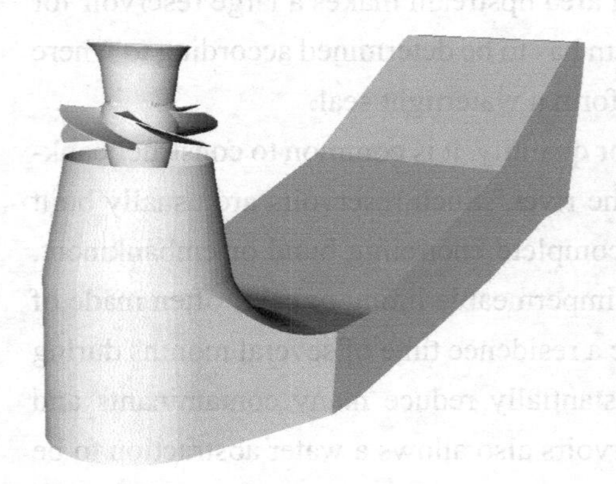

Fig.12.4 Draft Tube (http://www.tfd.chalmers.se)

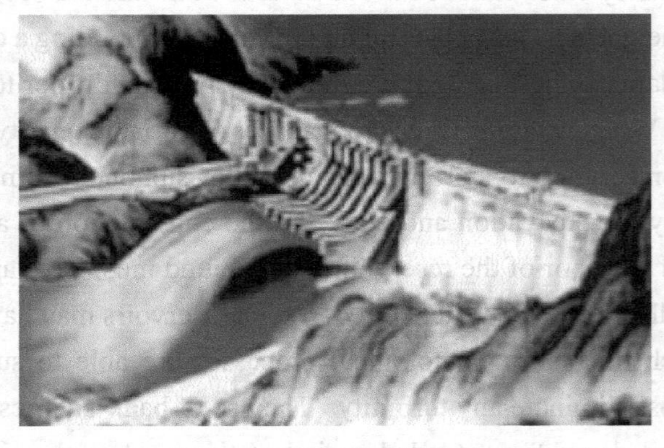

Fig.12.5 Discharge of Water in Longtan Hydropower Station

There are still some gantry cranes needed for hoisting device in the form of a bridge, as shown in Fig.12.6; it moves along rails. Hoisting device that travels along aboveground parallel rails is used to lift and carry heavy loads. Fig.12.7 shows the traveling crane.

Fig.12.6 Gantry Crane

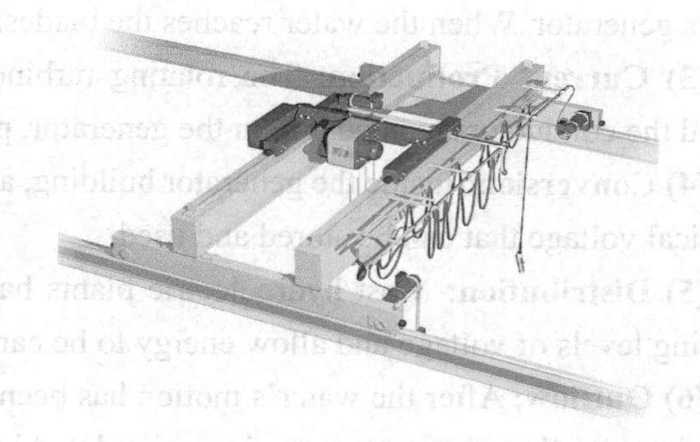

Fig.12.7 Traveling Crane (http://www.directindustry.com)

Inside the dam, there is necessary access gallery, which is underground passageway that provides access to various parts of the dam so that it can be inspected and maintained.

Part 4 Generating Process of Hydroelectric Power

Hydroelectric power plant produces electricity from energy generated by flowing water. In other words, hydroelectric power plant uses an energy source, here water, and converts it into electricity. Fig.12.8 shows the cross section of a typical hydroelectric power plant.

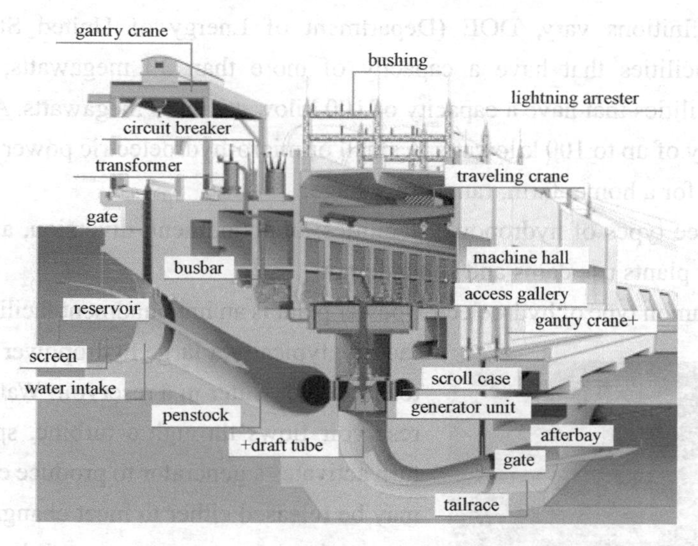

Fig.12.8 Cross Section of a Typical Hydroelectric Power Plant

Besides those parts described above, there are still some devices related to electrical production, such as busbar, transformer, circuit breaker and lightning arrester.

The stages in a typical hydroelectric power generating process include:

(1) Intake: When the dam opens its entrances the water flows into a penstock that channels it toward the turbines and builds up pressure as the water moves.

(2) Turbine Rotation: A turbine has vertical propeller blades set along a shaft linked to the plant's generator. When the water reaches the blades, it causes the turbine to turn along its axis.

(3) Current Production: The rotating turbine creates a corresponding rotation of magnets around the conductors located within the generator, providing an alternating current.

(4) Conversion: Inside the generator building, a transformer changes the alternating current into electrical voltage that can be stored and used.

(5) Distribution: Most hydroelectric plants have attached power lines that correspond to the differing levels of voltage and allow energy to be carried out of the plant.

(6) Outflow: After the water's motion has been harnessed, pipes carry it out of the plant where it continues to flow downstream or is re-circulated into the lower reservoir.

Part 5 Types of Hydropower Plants

Many dams were built for other purposes and hydropower was added later. In the United States, there are about 80 000 dams of which only 2400 produce power. The other dams are for recreation, farm ponds, flood control, water supply, and irrigation.

Hydropower facilities range in size from large power plants that supply many consumers with electricity to small and micro plants that individuals operate for their own energy needs or to sell power to utilities.

Although definitions vary, DOE (Department of Energy of United States) defines large hydropower as facilities that have a capacity of more than 30 megawatts, and defines small hydropower as facilities that have a capacity of 100 kilowatts to 30 megawatts. A micro hydropower plant has a capacity of up to 100 kilowatts. A small or micro-hydroelectric power system can produce enough electricity for a home, farm, ranch, or village.

There are three types of hydropower facilities: impoundment, diversion, and pumped storage. Some hydropower plants use dams and some do not.

The most common type of hydroelectric power plant is an impoundment facility. An impoundment facility, typically a large hydropower system, uses a dam to store river water in a reservoir. Water released from the reservoir flows through a turbine, spinning it, which in turn activates a generator to produce electricity. The water may be released either to meet changing electricity needs or to maintain a constant reservoir level.

Fig.12.9 The Tazimina Project in Alaska

A diversion, sometimes called run-of-river, facility channels a portion of a river through a canal or penstock. It may not require the use of a dam. The Tazimina project in Alaska, as shown in Fig.12.9, is an example of a diversion hydropower plant.

When the demand for electricity is low, a pumped storage facility stores energy by pumping water from a lower reservoir to an upper reservoir. During periods of high electrical demand, the water is released back to the lower reservoir to generate electricity.

* * * * * * * * * * * * * * * * * * Explanations * * * * * * * * * * * * * * * * *

* * * * * * * * * * * * * * * * * * New Words and Expressions * * * * * * * * * * * * * * * * *

1. afterbay ['ɑ:ftəbei] n. 下游池
2. chute [ʃu:t] n. 瀑布，斜道
3. conduit ['kəndit] n. 管道，导管，沟渠
4. diversion [dai'və: ʃən] n. 导流，分出，引出；转向，改道
5. headbay ['hedbei] n. [水] 上闸首，上游池
6. inclined [in'klaind]nadj. 倾斜的
7. installation [, instə'leiʃən] n.（整套）装置，设备
8. cement [si'ment] n. 水泥，结合剂
9. bund [bund] n. 急坡堤岸
10. cistern ['sistən] n. 蓄水池；储水器
11. impermeable [im'pə: miəbl] adj. 不渗透的
12. lining ['lainiŋ] n. 内层，衬套
13. seal [si:l] n. 封，密封
14. siphon ['saifən] vt. 虹吸
15. turbidity [tə:'biditi] n. 混浊，混乱
16. zooplankton [ˌzəuə'plæŋktən] n. 浮游动物
17. draft tube 尾水管
18. duct [dʌkt] n. 管，输送管
19. gantry crane 龙门起重机，高架移动起重机
20. hinder ['hində] vt. 阻碍，打扰
21. device 起重设备
22. intake ['inteik] n. 入口，进口
23. passageway ['pæsidʒwei] n. 过道，出入口
24. screen [skri: n] n. 筛子，掩蔽物
25. tailrace ['teilreis] n. （水轮，涡轮的）尾水，放水路
26. spiral ['spaiərəl] adj. 螺旋形的
27. alternating current 交流（电）
28. magnet ['mægnit] n. 磁体，磁铁
29. motion ['məuʃən] n. 运动，动作
30. power line 电力线，输电线
31. propeller blade 螺旋桨叶片
32. rotation [rəu'teiʃn] n. 旋转
33. shaft [ʃa:ft] n. 轴，杆状物
34. transformer [træns'fɔ:mə] n. 变压器
35. voltage ['vəultidʒ] n. 电压，伏特数
36. activate ['æktiveit] vt. 刺激，使活动
37. megawatt ['megəwɔt] n. 兆瓦（特）
38. utilities [ju: 'tilitiz] n. 公共事业，电力公司
39. capacity [kə'pæsiti] n. 容量，生产量
40. pumped storage 抽水蓄能
41. impoundment [im'paundmənt] n. 蓄水，积水，被坝所围住的水

* * * * * * * * * * * * * * * * * * Complicated Sentences * * * * * * * * * * * * * * * * * *

1. And the reservoir, basin formed by the construction of a dam, holds back a very large

volume of water so that the flow rate can be controlled.

【译文】水库（大坝拦河形成的盆地）汇集大量的水，从而使其流量可控。

【说明】句中 hold back 意思是“阻止”。英文中常用由两个逗号分隔的名词短语对前面出现的名词进行解释说明，两个逗号的作用相当于汉语中的括号。

2. The water discharging is controlled by spillway gate, a movable vertical panel; the reservoir's overflow is allowed to pass through when it is opened.

【译文】放水由泄洪道闸门（可移动的竖直平板）控制，当闸门其打开时，允许水库的水溢流通过。

【说明】a movable vertical panel 是对 spillway gate 的解释说明，用法参见上一条。overflow 表示水库中排放过量的水时形成的水流。when 后面的 it 指代 spillway gate。

3. And there are often log chutes structures that allow floating wood to travel from upstream to downstream of the dam.

【译文】常常还有运木通道，即允许漂浮的木材从大坝上游行进到下游的结构。

【说明】句中 that 引导定语从句，修饰 structures。

4. Reservoirs are often created by building a sturdy dam, usually out of concrete, earth, rock, or a mixture across a river or stream.

【译文】水库往往是通过修建横断河流的大坝（通常用混凝土、土料、岩石或其混合物）而形成。

【说明】句中 usually out of concrete, earth, rock, or a mixture 是对 dam 的补充说明，而句子的主体是 Reservoirs are often created by building a sturdy dam across a river or stream。

5. This is a design feature that allows particles and silts to settle out, as well as time for natural biological treatment using algae, bacteria and zooplankton that naturally live within the water.

【译文】这种设计风格允许颗粒和沉积物沉淀下去，也可让水中天然存在的藻类、细菌和浮游动物进行自然生物学处理。

【说明】as well as 表示并列，意思是“也，又”。

6. Where water is taken from a river of variable quality or quantity, it is common to construct bank-side reservoirs to store water pumped or siphoned from the river.

【译文】在从水质和水量会变化的河流中取水的场合，通常建造岸边水库来存储用水泵或虹吸管从河流中抽取的水。

【说明】to construct bank-side …不定式短语为真正主语，it 是形式主语。

7. The water stored in such reservoirs may have a residence time of several months during which time normal biological processes are able to substantially reduce many contaminants and almost eliminate any turbidity.

【译文】存储在这种水库中的水，可以有几个月的存留时间，期间正常的生物学过程能够持续减少很多污染物，并几乎消除任何混浊。

【说明】during which 引导定语从句，修饰前面的 months。

8. After the water has been used to drive the turinbe, the conduit named draft tube at the base of the turbine, as shown in Fig.12.4, is used to increase the runner's output by reducing the

pressure of the water as it exits.

【译文】水通过水轮机之后，就到了位于水轮机底部的尾水管（如图 12.4 所示），该管道用于在水流出时通过降低水压来增大叶轮的输出。

【说明】句子中的状语较多，翻译时需要合理地确定表达顺序。

*** * * * * * * * * * * * * * * * * * Summary of Glossary * * * * * * * * * * * * * * * * * ***

1. watercourse	水道，河道
2. spillway	溢洪道，泄洪道
3. penstock	压力水管，水渠，水道
4. training wall	导流墙
5. spillway chute	溢流坡道
6. diversion tunnel	导流隧洞
7. canal	运河，导管，槽，沟渠
8. log chute	原木通道
9. generator unit	发电机组
10. headbay	上游池
11. afterbay	下游池
12. reservoir	水库，蓄水池
13. cistern	水塔，蓄水池
14. bund	急坡堤岸
15. embankment	堤岸，路堤

*** * * * * * * * * * * * * * * * * * Abbreviations (Abbr.) * * * * * * * * * * * * * * * * * ***

DOE	Department of Energy	能源部 [美]

*** * * * * * * * * * * * * * * * * * * Exercises * * * * * * * * * * * * * * * * * * ***

(1) The _________, basin formed by the construction of a __________ , holds back ________ so that the flow rate can be controlled. Top of dam,_________, rises above the water level of the reservoir by several yards.

(2) The water discharging is controlled by ______________, a movable vertical panel; the reservoir's overflow is allowed to ___________ when it is ____________. And the _________ over which the reservoir's overflow discharge when the spillway gates are opened are called crests of spillway.

(3) Part of the reservoir immediately in front of the dam where the current originates is called __________. Channel that carries water under pressure to the power plant's turbines is called _________. And the area of the watercourse where water is discharged after passing through the turbines is called _________.

(4) The water stored in such reservoirs may ________________ during which time ________ are able to substantially reduce many contaminants and almost eliminate any turbidity.

*** * * * * * * * * * * * Word Building (12) im-; un- * * * * * * * * * * * * * ***

1. im- [前缀]，表示：否定的，相反的

balance	n./v.	imbalance	n.	不平衡，不均衡
material	n./adj	immaterial	adj.	不重要的，非本质的
mature	v./adj	immature	adj.	不成熟的，未完全发展的
palpably	adv.	impalpably	adv.	不能感知地，难以理解地
permeable	adj.	impermeable	adj.	不渗透性的

2. un- [前缀]，表示：否定，未

able	adj.	unable	adj.	不能的，不会的
acceptable	adj.	unacceptable	adj.	不可接受的，不合意的
avoidable	adj.	unavoidable	adj.	不可避免的
desirable	adj.	undesirable	adj.	不合需要的，不受欢迎的
predictable	adj.	unpredictable	adj.	不可预知的
stable	adj.	unstable	adj.	不稳定的
due	n./adj.	undue	adj.	不适当的
priced	adj.	unpriced	adj.	贵重的；无法估价的
desirable	adj.	undesirable	adj.	不受欢迎的
balanced	adj.	unbalanced	adj.	不均衡的；不平衡的
developed	adj.	undeveloped	adj.	不发达的，未开发的
protected	adj.	unprotected	adj.	无保护（者），无防卫的

*** Text Translation ***

第十二章 水电站

第一节 水电站枢纽

水电站枢纽包括水库和利用水能进行发电的整套设施。典型的水电站枢纽的横断面如图12.1所示。（单词个数：26）

根据水力发电的基本原理和功能，水电站枢纽可分为三个主要部分。（单词个数：18）

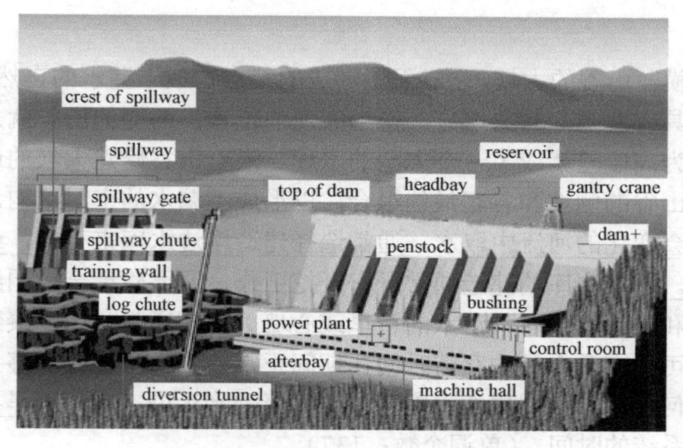

图 12.1 典型水电站综合建筑横断面

1. 大坝和水库

作为水电站枢纽里最大的建筑，大坝是为了蓄起发电用水而建造在河道上的挡水建筑物。水库（大坝拦河形成的盆地）汇集大量的水，从而使其流量可控。大坝的顶部，或其上部，要超出水库的水平面几码高。（单词个数：76）

2. 泄洪道

泄洪道是水库在汛期为避免水溢出大坝而排出多余水的通道。放水由泄洪道闸门（可移动的竖直平板）控制，在其打开时允许水库的水流通过。当泄洪道闸门打开，水库多余水量流过的水泥顶部称为泄洪道的堰顶。（单词个数：63）

放出的水流经的斜面称为溢流坡道。将坡道隔开的墙称为导流墙；它们用于引导水流。一些地下的管道称为导流隧洞，用于施工期分流。常常还有运木通道，即允许漂浮的木材从大坝上游输送到下游的结构。（单词个数：66）

3. 电站

安置发电机组的位置称为主机房。电力生产中需要的各种控制和监视装置所在的区域，称为控制室。（单词个数：33）

水库在大坝前面即水流上游的部分，称为上游池，把受压的水流导向电站水轮机的管道称为压力水管。流过水轮机的水汇集的河道称为下游池。（单词个数：49）

第二节 水 库

水电站枢纽的最大部分通常是大坝和水库。第十章介绍了大坝的详细内容。（单词个数：24）

水库是一个用于蓄水的人工湖。水库往往是通过修建横断河流的大坝（通常用混凝土、土料、岩石或其混合物）而形成。一旦大坝建成，水流就会流入水库。当水库主要由人工建造，而不是采用天然盆地时，可称为蓄水池。存储自然水的水库，不是简单地蓄水以备使用，它是水处理程序的第一步。水在泄放前的存储时间称为保持时间。这种设计允许泥沙和沉积物沉淀下去，也可让水中天然存在的藻类、细菌和浮游动物进行自然生物学处理。（单词个数：

136）

较为常见的是横断山谷的大坝，它靠天然形成的特点构成不漏水的自然环境。一般来说，工程师们寻找那些具有宽阔上游的地方，作为大坝的位置；两侧的山谷就是天然的墙壁，而广阔的上游地区因为高度差就成了一个大型的水库。沿着山谷修建大坝的最佳位置必须据此确定，即大坝可与山谷的侧坡和谷底最佳组合，以形成不漏水的密封结构。（单词个数：88）

从水质和水量会变化的河流中取水，通常选择建造岸边水库来存储，或用水泵、或虹吸管从河流中抽取水。这类水库的建造通常既有开挖工作，又要建造全包围结构的急坡堤岸或筑堤。水库的底层和堤岸都要有不渗水的衬里或心墙。它们通常是由混凝黏土构成。存储在这些水库中的水，可以有几个月的存留时间，期间正常的生物学过程能够持续减少很多污染物，并几乎消除任何混浊。使用岸边水库能够在河流受到污染，或是干旱导致水量减小的时候，延长关闭取水系统的时间。（单词个数：137）

第三节　水电站建筑的其他设施

水电站建筑里还有一些其他建筑设施。（单词个数：10）

引水口就是引导水从上游池进入压力水管的建筑物。压力水管是将受压的水导入水轮发电机的管道，如图 12.2 所示。入水口前通常装有称为滤网的栏杆，作用是阻隔所有会影响水轮机运行的杂物。（单词个数：58）

蜗壳常用来实现水轮机周围水的均匀分布，以使其平稳运行，蜗壳是一种螺旋梯形的管道，如图 12.3 所示。（单词个数：30）

图 12.2　压力水管

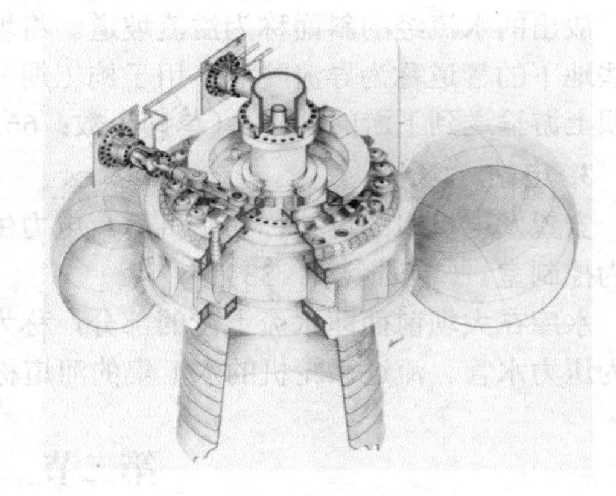

图 12.3　蜗壳

水流通过水轮机之后，就到了位于水轮机底部的尾水管（如图 12.4 所示），该管道用于在水流出时通过降低水压来增大叶轮的输出。（单词个数：43）

将水排进下游池以使其返回水源的管道称为泄水道。还有一些像闸门一样，控制向泄水道排水的可移动的垂直平板，如图 12.5 所示。（单词个数：41）

还有一些起重设备用的桥型龙门起重机，如图 12.6 所示，它沿铁轨移动。沿离地的水平轨道运行的起重设备用来吊起和移动重物。图 12.7 所示为一台移动式起重机。（单词个数：45）

大坝内部必须有交通廊道，它作为地下通道，为大坝的各部分之间提供通道，以实现大坝的检查和维护。（单词个数：29）

图 12.4　尾水管

图 12.5　水门

图 12.6　龙门起重机

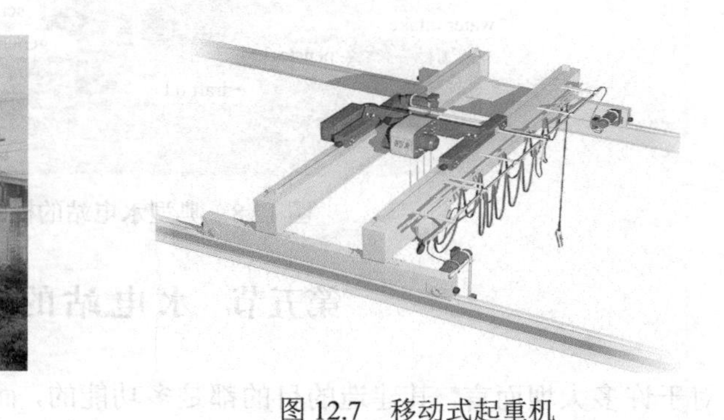

图 12.7　移动式起重机

第四节　水电站的发电过程

水电站利用流水的能量生产电能。换句话说，水电站将水的能量转化成了电能。图 12.8 所示为一个典型水电站的横断面。（单词个数：40）

除了上文描述的结构外，还有一些与电力生产相关的设施，如总线、变压器、断路器、避雷器。（单词个数：30）

典型水电站发电过程的阶段包括：（单词个数：10）

（1）进水口：当大坝引水口打开时，水流进了压力水管，压力水管通向水轮机，并给流水增加压力。（单词个数：27）

（2）水轮机转动：水轮机具有和发电机同轴相连的垂直的螺旋桨叶片。当水流接触到叶片时，推动水轮机绕其轴运动。（单词个数：32）

（3）电流产生：水轮机的旋转引起了磁场绕发电机中的导体相应旋转，于是产生了交流电。（单词个数：22）

（4）转换：在发电室内，变压器将交流电变为易于被存储和使用的电压值。（单词个数：20）

（5）配电：大多数水电厂都配备了对应不同电压等级的电力线，允许能量送出电厂。（单词个数：26）

（6）泄流：管道将利用后的水导出电站，然后水继续流入下游或重新流入下水库。（单词个数：28）

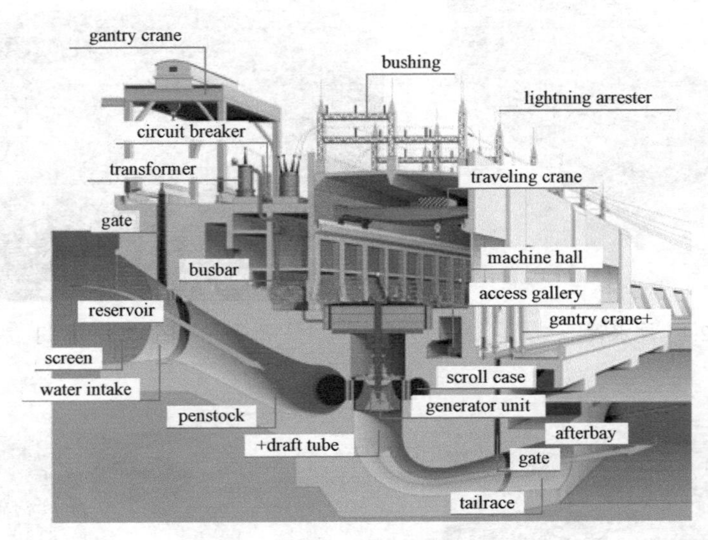

图 12.8　典型水电站的横断面

第五节　水电站的类型

对于许多大坝而言，其建造的目的都是多功能的，而水力发电是其中之一。在美国，大约有 80 000 座大坝，其中只有 2400 座发电。其他大坝的建设目的还有如风景旅游、养殖、防洪、供水以及灌溉等需要。（单词个数：41）

水电站的规模，包括从向许多消费者供电的大型水电站，到以满足自用或向电力公司售电为目的而独立运行的小型和微型水电站。（单词个数：34）

尽管定义各不相同，美国能源部（DOE）定义的大型水电站是具有 30MW 以上的容量的发电设施，而小型水电站是容量在 100kW 和 30MW 之间的发电设施。小水电站的容量上限是 100kW。一座小型或微型水电站可以产生足够一个家庭，农场，牧场，或村庄使用的电能。（单词个数：69）

水电站有三种类型：蓄水式，引水式和抽水蓄能电站。一些水电站有大坝，而另一些却没有。（单词个数：21）

水电站最普通的类型是蓄水式。蓄水式水电站，通常是利用大坝，将河水蓄在水库的大型水电系统，从水库放出的水流过水轮机，使其旋转，进而驱动发电机产生电能。放水的原因可能是为了满足发电需要，也可能是为了维持水库的蓄水量稳定。（单词个数：68）

引水式水电站，有时也称作径流式水电站，是利用沟渠或水道将河流分流引走而发电的电站。它可能不需要建造大坝。图 12.9 所示的阿拉斯加的塔齐米纳工程，就是一个引水式水电站的例子。（单词个数：43）

在电力需求低时，抽水蓄能电站通过将水从下游水库抽入上游水库的方式蓄能。在电力需求高时，再将水下泄到下游水库以产生电能。（单词个数：42）

图 12.9　阿拉斯加的塔齐米纳工程

Chapter 13 Water Environment

Part 1 Water Quality in Key River System

In the 1617 key sections under monitoring in the seven key river systems in 2016, the water quality of Grade I-III accounts for 67.8%. The water quality of Grade Ⅳ accounts for 16.8%. Grande Ⅴ and worse than Grade Ⅴ accounts for 15.5% (Fig.13.1). In the 184 national level controlling sections in the mainstreams of the seven key river systems, the sections with Grade I-III water quality account for 67.7%; Grade Ⅳ accounts for 19.2% and Grande Ⅴ and worse than Grande Ⅴ account for 13.1%. The water quality in the mainstreams of the river systems is better than that in the branches.

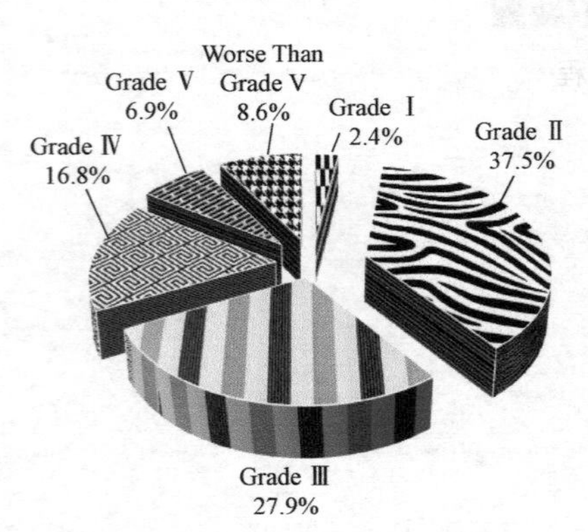

Fig.13. 1 Percentage of the Water Quality Grades of the Seven Key River Systems

The order of the pollution for serious level to light level in the seven key river systems in 2016 is: Hai River, Liao River, Huai River, Huang River, Song Hua River, Yangze River and Pearl River. Compared with the previous year, the order of the pollution level remained same and the pollution extent is quite close.

1. Yangtze River system

In the 510 sections for water quality monitoring, there are 59 mainstream sections, dominated by Grade Ⅱ water quality, and the percentage of the Grade Ⅰ, Ⅱ, Ⅲ and Ⅳ water quality is 2.7% , 53.5%, 26.1% and 9.6% (Fig.13.2). There are 451 sections in the major first level branches (Min River, Tuo River, Jialing River, Wu River, Han River, Xiang River, Zi River, Ruan River, Li River and Gan River), and the water quality of Grade Ⅰ-Ⅲ accounted for 82.3%. The major pollution indicators in the Yangtze River system are petroleum and ammonia nitrogen.

2. Yellow River system

There are 137 water quality monitoring sections, 31 of which are mainstream ones.[1] In 2016, the pollution in the Yellow River system was rather serious on the whole. The water quality of Grade V and worse than Grade V accounts for 20.5%. The water quality in the mainstream of the Yellow River is a little bit better. The percentage of the water of Grade Ⅱ, Ⅲ, Ⅳ, V and worse than V is 32.1%,24.8%,20.4%,6.6% and 13.9% (Fig.13.3). The major pollution indicators are dissolved oxygen, potassium permanganate, BOD, volatile phenol and petroleum. The pollution of the waters is more serious compared with the previous year.

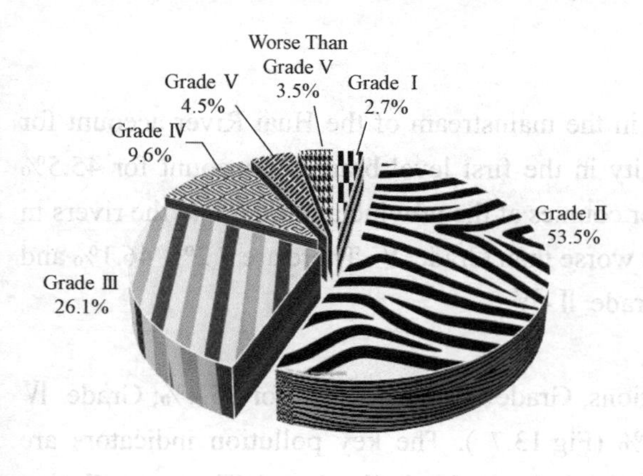

Fig.13. 2 Percentage of Water Quality Grades of the Yangtze River System

Fig.13.3 Percentage of Water Quality Grades of the Yellow River System

3. Pearl River

Among the 165 water quality monitoring sections, the ones with Grade Ⅱ-Ⅲ water account for 89.6%; Grade Ⅳ 4.8%; Grade Ⅴ 1.8% and worse than Grade Ⅴ 3.6% (Fig.13.4). The key pollution indicators are ammonia nitrogen and BOD.

The water quality of the Pearl River system in Guangxi is better than that in Guangdong. The water quality remains the same as that in the previous year.

4. Song Hua River

Among the 108 water quality monitoring sections, the ones with Grade Ⅱ-Ⅲ water account for 60.2%. Nen River attains the standards for Grade Ⅲ. Grade Ⅲ-Ⅳ water quality dominates the mainstream of the Songhua River in Jilin Province. Grade Ⅳ water quality dominates the mainstream of the Songhua River in Heilong Province (Fig.13.5). The rivers with serious pollution are Yitong River and An Zhaoxin River. The key pollution indicators are ammonia nitrogen, petroleum,

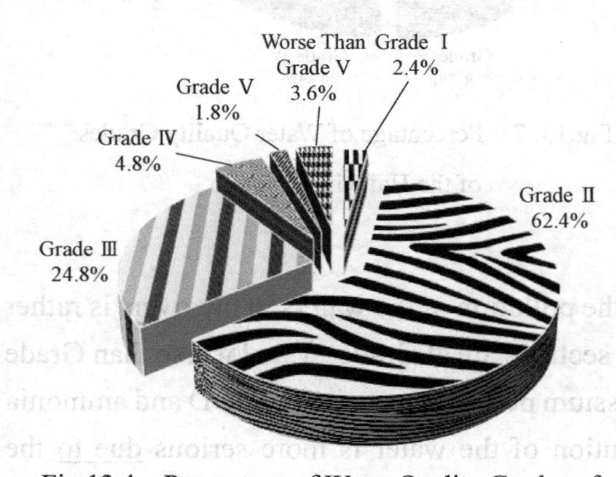

Fig.13.4 Percentage of Water Quality Grades of the Pearl River System

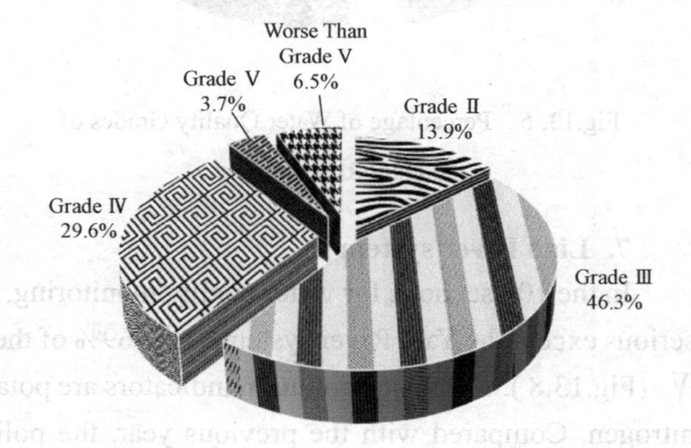

Fig.13. 5 Percentage of Water Quality Grades of the Songhua River System

potassium permanganate index and BOD. Pollution is more serious than in previous years because of the large reduction in water.

5. Huai River

The sections with Grade Ⅱ-Ⅲ water quality in the mainstream of the Huai River account for 53.3%. The sections with Grade Ⅱ-Ⅲ water quality in the first level branches account for 45.5% (Fig.13.6). The second and third branches, the river courses at the provincial border and the rivers in Shandong are dominated by water of Grade Ⅴ and worse than Grade Ⅴ. There are 7.2%, 46.1% and 23.9% of the sections attaining the standards for Grade Ⅱ-Ⅳ.

6. Hai River system

Among the 161 water quality monitoring sections, Grade Ⅱ-Ⅲ account for 35.4%; Grade Ⅳ 13%; Grade Ⅴ 8.7% and worse than Grade Ⅴ 41% (Fig.13.7). The key pollution indicators are ammonia nitrogen, petroleum, potassium permanganate index and volatile phenol. The key pollution in 77.7% of the section of Grade Ⅴ and worse than Grade Ⅴ is ammonia nitrogen. Compared with the previous year, the water quality remained basically the same.

There are 11 drinking water source reservoirs in the Hai River system. Except that the BOD in Huang Bi Zhuang Reservoir exceeded the standards, all other reservoirs attained or are better than the requirements of Grade Ⅲ water quality.

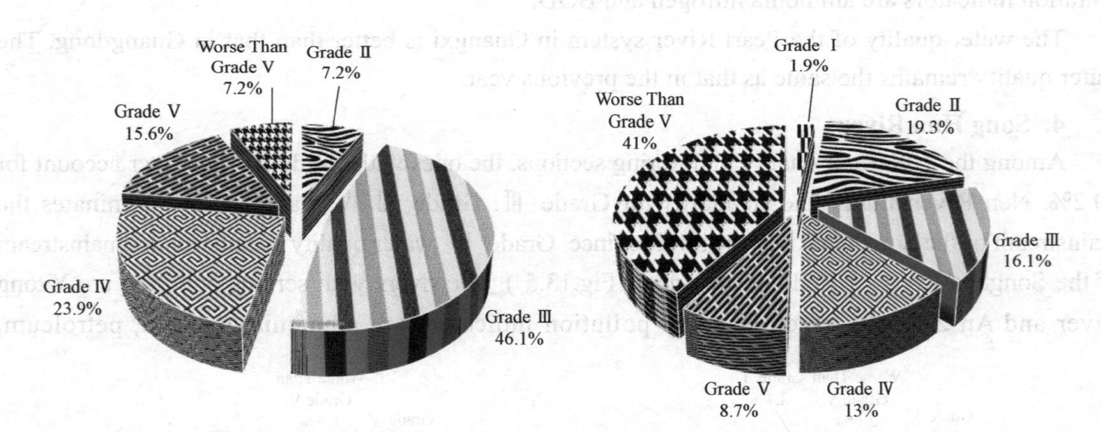

Fig.13. 6 Percentage of Water Quality Grades of the Huai River System

Fig.13. 7 Percentage of Water Quality Grades of the Hai River System

7. Liao River system

In the 106 sections for water quality monitoring, the pollution of the waters in the rivers is rather serious except the Yalu River system. Over 69% of the sections are at Grade Ⅴ and worse than Grade Ⅴ (Fig.13.8). The major pollution indicators are potassium permanganate index, BOD and ammonia nitrogen. Compared with the previous year, the pollution of the water is more serious <u>due to</u> the significant reduction of the water volume.

8. Rivers in Zhejiang and Fujian Provinces

The pollution is fairly light, and the sections attaining Grade Ⅰ account for 3.2%, Grade Ⅱ 53.6%, Grade Ⅳ 3.2% and Grade Ⅴ 2.4%. No sections are worse than Grade Ⅴ. The key pollution indicators are petroleum and ammonia nitrogen. The river course of Yong River in Ningbo is serious polluted.

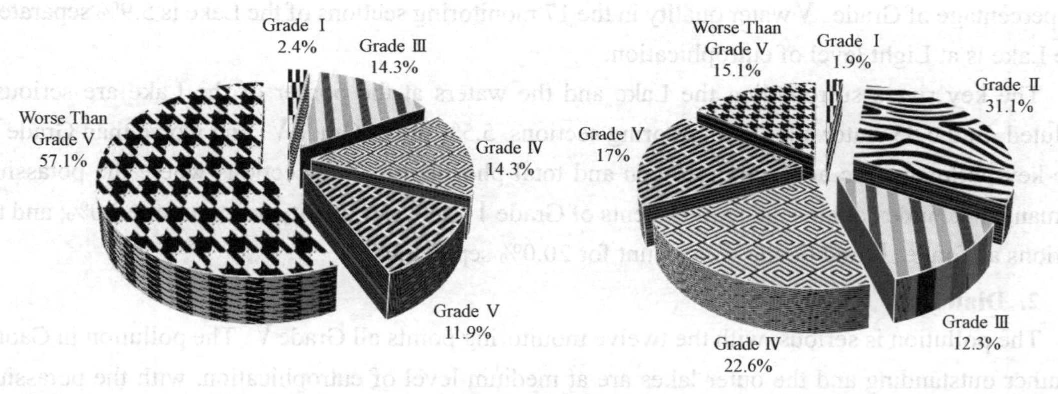

Fig.13. 8 Percentage of Water Quality Grades of the Liao River System

Fig.13. 9 Percentage of Water Quality Grades of Water Diversion Project (East Route)

9. Inland Rivers

The overall water quality is good. Grade Ⅰ sections account for 3.2%; Grade Ⅱ 77.6%; Grade Ⅲ 11.2% and Grade Ⅳ 6.35%.

10. Water Diversion Project (East Route)

In the 44 sections for water quality monitoring, the water quality of the sources of the Water Diversion Project–Jiangdu (Zhenjiang) of the Yangze River, Hongze lake, Dongping Lake and Bei Dagang Reservoir is good, attaining Grade Ⅲ (Fig.13.9), with the rest river courses polluted to various extents. The key pollution indicators are ammonia nitrogen, potassium permanganate index and petroleum.

Part 2 Water Quality in Lakes and Reservoirs

The eutrophication in the major lakes is still rather serious. In 2016, Tai Lake and the outer lakes of Dian lake were at mid eutrophication level and Chao Lake was at light eutrophication level (shown in Table 13.1).

Table 13.1 Index of Eutrophication of Three lakes in 2016

Lakes	Index of Eutrophication	Extent Eutrophication
Tai Lake	54.6	Light
Chao Lake	54.5	Light
Dian Lake (outer lakes)	62.5	Medium

1. Tai Lake

In the 142 water quality monitoring points, there are 17 ones in the lake, 55 in the key rivers surrounding the lake.[2] Compared with the previous year, the water quality remains almost the same. The potassium permanganate index of the Lake is at Grade Ⅲ. The total phosphorus is at Grade Ⅴ and the total nitrogen worse than Grade Ⅴ. Due to the serious pollution of nitrogen and phosphorus, the percentage of Grade Ⅴ water quality in the 17 monitoring sections of the Lake is 5.9% separately. The Lake is at Light level of eutrophication.

The key rivers surrounding the Lake and the waters at the border of the Lake are seriously polluted. In the 55 water quality monitoring sections, 5.5% are at Grade Ⅴ and worse than Grade Ⅴ. The key pollutants are ammonia nitrogen and total phosphorus. The sections where the potassium permanganate index attains the requirements of Grade Ⅰ-Ⅲ water quality account for 80.0%; and the sections at Grade Ⅳ and Grade Ⅴ account for 20.0% separately.

2. Dian Lake

The pollution is serious, with the twelve monitoring points all Grade Ⅴ. The pollution in Caohai is rather outstanding and the outer lakes are at medium level of eutrophication, with the potassium permanganate basically attaining the requirements of Grade III.[3] Caohai is at high eutrophication level and the pollution is obviously more serious than the outer lakes. Compared with the previous year, the water quality remains almost the same.

3. Chao Lake

Due to the serious pollution of total nitrogen and total phosphorus, the 8 monitoring points in the Lake are all Grade Ⅴ. The potassium permanganate index attains the standards for Grade Ⅲ water quality. The western half of Chao Lake is at medium eutrophication level. The average figure of the entire Lake is at light eutrophication level. The pollution in the western half of Chao Lake is obviously more serious than that in the eastern half. Compared with the previous year, the pollution of the water is more serious.

4. Other Large Lakes

In the ten fresh water lakes—Dongting Lake, Dalai Lake, Hongze Lake, Xingkai Lake, Nansi Lake, Bositeng Lake, Baiyangdian Lake, Erhai Lake, Jingpo Lake and Hong Lake, the water quality of Erhai Lake, Xinkai Lake, Bositeng Lake and Hong Lake is good, attaining standards for Grade Ⅲ; the water quality of Dongting Lake and Jingpo Lake attains standards of Grade Ⅳ; and Baiyangdian Lake, Dalai Lake and Nansi Lake are seriously polluted, all at Grade Ⅴ.

5. Urban Lakes

The Kunming Lake in Beijing is at Grade Ⅲ, West Lake in Hangzhou is at Grade Ⅳ. Xuanwu Lake in Nanjing, East Lake in Wuhan and Daming Lake in Jinan are all at Grade Ⅴ.

6. Large Reservoirs

In the ten large-sized reservoirs—Miyun in Beijing, Dahuofang in Fushun, Songhua Lake in Jilin, Yuqiao in Tianjin, Danjiangkou in Hubei, Dongpu in Hefei, Laoshan in Qingdao, Menlou in Yantai, Shimen in Hankou and Qiandao Lake Hangzhou, the water quality of the Qiandao lake and

Danjiankou Reservoir is at Grade Ⅰ; Yuqiao Reservior and Songhua Lake are at Grade Ⅲ; and the other six ones at Grade Ⅱ. The overall water quality of the large reservoirs is good.

Part 3 Water Quality in Ground Water

Among the 218 major ground water table monitoring cities and regions in the entire country, the ground water table in 75 cities and regions has risen to certain extent, accounting for 34% of the cities and regions same as the previous year.[4] However, the ground water table tends to descend, accounting for 50% of the cities and regions.

Overall speaking, the quality of the ground water in most of the cities and regions in the entire country is very good, with point and non-point pollution in certain regions to certain extent and certain indicators exceeding the standards. The key pollution indicators include mineralization degree, total hardness, nitrate, sub-nitrate, ammonia nitrogen, iron, manganese, chloride, sulfate, fluoride and pH value. The pollution of the ground water in the northern cities is more serious than that in the south, with high unattainment rate. The pollution of nitrogen is rather significant in various regions in the whole country. The unattainment of mineralization and total hardness is mainly located in northeast, north China, north-west and south-west regions. The unattainment of iron and manganese is mainly located in northeast and south regions.

Part 4 Where Does Wastewater Come from?

Wastewater comes from homes, and includes human and household wastes from toilets, sinks, baths and drains. It also comes from industries, schools and business. This wastewater is collected in a sanitary sewer system and conveyed to the wastewater treatment plant for treatment. Water that collects in street drains during a rainstorm is not considered wastewater but rather storm water.[5] This water is collected in a network of storm sewers and conveyed to the local rivers with no treatment.

1. Why is it necessary to treat wastewater?

The federal Clean Water Act requires municipalities like the City of Moline to treat its wastewater. In Moline, treated wastewater is discharged to either the Mississippi or Pock Rivers. Treatment helps protect aquatic life as well as keep waters safe for fishing and recreational uses. Our wastewater treatment plants remove suspended and dissolved solids from water, and reduce organic matter and pollutants. They also **disinfect** the wastewater to ensure that discharged water does not contain pathogenic bacteria life.

2. How does a wastewater treatment plant work?

There are two steps to treating wastewater at either of the City of Moline plants. The first step, called primary treatment, removes about half of the solids from wastewater. Primary treatment first involves Screens to collect trash as water passes. Next water flow is slowed to allow heavy solids to settle and lighter greases to rise where each can be removed.

In the second stage, or secondary treatment, 85%-90% of pollutants are removed. Air is mixed with wastewater, along with bacteria and other microorganisms that are grown and maintained at the treatment plant. These organisms consume harmful organic matter. A sedimentation tank allows the microorganisms and the solid wastes which they have converted to settle and separate from wastewater. Chlorine is added to wastewater as a **disinfectant** before it is discharged.

The solids that were removed from both primary and secondary treatment are sent for further processing in digestion tanks where they decompose. [6] When stabilized, these bio-solids meet all local, state and federal regulations for utilization for beneficial land application reuse as fertilizer.

* * * * * * * * * * * * * * * * * **Explanations** * * * * * * * * * * * * * * * * *

* * * * * * * * * * * * * * * * * **New Words and Expressions** * * * * * * * * * * * * * * *

1. section ['sekʃən] n. 部分，部件，节，项，区，地域，截面
2. controlling section 监测断面
3. mainstream ['meinstri:m] n. 主流，干流
4. pollution extent 污染程度
5. branch [bra: ntʃ] n. 支流，分枝，树枝 v. 分枝
6. dissolve [di'zɔlv] v. 溶解，解散，分解
7. petroleum [pi'trəuliəm] n. 石油
8. ammonia [ə'məunjə] n. 氨水，阿摩尼亚
9. nitrogen ['naitrədʒən] n. 氮
10. monitor ['mɔnitə] v. 监测 n. 班长，监听器，监视器，监控器
11. course [kɔ: s] n. 路线，过程，课程，讲座，一道（菜）
12. provincial [prə'vinʃəl] adj. 省的，地方的 n. 乡下人，地方人民
13. overall ['əuvərɔ: l] adj. 总体的，全部的，全体的，一切在内的 adv. 总的来说，全部地
14. Water Diversion Project 南水北调工程
15. permanganate [pə: mæŋgəneit] n. [化] 高锰酸
16. potassium [pə'tæsjəm] n. [化] 钾（19 号元素，符号 K）
17. pollutant [pə'lu:tənt] n. 污染物质
18. monitoring point (section) 监测点位（断面）
19. outstanding [aut'stændiŋ] adj. 突出的，显著的
20. eutrophication [ju,trɔfi'keiʃən] n. 营养化，超营养作用
21. phosphorus ['fɔsfərəs] n. 磷
22. ground water table 地下水水位
23. descend [di'send] v. 下降，传
24. indicator ['indikeitə] n. 污染指标，指示器，指示剂；[计算机] 指示符
25. nitrate ['naitreit] n. [化] 硝酸盐，硝酸钾
26. iron ['aiən] n. 铁
27. manganese [,mæŋgə'ni:z] n. 锰
28. sulfate ['sʌlfeit] n. [化] 硫酸盐 v. 以硫酸或硫酸盐处理，使变为硫酸盐
29. fluoride ['flu:əraid] n. 氟化物
30. drain [drein] vt. 排出沟外，喝干，耗尽 vi. 排水，流干 n. 排水沟

31. rainstorm ['reinstɔ:m] n. 暴风雨
32. aquatic [ə'kwætik] n. 水生动物，水草
33. recreational [ˌrekri'eiʃənəl,-kri:-] adj. 娱乐的，休养的
34. disinfect [ˌdisin'fekt] vt. 消毒
35. pathogenic [ˌpæθə'dʒenik] adj. 致病的，病原的，发病的
36. treatment ['tri: tmənt] n. 处理
37. wastewater ['weistwɔ:tə] n. 废水（污水）
38. grease [gri:s] n. 油脂，贿赂 vt. 涂脂于，[俗] 贿赂
39. microorganism [maikrəʊ'ɔ:gəniz(ə)] n. 微生物
40. organism ['ɔ:gənizəm] n. 生物体，有机体

****************** **Complicated Sentences** ******************

1. There are 137 water quality monitoring section, 31 of which are mainstream ones.

【译文】（黄河流域）水质监测断面有 137 个，其中 31 个为干流断面。

【说明】句中 which 和 ones 都是指 water quality monitoring section。

2. In the 142 water quality monitoring points, there are 17 ones in the Lake, 55 in the key rivers surrounding the Lake and 52 in the waters at the border of the Lake.

【译文】142 个监测断面中，位于湖内的 17 个，环湖主要河流上 55 个，环湖交界水体中 52 个。

【说明】句中 ones 指代 water quality monitoring pointes，55 后面省略了 ones。

3. The pollution in Caohai is rather outstanding and the outer lakes are at medium level of eutrophication, with the potassium permanganate basically attaining the requirements of Grade III.

【译文】滇池草海污染相当严重。外海属中度富营养状态，高锰酸盐指数基本符合Ⅲ类水质要求。

【说明】句中 with the potassium permanganate basically attaining the requirements of Grade Ⅲ修饰 at medium level of eutrophication。

4. Among the 218 major ground water table monitoring cities and regions in the entire country, the ground water table in 75 cities and regions has risen to certain extent, accounting for 34% of the cities and regions same as the previous year.

【译文】全国 218 个主要地下水水位监测城市和地区中，有 75 个城市和地区水位有所回升，回升区所占比例为 34%，与上年一样。

【说明】此句的主体结构就是“主语+谓语+宾语”的简单句，主语 the ground water table，谓语宾语 has risen to certain extent。Among 引导的短语作地点状语，accounting for 34% of the cities and regions same as the previous year. 进一步解释 certain extent。

5. Water that collects in street drains during a rainstorm is not considered wastewater but rather storm water.

【译文】下暴雨时积聚在街道排水沟里的水不是废水，而是雨水。

【说明】but 表示转折，rather 用于强调，大致意思为“更是…”。

6. The solids that were removed from both primary and secondary treatment are sent for further processing in digestion tanks where they decompose.

【译文】将初步处理和二次处理所去除的固体送到分解消化箱，等待进一步处理。

【说明】该句中 that 引导的定语从句修饰主语 The solids，both…and 连接两个宾语 primary 和 secondary treatment，后含有 where 引导的状语从句修饰 digestion tanks。

******************** Summary of Glossary ********************

1. mainstream	干流
2. branch	支流
3. wastewater	废水
4. water quality	水质
5. key river ststem	主要水系
6. Pearl River	珠江
7. sub-nitrate	亚硝酸盐
8. Yangtze River	长江
9. ammonia nitrogen	氨氮
10. pollution index	污染指标
11. total phosphorus	总磷
12. mineralization degree	矿化度
13. pH value	pH 值
14. wastewater treatment plant	废水处理厂
15. in the mainstream	干流上
16. Water Diversion Project	南水北调工程
17. pollution indicator	污染指标
18. monitoring point (section)	监测点（断面）
19. secondary treatment	二次处理
20. solid wastes	固体废物

******************** Abbreviations (Abbr.) *********************

BOD	Biochemical Oxygen Demand	生化需氧量

********************** Exercises **********************

(1) The water quality in the ______ of the river systems is better than that in the ______.

(2) Compared with the previous year, the order of the ______ level remained same and the pollution ______ is quite close.

(3) The key ________ indicators are _______, ________, potassium permanganate index

and ________.

************* Word Building (13) -logy ; re- *************

1. -logy [词尾]，表示：学，科学

hydro	n.	hydrology	n.	水文学
bio-		biology	n.	生物学
meteor	n.	meteorology	n.	气象学
geo-		geology	n.	地质学

2. re- [前缀]，表示：重复，再次

arrange	v.	rearrange	v.	重安排，再设置
cover	n./v.	recover	v.	复原，恢复
establish	v.	reestablish	v.	重建，恢复
locate	v.	relocate	v.	重新部署

***************** Text Translation *****************

第十三章　水　环　境

第一节　主要水系的水质

2016 年，七大水系 1617 个重点监测断面中，有 67.8%的断面属Ⅰ～Ⅲ类水质，16.8%的断面属Ⅳ类水质，15.5%的断面属Ⅴ类和劣Ⅴ类水质。其中，七大水系干流的 184 个国家重点监控断面中，Ⅰ～Ⅲ类水质断面占 67.7%，Ⅳ类水质断面占 19.2%，Ⅴ类和劣Ⅴ类水质断面占 13.1%（见图 13.1），而且各水系干流水质均好于支流水质。（单词个数：56）

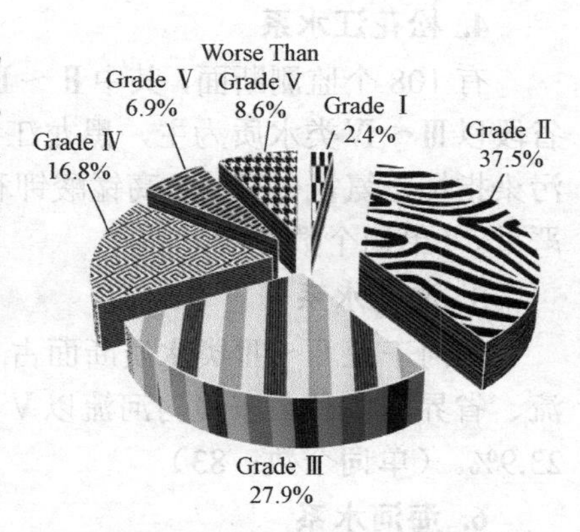

图 13.1　七大流域水质等级百分比

2016 年七大水系污染程度由重到轻依次为：海河、辽河、淮河、黄河、松花江、长江、珠江。与往年相比，污染程度的次序一样，污染程度也很接近。（单词个数：99）

1. 长江水系

长江流域有 510 个监测断面，其中干流监测断面有 59 个。干流以Ⅱ类水质为主，Ⅰ、Ⅱ、Ⅲ、Ⅳ类水质的比例分别为 2.7%、53.5%、26.1%和 9.6%。主要的一级支流（岷江、沱江、嘉陵江、乌江、汉江、湘江、资水、沅江、澧江和赣江）有 451 个监测断面，Ⅰ~Ⅲ类水质断面占 82.3%。长江水系的主要污染指标是石油和氨氮。（单词个数：95）

2. 黄河水系

有 137 个监测断面，其中 31 个为干流断面。2016 年黄河水系整体污染相当严重。Ⅴ类和劣Ⅴ类水体占 20.5%。黄河干流水质稍好些。Ⅱ、Ⅲ、Ⅳ、Ⅴ类和劣Ⅴ类水质比例分别为 32.1%、24.8%、20.4%、6.6%和 13.9%。黄河水系主要污染指标为溶解氧、高锰酸钾、BOD、挥发性苯酚和石油。与上年相比，水污染更加严重。（单词个数：104）

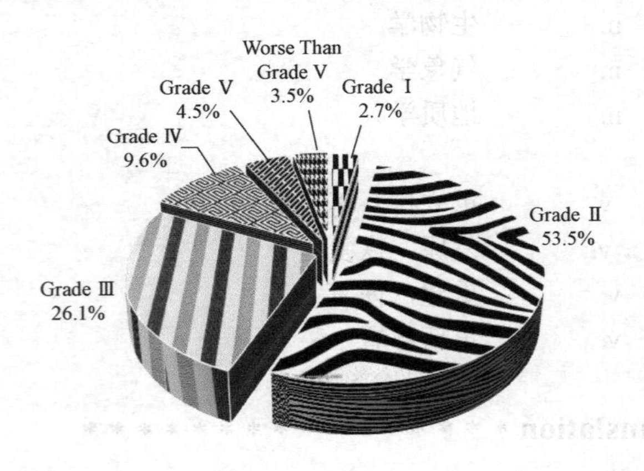

图 13.2 长江流域水质等级百分比七大流域水质等级百分比

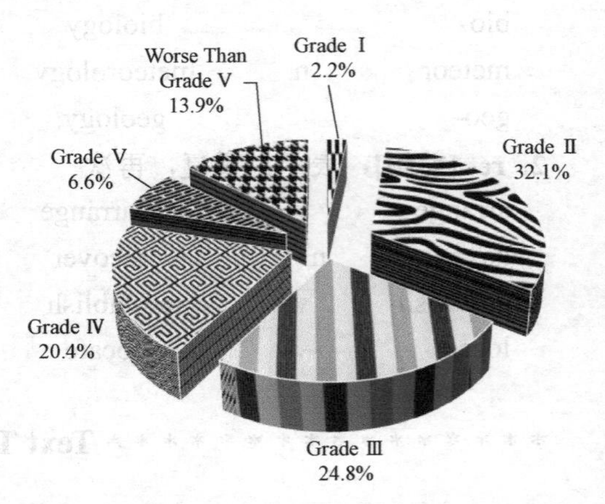

图 13.3 黄河流域水质等级百分比

3. 珠江水系

有 165 个监测断面，其中 89.6%的断面为Ⅱ～Ⅲ类水质，4.8%的断面属Ⅳ类水质，1.8%的断面属Ⅴ类水质，3.6%的断面属劣Ⅴ类水质。主要污染指标是氨氮和 BOD。珠江水系广西境内水质比广东境内水质好。水质与往年一样。（单词个数：64）

4. 松花江水系

有 108 个监测断面，其中Ⅱ～Ⅲ类水质断面占 60.2%。嫩江为Ⅲ类水质。松花江干流吉林省段以Ⅲ～Ⅳ类水质为主，黑龙江省段以Ⅵ类为主。污染严重的是伊通河和安肇新河。主要污染指标是氨氮、石油、高锰酸钾和 BOD 指数。由于水量减少很多，与往年相比，污染更加严重。（单词个数：84）

5. 淮河水系

淮河干流Ⅱ～Ⅲ类水质断面占 53.3%，一级支流Ⅱ-Ⅲ类水质断面占 45.5%。二、三级支流、省界河段和山东境内河流以Ⅴ类和劣Ⅴ类水质为主。Ⅱ-Ⅳ类水质断面占 7.2%、46.1%、23.9%。（单词个数：83）

6. 海河水系

有 161 个监测断面，其中Ⅱ～Ⅲ类水质断面占 35.4%；Ⅳ类水质断面占 13%；Ⅴ类水质断面占 8.7%；劣Ⅴ水质断面占 41%。主要污染指标为氨氮、石油、高锰酸钾和挥发性苯酚。与往年相比，水质基本相同。

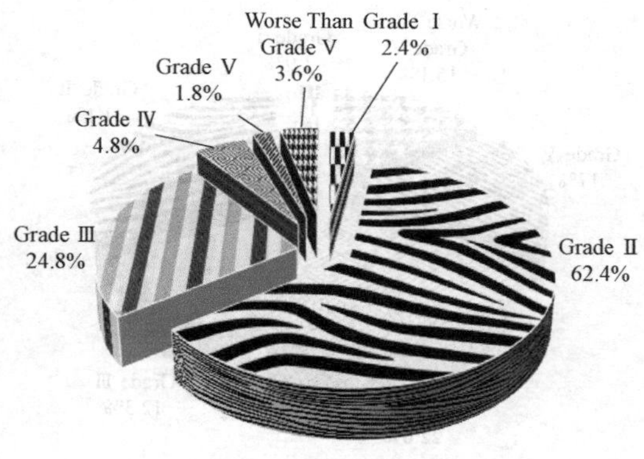

图 13.4 珠江流域水质等级百分比

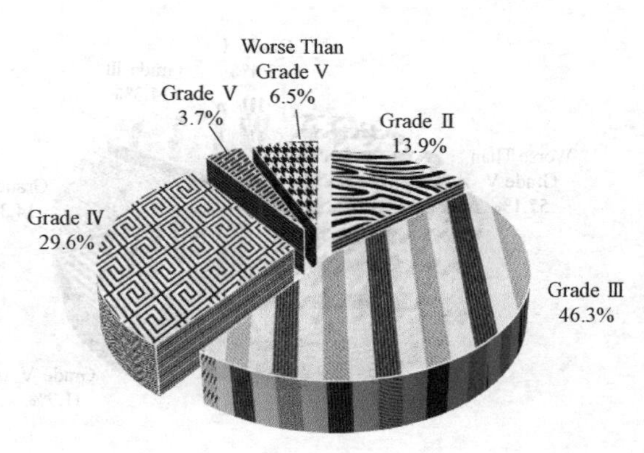

图 13.5 松花江流域水质等级百分比

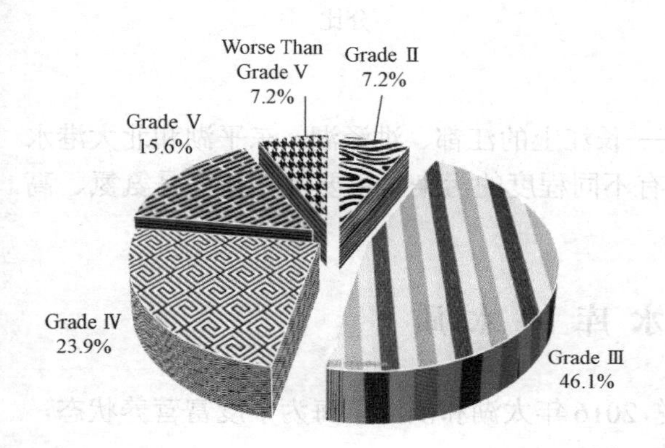

图 13.6 淮河流域水质等级百分比

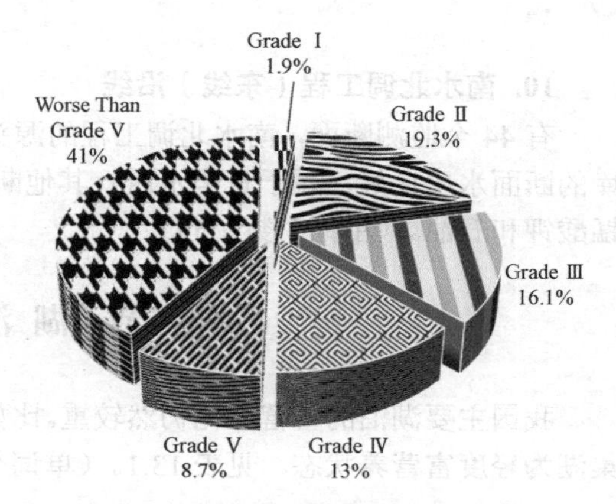

图 13.7 海河流域水质等级百分比

海河水系有 11 个饮用水源。除黄壁庄水库的 BOD 超标之外，其他所有水库水质都为Ⅲ类水质或好于Ⅲ类水质。(单词个数：104)

7. 辽河水系

有 106 个监测断面，除了鸭绿江水系外，其他水系的污染相当严重。69%以上属于Ⅴ类和劣Ⅴ类断面。主要污染指标是高锰酸钾、BOD 和氨氮。由于水量减少很多，与往年相比，污染更加严重。(单词个数：71)

8. 浙闽片河流

污染程度非常轻。Ⅰ类水质断面占 3.2%。Ⅱ、Ⅳ和Ⅴ类水质断面分别为 53.6%、3.2%、2.4%。无劣Ⅴ类水质断面。主要污染指标是石油和氨氮。污染较重断面集中在宁波的甬江。(单词个数：45)

9. 内陆河流

内陆河流水质总体良好。满足Ⅰ类水质的断面占 3.2%，Ⅱ、Ⅲ和Ⅳ类分别占 77.6%、11.2%和 6.35%。(单词个数：20)

Grade Ⅰ
2.4%
Grade Ⅲ
14.3%
Worse Than
Grade Ⅴ
57.1%
Grade Ⅳ
14.3%
Grade Ⅴ
11.9%

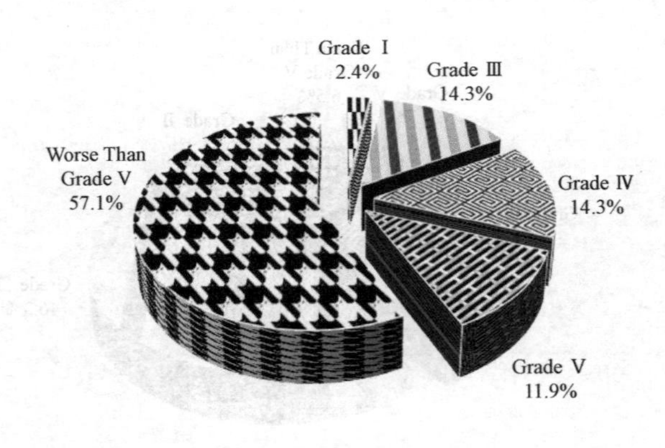

图 13.8 辽河流域水质等级百分比

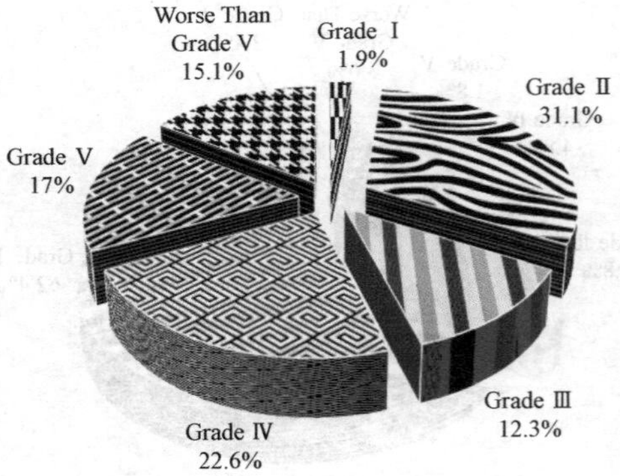

Worse Than
Grade Ⅴ
15.1%
Grade Ⅰ
1.9%
Grade Ⅱ
31.1%
Grade Ⅴ
17%
Grade Ⅲ
12.3%
Grade Ⅳ
22.6%

图 13.9 南水北调工程（东线）沿线水质百分比

10. 南水北调工程（东线）沿线

有 44 个监测断面，南水北调工程的源头——长江上的江都、洪泽湖、东平湖和北大港水库的断面水质良好，属于Ⅲ类水质。其他断面有不同程度的污染。主要污染指标是氨氮、高锰酸钾和石油。（单词个数：59）

第二节 湖泊水库的水质

我国主要湖泊的富营养化仍然较重。比如说，2016 年太湖和滇池草海为中度富营养状态，巢湖为轻度富营养状态，见表 13.1。（单词个数：34）

表 13.1 三个湖泊富营养情况

湖泊	氮、磷污染指数	污染程度
太湖	54.6	轻度
巢湖	54.5	轻度
滇池	62.5	中度

1. 太湖

有 72 个水质监测点，其中湖中 17 个，环湖主要河流 55 个。与往年相比水质几乎未变。（单词个数：35）

太湖高锰酸钾指标为Ⅲ类，总磷指标为Ⅴ类，总氮指标为劣Ⅴ类。太湖的 17 个监测点位中，由于受到氮和磷的严重污染，Ⅴ类水质的监测断面数的比例为 5.9%。太湖湖体处于轻度富营养状态。（单词个数：58）

环湖主要河流和环湖交界水体污染严重，55 个水质监测断面中，有 5.5%为Ⅴ类水质。主要污染指标为氨氮和总磷指标。高锰酸钾指数符合Ⅰ～Ⅲ类水质要求的断面占 80.0%。Ⅳ类和Ⅴ类水质的断面比例为 20.0%。（单词个数：72）

2. 滇池

污染严重，12 个水质监测点全都为Ⅴ类水质。滇池草海污染相当严重。外湖属于中度富营养状态，高锰酸盐指数基本符合Ⅲ类水质要求。草海属重度富营养状态，污染明显比外海严重。同往年相比，水质几乎一样。（单词个数：68）

3. 巢湖

由于总氮和总磷污染严重，湖面 8 个监测点全部为Ⅴ类水质。湖体高锰酸盐指数达到Ⅲ类水质标准，西半湖区属于中度富营养区。全湖平均营养状态属于轻度富营养状态。巢湖西半湖污染明显重于东半湖，水污染更加严重。（单词个数：89）

4. 其他大型湖泊

在洞庭湖、达赉湖（呼伦湖）、洪泽湖、兴凯湖、南四湖、博斯腾湖、白洋淀、洱海、镜泊湖和洪湖这 10 个淡水湖泊中，洱海、兴凯湖、博斯腾湖和洪湖水质良好，湖体水质为Ⅲ类水质标准；洞庭湖和镜泊湖的水质为Ⅳ类水质标准；白洋淀、达赉湖和南四湖湖体污染较重，水质均为Ⅴ类。（单词个数：77）

5. 城市内湖

北京昆明湖水质为Ⅲ类水质，杭州西湖为Ⅳ类，南京玄武湖、武汉东湖和济南大明湖水质均为Ⅴ类。（单词个数：36）

6. 大型水库

10 座大型水库——北京密云水库、抚顺大伙房水库、吉林松花湖（丰满水库）、天津于桥水库、湖北丹江口水库、合肥董铺水库、青岛崂山水库、烟台门楼水库、汉口石门水库和杭州千岛湖水库中，其中千岛湖水库和丹江口水库水质达到Ⅰ类水质标准；于桥水库和松花湖达到Ⅲ类水质标准；其他都达到Ⅱ类水质标准。所以说，大型水库整体水质较好。（单词个数：77）

第三节 地下水水质

全国 218 个主要地下水水位监测城市和地区中，有 75 个城市和地区水位有所回升，回升区所占比例为 34%，与上年相同。而地下水位仍以下降为主，所占比例为 50%。（单词个数：59）

总的来说，全国大部分城市和地区地下水水质总体较好，局部受到一定程度的点或面污染，部分指标超标。主要污染指标有矿化物、总硬度、硝酸盐、亚硝酸盐、氨氮、铁、锰、氯化物、硫酸盐、氟化物和 pH 值等。北部城市地下水污染比南部城市严重，不可使用率很高。全国不同地区氮污染程度相当严重。矿化度和总硬度超标主要分布在东北、华北、西北和西南地区，铁和锰超标主要分布在东北和南方地区。（单词个数：129）

第四节 废水从何而来?

废水一方面来自家庭，包括卫生间、洗涤池、洗澡和排水所放出的人用和家用废弃物。另一方面来自工业、学校和企业。废水积聚在卫生排污系统并送到废水处理厂进行处理。下暴雨时积聚在街道排水沟里的水不是废水，而是雨水。这些水通过暴雨排水沟网进行汇集，再输送到附近的河流中，不需要处理。（单词个数：78）

1. 为何需要处理废水？

联邦净水法案要求像莫林这样的大都市对其废水进行处理。在莫林，处理过的废水被排放到密西西比河或泡克河。对废水进行处理有助于保护水生动植物，同时，保证了鱼类的用水需求和水的循环再利用。废水处理厂去除水中悬浮和溶解的固体，减少有机物质和污染物质的含量。同时，给废水消毒可以确保排出的水不含致病的细菌。（单词个数：78）

2. 废水处理厂如何工作？

莫林处理厂废水处理有两个步骤。第一步，叫作初步处理，将废水中半数以上的固体去除。初步处理首先将水流动时的废物积聚到过滤网。接着，随着水流变慢，重些的固体停滞下沉，而轻些的油脂将会漂浮到水面，从而可以将上、下两层的污染物全部去除。（单词个数：61）

第二步，叫作二次处理，可去除掉85%～90%的污染物质。空气中掺杂着污水、细菌以及生长并停留在处理厂的其他微生物。这些微生物消耗有害的有机物质。沉淀物箱可以让微生物和转化成的下沉的固体废物从废水中分离。在排放废水前将氯作为消毒剂加入废水中。（单词个数：71）

将初步处理和二次处理所去除的固体送到分解消化箱待进一步处理。处理完之后，这些符合地方、州和联邦规定的生物固体可作为肥料用于可耕作的土地进行再利用。（单词个数：42）

Chapter 14 Climate Change

Part 1 Introduction

Climate change is any long-term change in the patterns of average weather of a specific region or the earth as a whole. Climate change reflects abnormal variations to the earth's climate and subsequent effects on other parts of the earth, such as in the ice caps over durations ranging from decades to millions of years.

In recent usage, especially in the context of environmental policy, climate change usually refers to changes in modern climate (see global warming). For information on temperature measurements over various periods, and the data sources available, see temperature record. For attribution of climate change over the past century, see attribution of recent climate change.

Evidence is mounting that we are in a period of climate change brought about by increasing atmospheric concentration of greenhouse gases. [1] Global mean temperatures have risen 0.3-0.6℃ since the late 19th century and global sea levels have risen between 10 and 25 cm (Intergovernmental Panel on Climate Change, IPCC, 1995) the IPCC has reported that the expected global rise in temperature over the next century would probably be greater than observed in the last 10 000 years. As a direct consequence of warmer temperatures, the hydrologic cycle will undergo significant impact with accompanying changes in the rates of precipitation and evaporation. Predictions include higher incidences of severe weather events, a higher likelihood of flooding, and more droughts.

Part 2 Climate Change Factors

Climate change is the result of a great many factors including the dynamic processes of the earth itself, external forces including variations in sunlight intensity, and more recently by human activities. External factors that can shape climate are often called climate forcings and include such processes as variations in solar radiation, deviations in the earth's orbit, and the level of greenhouse gas concentrations. [2] There are a variety of climate change feedbacks that will either amplify or diminish the initial forcing.

Most forms of internal variability in the climate system can be recognized as a form of hysteresis, where the current state of climate does not immediately reflect the inputs. Because the earth's climate system is so large, it moves slowly and has time-lags in its reaction to inputs. For example, a year of dry conditions may do no more than to cause lakes to shrink slightly or plains to dry marginally. In the following year however, these conditions may result in less rainfall, possibly leading to a drier year the next. When a critical point is reached after "x" number of years, the entire system may be altered inexorably. In this case, would result in no rainfall at all. It is this hysteresis that has been mooted to be the possible progenitor of rapid and irreversible climate change. [3]

1. Plate tectonics

On the longest time scales, plate tectonics will reposition continents, shape oceans, build and tear down mountains and generally serve to define the stage upon which climate exists.[4] During the Carboniferous period, plate tectonics may have triggered the large-scale storage of carbon and increased glaciation. More recently, plate motions have been implicated in the intensification of the present ice age, when approximately 3 million years ago, the North and South American plates collided to form the Isthmus of Panama and shut off direct mixing between the Atlantic and Pacific Oceans.

2. Solar output

Variations in solar activity during the last several centuries based on observations of sunspots and beryllium isotopes. The sun is the source of a large percentage of the heat energy input to the climate system. Lesser amounts of energy are provided by the gravitational pull of the moon (manifested as tidal power), and geothermal energy. The energy output of the sun, which is converted to heat at the earth's surface, is an integral part of the earth's climate. Early in earth's history, according to one theory, the sun was too cold to support liquid water at the earth's surface, leading to what is known as the faint young sun paradox. Over the coming millennia, the sun will continue to brighten and produce a correspondingly higher energy output; as it continues through what is known as its "main sequence", and the earth's atmosphere will be affected accordingly.

On more contemporary time scales, there are also a variety of forms of solar variation, including the 11-year solar cycle and longer-term modulations. However, the 11-year sunspot cycle does not appear to manifest itself clearly in the climatological data. Solar intensity variations are considered to have been influential in triggering the little Ice Age, and for some of the warming observed from 1900 to 1950. The cyclical nature of the sun's energy output is not yet fully understood; it differs from the very slow change that is happening within the sun as it ages and evolves, with some studies pointing toward solar radiation increases from cyclical sunspot activity affecting global warming.

3. Orbital variations

In their effect on climate, orbital variations are in some sense an extension of solar variability, because slight variations in the earth's orbit lead to changes in the distribution and abundance of sunlight reaching the earth's surface. These orbital variations, known as Milankovitch Cycles, directly affect glacial activity. Eccentricity, axial tilt, and precession comprise the three dominant cycles that make up the variations in earth's orbit. The combined effect of the variations in these three cycles creates changes in the seasonal reception of solar radiation on the earth's surface. As such, Milankovitch Cycles affecting the increase or decrease of received solar radiation directly influence the earth's climate system, and influence the advance and retreat of earth's glaciers. Subtler variations are also present, such as the repeated advance and retreat of the Sahara Desert in response to orbital precession.

4. Volcanism

Volcanism is the process of conveying material from the depths of the earth to the surface, as part of the process by which the planet removes excess heat and pressure from its interior.[5] Volcanic eruptions, geysers and hot springs are all part of the volcanic process and all release varying levels of particulates into the atmosphere.

A single eruption of the kind that occurs several times per century can affect climate, causing cooling for a period of a few years or more. The eruption of Mount Pinatubo in 1991, for example, produced the second largest terrestrial eruption of the 20th century (after the 1912 eruption of Novarupta) and affected the climate substantially, with global temperatures dropping by about 0.5°C (0.9°F), and ozone depletion being temporarily substantially increased. Much larger eruptions, known as large igneous provinces, occur only a few times every hundred million years, but can reshape climate for millions of years and cause mass extinctions. Initially, it was thought that the dust ejected into the atmosphere from large volcanic eruptions was responsible for longer-term cooling by partially blocking the transmission of solar radiation to the earth's surface. However, measurements indicate that most of the dust hurled into the atmosphere may return to the earth's surface within as little as six months, given the right conditions.

Volcanoes are also part of the extended carbon cycle. Over very long (geological) time periods, they release carbon dioxide from the earth's interior, counteracting the uptake by sedimentary rocks and other geological carbon dioxide sinks. According to the US Geological Survey, however, estimates are that human activities generate more than 130 times the amount of carbon dioxide emitted by volcanoes.

5. Ocean variability

On a timescale often measured in decades or more, climate changes can also result from the interaction between the atmosphere and the oceans. Many climate fluctuations, including the El Niño Southern oscillation, the Pacific decadal oscillation, the North Atlantic oscillation, and the Arctic oscillation, owe their existence at least in part to the different ways that heat may be stored in the oceans and also to the way it moves between various "reservoirs ". On longer time scales (with a complete cycle often taking up to a thousand years to complete), ocean processes such as thermohaline circulation also play a key role in redistributing heat by carrying out a very slow and extremely deep movement of water, and the long-term redistribution of heat in the oceans.

6. Human influences

Anthropogenic factors are human activities that change the environment. In some cases, the chain of causality of human influence on the climate is direct and unambiguous (for example, the effects of irrigation on local humidity), whilst in other instances it is less clear. Various hypotheses for human-induced climate change have been argued for many years though, generally, the scientific debate has moved on from scepticism to a scientific consensus on climate change that human activity is the probable cause for the rapid changes in world climate in the past several decades. Consequently, the debate has largely shifted onto ways to reduce further human impact and to find ways to adapt to change that has already occurred.

Of most concern in these anthropogenic factors is the increase in CO_2 levels due to emissions from fossil fuel combustion, followed by aerosols (particulate matter in the atmosphere) and cement manufacture. Other factors, including land use, ozone depletion, animal agriculture and deforestation, are also of concern in the roles they play-both separately and in conjunction with other factors-in affecting climate.

7. Sea level change

Climate models for the substantiation of theories regarding global warming rely heavily on the measurement of long-term changes in global average sea level. Global sea level change for much of the last century has generally been estimated using tide gauge measurements collated over long periods of time to give a long-term average. More recently, altimeter measurements—in combination with accurately determined satellite orbits—have provided an improved measurement of global sea level change.

* * * * * * * * * * * * * * * * * * Explanations * * * * * * * * * * * * * * * * * *

* * * * * * * * * * * * * * * * New Words and Expressions * * * * * * * * * * * * * * * *

1. pattern [ˈpætən] vt. 仿制
   n. 模型，模式；花样，图案
2. abnormal [æbˈnɔ:məl] adj.反常的，异常的
3. usage [ˈju: sidʒ] n. 使用，用法；惯用法
4. policy [pɔləsi] n. 政策，方针；保险单
5. atmospheric [ˌætməsˈferik] adj. 大气的；有…气氛的
6. greenhouse [ˈgri:nhaus] n. 温室，暖房
7. global [ˈgləubəl] adj. 全球的，全世界的；总的，完整的
8. observe [əbˈzə:v] vt. 注意到；观察；评论；遵守，奉行
9. deviation [ˌdi:viˈeiʃən] n. 背离，偏离；偏差，偏向；离题
10. feedback [ˈfi:dbæk] n. 反馈，反馈信息
11. diminish [diˈminiʃ] vt. 减少，减小，降低
    vi. 变少，变小，降低
12. immediately [iˈmi:diətli] adv. 立即，马上；直接地，紧接着地
13. time-lag 时间滞后
14. shrink [ʃriŋk] vi. 起皱，收缩；畏缩
    vt. 使起皱，使收缩
15. critical [ˈkritikəl] adj. 决定性的，关键性的，危急的；批评(判)的
16. inexorably [inˈeksərəbəli] adv. 不为所动的，坚决不变的
17. Carboniferous [ˌkɑ:bəˈnifərəs] n. 石炭纪（时期）
18. intensification [inˌtensifiˈkeiʃən] n. 增强，加剧
19. isotope [ˈaisətəup] n. 同位素
20. geothermal energy 地热能
21. paradox [ˈpærədɔks] n. 似乎矛盾却正确的说法；自相矛盾的人（物）
22. contemporary [kənˈtempərəri]
    n. 同代人，当代人
    adj. 当代的；同时代的
23. modulation [ˌmɔdjuˈleiʃən] n.（声音之）抑扬；变调

24. trigger ['trigə] vt. 触发，引起

n. 扳机；引起反应的行动

25. Ice Age 冰河世纪

26. volcanism ['vɔlkənizəm] n. 火山活动，火山作用

27. geyser ['gi: zə] n.天然热喷泉

28. Mount Pinatubo 皮纳图博火山

29. Novarupta 诺瓦拉普塔火山

30. ozone ['əuzəun] n. 臭氧

31. igneous ['igniəs] adj. 火的，火成的

32. mass [mæs] adj. 大规模的

v. 集中

33. transmission [trænz'miʃən] n. 播送，发射；传送，传递，传染

34. uptake ['ʌpteik] n. 摄取，领会

35. El Niño Southern oscillation 厄尔尼诺南方涛动

36. unambiguous ['ʌnæm'bigjuəs] adj. 不引起歧义的，清楚的，清晰的

37. tide [taid] n. 潮，潮汐；潮流，趋势

38. cement [si'ment] vt. 水泥，结合剂

****************** **Complicated Sentences** ******************

1. Evidence is mounting that we are in a period of climate change brought about by increasing atmospheric concentration of greenhouse gases.

【译文】事实证明，温室气体的排放使得大气浓度增加，从而导致目前的气候变化。

【说明】复合宾语从句，句中 increasing ...为介词短语做宾语。

2. External factors that can shape climate are often called climate forcings and include such processes as variations in solar radiation, deviations in the earth's orbit, and the level of greenhouse gas concentrations.

【译文】可以改变气候的外部因素通常称为气候力，包括太阳辐射变化、地球轨道偏差以及温室气体集中程度等作用。

【说明】and 后面继续解释气候力的具体情况。

3. It is this hysteresis that has been mooted to be the possible progenitor of rapid and irreversible climate change.

【译文】正是这样的滞后的变化使得气候变化已成为迅速和不可逆转的现象。

【说明】hysteresis 意思是“滞后”。Progenitor 意思是“祖先，起源”。Moot 意思是“提出……供讨论”。

4. On the longest time scales, plate tectonics will reposition continents, shape oceans, build and tear down mountains and generally serve to define the stage upon which climate exists.

【译文】在最长的时间尺度上，由于我们赖以生存的气候因素的变化将导致大陆板块重组、海洋形状重塑以及山岭海拔发生变化。

【说明】which 引导条件状语从句。

5. Volcanism is the process of conveying material from the depths of the earth to the surface, as part of the process by which the planet removes excess heat and pressure from its interior.

【译文】火山活动是将地球深处的材料输送到地球表面的过程，伴随着这个过程的发生，地球也将除去其内部多余的热量和一部分的压力。

【说明】原文为 which 引导定语从句，which 指代前面的 the process。

*** * * * * * * * * * * * * * * * * * Summary of Glossary * * * * * * * * * * * * * * * * * ***

1. climate change 气候变化
2. climate forcing 气候强制力
3. greenhouse gas 温室气体
4. solar radiation 太阳辐射
5. volcanism 火山作用
6. Ice Age 冰河世纪

*** * * * * * * * * * * * * * * * * * Abbreviations (Abbr.) * * * * * * * * * * * * * * * * * ***

IPCC Intergovernmental Panel on Climate Change 政府间气候变化小组

*** Exercises ***

(1) Predictions include higher incidences of severe weather events, a higher likelihood of flooding, and more ________.

(2) As a direct _______ of warmer temperatures, the hydrologic cycle will undergo significant ________ with accompanying changes in the rates of ______ and evaporation.

(3) There are a variety of climate change ________ that will either amplify or diminish the initial forcing.

(4) In their effect on climate, ______ variations are in some sense an _____ of solar variability, because slight variations in the earth's orbit lead to changes in the distribution and abundance of sunlight reaching the earth's surface.

(5) Global sea level change for much of the last century has generally been estimated using _______ gauge measurements_________ over long periods of time to give a long-term average.

*** * * * * * * * * * * * * * * Word Building (14) ir-; in- * * * * * * * * * * * * * * ***

1. ir- [前缀]，表示：否定，与……相反（用于 r 之前）

respective	adj.	irrespective	adj.	不顾的，不考虑的，无关的
rational	n./adj.	irrational	adj.	无理性的，失去理性的
realizable	adj.	irrealizable	adj.	不能实现的，不能达到的
reconcilable	adj.	irreconcilable	adj.	不能协调的，矛盾的
recoverable	adj.	irrecoverable	adj.	无可挽救的
regular	n./adj.	irregular	adj.	不规则的，无规律的
relative	n./adj.	irrelative	adj.	无关系的
relevant	adj.	irrelevant	adj.	不相关的，不切题的

2. in-[前缀]，表示：否定，与……相反，可译为：不，非

dependent	adj.	independent	adj.	不受约束的；独立自主的
evitable	adj.	inevitable	adj.	不可避免的
numerable	adj.	innumerable	adj.	无数的，数不清的
legal	adj.	illegal	adj.	违法的，不合规定的
exact	vt./adj.	inexact	adj.	不精确的
regular	n./ad.	irregular	adj.	不规则的，不规律的
convenient	adj.	inconvenient	adj.	不便的，有困难的
exhaustible	adj.	inexhaustible	adj.	无穷尽的，用不完的
flexible	adj.	inflexible	adj.	不灵活的；不可变更的

******************* Text Translation *******************

第十四章 气候变化

第一节 概 述

气候变化是指任何一个特定地区或整个地球的平均天气模式的长期变化。气候的异常变化会对地球的气候和地球其他地区造成的影响，例如在过去的几十年到几百万年不等时期的冰盖层厚度的变化。（单词个数：55）

按照最新用法，特别是在环境政策实施的背景下，气候变化通常是指现代气候变化（如全球变暖等）。可以通过查阅气温记录，而得到在不同时期的有关温度测量的资料和数据。我们可以通过最近的气候变化起因，来探求在过去一个世纪里的气候变化的起因。（单词个数：53）

事实证明，温室排放的气体使得大气浓度增加。我们目前正处于一个气候变暖的时期。自从19世纪后期以来，全球平均温度上升了0.3～0.6℃，全球海平面上升10～25cm（专门针对全球气候变化的政府间气候变化组织，IPCC，1995）。据IPCC报道，预计下世纪全球温度上升幅度要比过去10 000年实测温度上升幅度大。由于温度的上升，水文循环也会发生较大的变化，降水量和蒸发量也会发生相应的变化。恶劣的气象事件，如洪水暴发、干旱事件都会增大发生的可能性。（单词个数：116）

第二节 气候变化因素

造成气候变化的因素包括地球本身，以及在阳光的强度变化的外部环境，和最近由人类活动所引起的动态过程等。这些能够对气候造成影响的外部因素通常称为气候因子，这些因子包括太阳辐射的变化、地球轨道上的偏差以及温室气体的浓度大小等。气候变化的各种反馈，有可能扩大或缩小因气候而造成影响。（单词个数：81）

气候系统内部变异大多数形式可以视为一个滞后的形式，即当前的气候状态并不立即反映出来。由于地球气候系统是如此之大，其步伐是如此缓慢，这当然也有其气候变化输入的时间滞后方面的影响。例如，今年的干旱情况，可能只不过是导致湖泊轻微的萎缩，或轻微的平原干旱。然而在第二年，这些条件可能会导致降水减少，从而可能导致下一年干旱加剧。当达到临界点“*x*”值后面的数年，整个系统可能会发生较大地改变。在这种情况下，则会导致不再降雨。正是这样的变化，使得气候变化已成为迅速和不可逆转的滞后的变化。（单词个数：139）

1. 板块构造

在最长的时间尺度，由于我们赖以生存的气候因素的变化将使大陆板块重组，海洋形状的重塑以及山岭海拔的变化。在石炭纪时期，板块构造可能引发的碳含量和冰川大规模存储增加。最近，板块运动的强度比冰河时期加强，大约 300 万年前，北美和南美洲板块相撞形成的巴拿马地峡阻止了大西洋和太平洋接洽。（单词个数：91）

2. 太阳能辐射输出

过去几个世纪对太阳活动的变化观察，主要是通过太阳黑子的活动和对同位素铍的观测。气候系统中的热能来源大部分是由太阳热能辐射输出所提供的。当太阳的能源总量较少时会对月球的引力产生的影响（如潮汐能表现），或产生地热能。太阳的辐射输出能量，通过热传递到地球表面，是影响地球气候的重要组成部分。在地球早期的历史上，一种理论认为太阳不够热，从而使得地球表面的液态水得以存在，虽然众所周知，但确是一种似乎矛盾又正确的说法。在接下来的数千年，太阳将通过所谓的“主序”继续照亮，并产生相应的更高的能量输出，从而相应地影响了地球大气层。（单词个数：147）

在最近一段时期，也有一些太阳变化的不同形式，包括 11 年的太阳周期和较长期的调节。然而，11 年的太阳黑子周期似乎没有明确的气候资料表现出来。从冰河世纪时期，以及一些从 1900 年至 1950 年观察到气候变暖的现象，我们认为太阳能辐射强度的变化已经对气候产生影响。太阳的辐射能量输出的周期性尚未被充分认识和理解，它是从太阳内部很缓慢的变化开始的，因为它的成熟和演变产生了一些对太阳黑子的周期性活动，而一些研究发现太阳黑子的活动影响着全球气候变暖。（单词个数：111）

3. 轨道变化

轨道变化，从某种意义上是对太阳变化的延伸，因为地球轨道上的变化，将导致到达地球表面的太阳光的分配和强度略有不同。这些轨道的变化，被人们称为米兰科维奇周期，直接影响冰川活动。地球轨道变化的米兰科维奇周期的三个主要组成包括偏心、轴向倾斜和岁差。在这周期的三个变化产生的综合效应下，太阳辐射到地球表面的光线，也季节性地变化着。因此，米兰科维奇循环影响着收到的太阳辐射的增加或减少，直接影响地球的气候系统，并影响和推动地球的冰川减少。更微妙的一些变化也开始出现了，比如轨道岁差会造成哈拉沙漠的延展或收缩。（单词个数：138）

4. 火山

火山是将地球深处的某些物质输送到地球表面的过程，伴随着这个过程的发生，地球也将释放其内部多余的热量和一部分的压力。火山爆发、间歇泉、温泉都是火山过程的一部分，

都不同程度释放微粒到大气中。（单词个数：57）

那种每个世纪都发生几次的可以影响气候的单喷发，需要数年或更长时间的冷却。例如：在 1991 年皮纳图博火山喷发，产生了 20 世纪第二大地面喷发（1912 年的诺瓦拉普塔火山喷发后）和气候的影响大幅增加，导致了全球气温下降了约 0.5℃（0.9℉），以及臭氧消耗暂时性的大幅度增加。更大的爆发，就如大火成岩范围，已知出现的概率，只有每亿年几次，但可以重塑了数百万年的气候，并导致大规模的物种灭绝。起初，人们认为由于大规模的火山喷发而抛射入大气层的尘埃，长期挡住了太阳辐射传输到地球表面，而导致长时间气温下降冷却的。然而，测量结果显示，在适当的条件下，大部分被投掷而进入大气层内的细小颗粒灰尘，经过 6 个月的时间就可能返回到地球表面。（单词个数：161）

火山也是延续的碳循环的一部分。在很长的（地质）时期里，他们释放了地球内部的二氧化碳，抵消由沉积岩等地质吸收了的二氧化碳。美国地质调查局的调查报告称，人类活动产生的二氧化碳估计是火山的二氧化碳排放量的 130 多倍。（单词个数：60）

5. 海洋变化

在长达几十年或更长时间的观测过程中，我们发现气候变化也可能来自大气和海洋之间的相互作用。许多气候波动的影响，包括厄尔尼诺现象、太平洋洋流、北大西洋洋流和北极洋流，至少有一部分热量可能会储存在海洋中，也可能以不同的方式方法穿梭于各大洋形成的巨型“水库”。在较长（一个完整的周期往往需要到上千年才能完成）的时间里，海洋过程，如温盐环流中热发挥关键作用，通过开展十分缓慢和极深水运动热的再分配，以及热在海洋中的长期的再分配。（单词个数：125）

6. 人类活动的影响

人为因素是人类活动，改变了环境。在某些情况下，关于人类对气候的影响是直接的和明确的（例如，灌溉对当地湿度的影响）。而其他情况，现在还不太清楚。各种假设人类活动引起的气候变化已经争论了多年，总体而言，科学辩论也已经从怀疑转向到对气候变化的科学共识，过去几十年，人类活动很可能是造成全球气候迅速变化的原因。因此，辩论在很大程度上转移到如何进一步降低人类的影响，并找到方法来适应已经发生了的变化。（单词个数：116）

大多数这些人为气候变化的原因，是由于化石燃料燃烧排放以及空气中悬浮颗粒（大气中的颗粒物质），以及胶结材料制造而导致二氧化碳气体的增加。其他因素，包括土地利用、臭氧层破坏、动物农业和森林砍伐，同时，为人类自己扮演的角色而备受关注——单独或与其他因素结合起来，也在影响着气候变化。（单词个数：64）

7. 海平面变化

对全球变暖问题的理论证据，来源于气候模式在很大程度上依赖于全球平均海平面长期变化的观测。通过潮位观测，一般估计和与过去很长一段时间的平均水位值相比，上个世纪的全球海平面变化很大。最近，与确定卫星轨道相结合的测高仪观测，已经提供了一个观测全球海平面变化的方法。（单词个数：73）

Chapter 15 Water Use Efficiency and Water Conservation: Definitions

Part 1 Water Use Efficiency

Water use efficiency and water conservation are commonly used terms applied to irrigated agriculture. However, confusion and misunderstanding can arise because these terms have multiple definitions. Water use efficiency is calculated by a ratio of terms which is then multiplied times a hundred to report the efficiency as a percentage. Different measurements may be used in the ratio resulting in different numbers and yet all may be referred to as water use efficiency. For example, water use efficiency can be defined as a ratio of beneficial water use to the Applied Water (AW) to the field. However, beneficial use sometimes is defined as ET (Evapotranspiration) and at other times it is ET plus the amount of water required for leaching salts from the root zone. Some individuals include all of the water delivered to the field as AW, whereas others might only consider the infiltrated water that would be available for crop use. For example, in a surface irrigated system, the runoff may be considered as part of the AW, or it may be subtracted from the AW. Obviously different numbers result for different combinations of these terms. Furthermore, the computations can be made on different area sizes. For example, the ratio can be calculated for a field, the total farm, or the total basin. Different numbers result depending on which is selected. Possibly the biggest problem however is the common belief that a higher efficiency number is always better than a lower number. Any number less than 100 percent is considered to have some degree of inefficiency. As will be discussed later in this paper, this is not usually the case.

Water conservation, likewise, is subject to different definitions. One definition is to use less water. This can be accomplished by various means, each with a specific consequence. A farmer can use less water by not growing a crop. Or, a farmer can grow a crop and apply a small amount of water resulting in very low crop production. Farmers can also grow a crop for high production and either eliminate runoff or capture runoff and use it as part of the irrigation supply.

The main point is that in using these terms, the definition must be clearly specified and the consequences of the action properly evaluated. A further complication is that all of the definitions are based on water quantity without reference to water quality and water quality cannot be ignored in a management scheme.[1]

A distinction between water use and water consumption is required to properly assess water conservation practices. Water consumption is water that is lost for future use. For example, ET is water consumption. It is water that is lost and not available until it is returned again in the form of precipitation, usually at a different location. For a non-pressurized irrigation system with runoff from the field, part of the water is consumed through ET and part is used but runs off and is available for other uses. Water that percolates beyond the root zone might be consumed or used depending upon the fate of that water. If the water migrates to groundwater or a stream, it is still available for use and therefore not consumed. Conversely, if it migrates to a location where it cannot be retrieved, it should be considered as consumed.

Part 2 Economic Irrigation Efficiency

Because of deficiencies in the traditional water use efficiencies, as stated above, a different criterion must be used to characterize the optimal irrigation management. Since agriculture production is a business operation, maximizing economic irrigation efficiency would be an appropriate goal. Economic irrigation efficiency is defined at the farm level as the irrigation management that maximizes profits. In a broader context economic irrigation efficiency could be defined as irrigation management that maximizes net social benefits. The difference between the two definitions is the result of externalities. An externality arises when some of the costs or benefits of irrigation agriculture accrue to society as a whole and the costs (as reflected in market prices) are not borne by the farmers or the consumers of their products.[2] Externalities can be positive or negative. An example of a positive externality occurs when water purchased by a farmer runs off his farm and serves some beneficial societal use. However, if the water is polluted it can impose a cost to society and create a negative externality.

Benefit curves by multiplying the yield by the market price for the crop are depicted in Fig.15.1 as a hypothetical case representing two irrigation uniformities. The total benefit (*TB*) in $ ha-1 for a

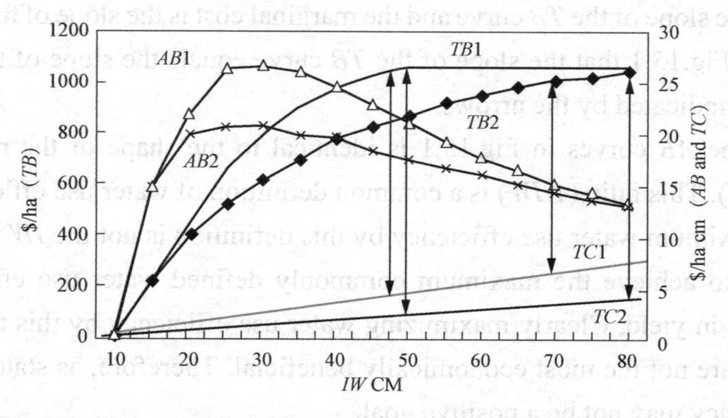

Figure15.1 The total benefit (*TB*) and average benefit (*AB*) as a function of infiltrated water (*IW*) for two levels of irrigation uniformity, when irrigation uniformity is more uniform for case 1 than case 2. Also presented is the total cost for water (*TC*) as a function of *IW* when the cost for water is higher for case 1 than case 2. The arrows represent *IW* that maximize profits.

given IW is higher for the more uniform (*TB*1) than the less uniform irrigation (*TB*2). The total cost of water (*TC*) is also depicted in Fig.15.1 where the price of water for case 1 (*TC*1) is greater than for case 2 (*TC*2).

The highest profit is achieved where the differences between *TB* and *TC* is the greatest. These points are identified by arrows in Fig.15.1. Some general conclusions can be derived from the information depicted in Fig.15.1. The economically optimal (profit maximizing) level of *IW* (Infiltrated Water) depends on the shape of the crop-water production function and the price of the water. Improving the uniformity of irrigation results in a decrease in the value of *IW* that achieves economic efficiency. Also raising the price of water lowers the value of *IW* that achieves economic efficiency. Raising the price of water has a greater effect on decreasing the economically efficient *IW* value under the nonuniform irrigation system as compared to the more uniform irrigation system.[3] Indeed, raising the price of water had relatively little effect on changing the economically efficient level of *IW* for the most uniform system.

A shift in irrigation technology or management to achieve more uniform irrigation usually imposes a cost. The increased cost may not be offset by the increased benefits associated with improved irrigation uniformity to justify the investment. This factor must be evaluated on a case by cases basis.

The Average Benefit (*AB*) as a function of *IW* is also depicted in Fig.15.1 for the two irrigation uniformities. *AB* is calculated as *TB* divided by *IW* and has the units of $ (ha cm)-1. *AB* is the average dollar return per hectare-centimeter of *IW*. Note that the maximum *AB* value occurs at a lower *IW* value than the economically optimal quantity. Even though the average return decreases, the cost for additional irrigation water is exceeded by additional return associated with an incremental increase in water. Economically optimal input is at the point where the marginal benefit equals the marginal costs. The marginal benefit is defined as the benefit associated with the next incremental increase in input and likewise the marginal cost is the increase in cost associated with the next incremental input.[4] The marginal benefit is the slope of the *TB* curve and the marginal cost is the slope of the *TC* curve depicted in Fig.15.1. Note in Fig.15.1 that the slope of the *TB* curve equals the slope of the *TC* curve for the optimal irrigation as indicated by the arrows.

The shape of the *AB* curves in Fig.15.1 is identical to the shape of the ratio of yield (Y) to Infiltrated Water (*IW*). This ratio (*Y*/*IW*) is a common definition of water use efficiency. Note that the *IW* that achieves maximum water use efficiency by this definition is not the *IW* that is economically efficient. Irrigating to achieve the maximum commonly defined water use efficiency results is a significant reduction in yield. Clearly maximizing water use efficiency by this traditional definition leads to results that are not the most economically beneficial. Therefore, as stated above, increasing the water use efficiency may not be a positive goal.

A shift in production function from less uniform to more uniform irrigation does result in higher water use efficiency for a given value of *IW*. Therefore increasing the numerical value of water use efficiency by a change in management that entails a change in production function is positive. However, it is not necessarily economical. It is not obvious that the shift in production function from the less uniform to the more uniform irrigation is economically efficient. The main conclusion is that generalizations can not be made, and each situation has to be thoroughly evaluated from a production and economic consideration.

Part 3 Water Use Monitoring

Irrigation scheduling refers to the time, duration, and quantity of an irrigation. Although crop water production functions as depicted in the figures provide the scientific and economic basis for optimizing irrigation, farmers do not have such complete detailed information available to guide their irrigation management. Nevertheless, the general principles still apply. Since the purpose of irrigation is to replace the water lost from the storage zone between irrigations, knowing the amount of ET that has occurred since the last irrigation is important. Alternatively, the farmer could monitor the soil water content as a function of time to determine when the soil is sufficiently dry to warrant recharge.

Climatic conditions drive ET, therefore monitoring the potential ET by an evaporation pan or from other climatological data. In California several weather stations have been established throughout the state to form a California Irrigation Management Information System (CIMIS). Farmers with computer systems can get daily information from the weather station located nearest to their farms.

The climatological data identify the potential ET. The crop ET is not always equal to the potential ET. For example, during the early part of the season for annual row crops, the plant is small, and the crop ET is much less than the potential ET. As the crop grows and the canopy cover increases, the crop ET approaches potential ET. Thus, to estimate crop ET the potential ET is multiplied by a crop coefficient (K_c), which must be empirically determined for each crop as a function of time. Studies have established the crop coefficients for several crops in California. Results from these studies can be used to guide irrigation management. Nevertheless, the study results are not absolute and the farmer must use judgment and make observations in the field to be sure that his irrigation is appropriate.[5]

Monitoring the soil water content as a function of time requires instrumentation. The neutron probe can be used to measure the water content in the soil profile, but this method is labor intensive. It requires the installation of neutron probe access tubes and then measurements on some predetermined schedule. Other instruments such as tensiometers can be installed at various depths and require reading on a timely basis. Some of the instruments have electrical signals that can be connected to a recorder for continuous monitoring with minimal labor input. Soil moisture monitoring requires capital investment and then some level of operational expense.

* * * * * * * * * * * * * * * * * Explanations * * * * * * * * * * * * * * * * *

* * * * * * * * * * * * * * * * New Words and Expressions * * * * * * * * * * * * * * * *

1. leaching [ˈliːtʃɪŋ] n. 沥滤　　v. 浸出，淋洗；滤取；滤去

2. infiltrate [ɪn'filtret] vt. 使潜入；使渗入

vi. 渗入

n. 渗透物

3. consequence['kɑnsɪkwəns] n. 结果；重要性；推论

4. eliminate [ɪ'lɪmɪnet] vt. 消除；排除

5. percolate['pə:kəleɪt] vt. 过滤；渗出；浸透

vi. 使渗出；使过滤

n. 滤过液；渗出液

6. retrieve [rɪ'triv] vt. 检索；恢复；重新得到

vi. 找回猎物

n. 检索；恢复，取回

7. externality [ˌɛkstər'næləti] n. 外在性；外形；外部事物

8. accrue [ə'kru] vt. 产生；自然增长或利益增加

9. climatological [ˌklaɪmətə'lɒdʒɪkəl] adj. 与气候学有关的

10. tensiometer [ˌtensɪ'ɒmɪtə] n. 张力计；表面张力计；土壤湿度计

*** * * * * * * * * * * * * * * * Complicated Sentences * * * * * * * * * * * * * * * ***

1. A further complication is that all of the definitions are based on water quantity without reference to water quality and water quality cannot be ignored in a management scheme.

【译文】另外一个复杂化的因素是所有关于节水的定义都是基于水量而没有涉及水质，而水质在管理方案中不可忽视。

【说明】that 引导的长从句直至句尾，从句的主语又由两个短句构成，短句用 and 连接。前一短句表示定义只考虑水量没有考虑水质，后一短句表示水质是不能忽视的，两短句表达转折的含义。

2. An externality arises when some of the costs or benefits of irrigation agriculture accrue to society as a whole and the costs (as reflected in market prices) are not borne by the farmers or the consumers of their products.

【译文】当灌溉农业的一些成本或收益产生于整个社会，并且该成本（反映为市场价格）不由农民或消费者承担时，就产生了正外部效应。

【说明】externality 此处意为外部效应。accrue to 意为归属、融合。

3. Raising the price of water has a greater effect on decreasing the economically efficient IW value under the nonuniform irrigation system as compared to the more uniform irrigation system.

【译文】同样提高水的成本也会降低入渗水的经济效率。比起更均匀的灌溉系统，不均匀系统提高水的成本对降低渗水的经济效率有更大的影响。

【说明】has a greater effect on 表示"对……有重大影响"，nonuniform 意为不均匀的。

4. The marginal benefit is defined as the benefit associated with the next incremental increase in input and likewise the marginal cost is the increase in cost associated with the next incremental input.

【译文】边际收益指的是每增加一件产品获得的收益，同样边际成本指的是每增加一件产品增加的成本。

【说明】marginal 意为边缘的、边际的，incremental 意为增加的、增值的。

5. Nevertheless, the study results are not absolute and the farmer must use judgment and make observations in the field to be sure that his irrigation is appropriate.

【译文】然而，研究结果并不是绝对的，农民必须在农田灌溉过程中做出判断并观察，以确保灌溉的合理性。

【说明】to be sure 意为确保、表示目的，that 后面引导从句补充说明“确保”的内容。absolute 意为完全的、绝对的。

********************* Summary of Glossary *********************

1. be subject to	受限于
2. without reference to	不管；与……无关
3. marginal benefit	边际效益
4. as stated above	如前文所述
5. irrigation scheduling	灌溉制度
6. neutron probe	中子探测器
7. labor intensive	劳动密集型
8. accrue to	归于
9. be identical to	与……相同

********************* Abbreviations (Abbr.) *********************

CIMIS California Irrigation Management Information System
加利福尼亚灌溉管理信息系统

*********************** Exercises ***********************

(1) Benefit use has two calculation methods: 1. Single ______ and 2. ______ plus the amount of water required for ______ from the root zone.

(2) Usually, maximum average benefit value occurs at a lower infiltrated water value than the ______ quantity. Economically optimal input is at the point where the ______ equals the ______.

(3) Irrigation scheduling refers to the ______, ______, and ______ of an irrigation.

(4) ______ can be used to measure the water content in the soil profile, but this method is ______.

****************Word Building (15) pre-;mis-****************

1. pre- [前缀]，表示：提前, 之前

determined	adj.	predetermined	adj.	预设的
bid	n.	pre-bid	adj.	投标前的
school	n.	preschool	adj.	入学前的
paid	adj	prepaid	adj.	预付的

2. mis-[前缀]，表示：相反

understanding	n.	misunderstanding	n.	误解
aligned	adj.	misaligned	adj.	没有对准的
fortune	n.	misfortune	n.	不幸，灾祸

**************** **Text Translation** ****************

第十五章　水利用率与节水

第一节　水　利　用　率

水利用率和节水是灌溉农业常用的术语。然而，由于它们定义很多，就产生了困惑和误解。水利用率是用一个比例乘以100后，以百分比的形式呈现的。不一样的计量会得出不一样的数值，但所有的都被认为是水利用率。例如，水利用率是指有效利用的水量与灌溉水总量的比例。然而，有效利用水量有时指蒸腾所需的水，有时又指蒸腾水量加稀释根区盐分所需的水量。有些人把输入农田的所有水量称为灌溉水，但另外一些人只把庄稼利用的渗入水认为是灌溉水。例如，农田表面上漫流的水被认为是灌溉水的一部分，或者不是其中的一部分。很显然，不同的计量方法会导致不同的数值。此外，不同的田地大小也是可以计算的。例如，一个田地水分利用率可以是计算整块田，或者是凹下盆状地带。不同的数值结果取决于计量范围的大小。也许最大的问题就是人们普遍认为高水分利用率比低水分利用率好。任何低于 100%的数值都被认作是某种程度的低效率。接下来的讨论并不是以上所说的通常情况。(单词个数：271)

节水同样也有不同的定义，其中一种说法是少用水。节水可以通过各种方法实现，每一种方法都有其特定的结果。农民可以不种庄稼来节水。或者种庄稼，但缺乏灌溉就会导致低产。农民还可以通过降低流量或者充分利用流量作为灌溉水量，从而提高产量。(单词个数：83)

在使用这些术语时，最重要的是必须清楚地区分每个定义，合理地评估每个节水行为将产生的后果。另外一个复杂化的因素，是所有关于节水的定义都是基于水量没有涉及水质，而水质在管理方案中不可忽视。(单词个数：52)

合理评定节水实践必须区分水利用量和耗水量。耗水量指未来消耗的水分。例如，蒸腾掉的水是耗水量。这种消耗的水分只有等到它在不同的地点重新以降水的形式回归大地时才可利用。无压系统灌溉方式，灌溉水在田间径流，一部分水分通过蒸腾被消耗，另一部分径流的水分可做其他用处。渗透进根区的水分可能被消耗利用了，但这要看它们各自的命运了。如果水分转移到地下水或溪流中，那么仍旧可以利用不至于消耗了。相反地，如果转移到不能重新取回的地方，那么就是被消耗了。（单词个数：143）

第二节 经济灌溉效率

正如上面所述的那样，由于传统的水利用效率存在不足，必须用一个不同的标准来描述最佳灌溉管理的特征。农业生产是一个商业运营活动，扩大灌溉效率是个合适的目标。经济灌溉效率在农场级别被定义为扩大经济效益的灌溉管理。从更广的层面来说，经济灌溉效率指扩大净效益的灌溉管理。两种定义的差异是由外部性引起的。当部分农业灌溉的成本或效益归于整个社会，而成本（市场价格所反映的）不由农民或者产品消费者所负担时，就产生了外部性。外部性分为正外部性和负外部性。当灌溉农业的一些成本或收益产生于整个社会，并且该成本（反映为市场价格）不由农民或消费者承担时，就产生了正外部效应。而当灌溉水被污染时，就会给社会造成负担，产生消极的外部性。（单词个数：171）

图 15.1 所示的作物水分生产函数可以转变为收益曲线，通过作物的市价来增产。图 15.1 以假定的两种灌溉均匀性来描述收益曲线。在一定的入渗水的情况下，灌溉均匀性高的整体效益（*TB*1）比灌溉均匀性低的整体效益（*TB*2）要高。图 15.1 描述了水的总成本，其中案例 1（*TC*1）比案例 2（*TC*2）高。（单词个数：78）

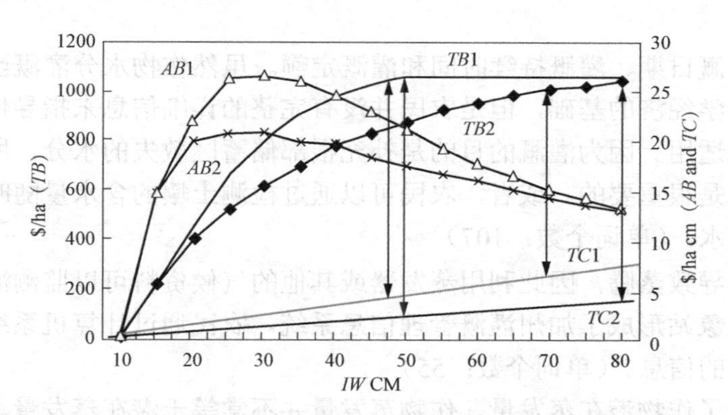

图 15.1 当案例 1 灌溉水的均匀性高于案例 2 时，两种灌溉均匀性渗入水整体效益和平均效益的曲线函数。同时描述了当案例 1 的水成本高于案例 2 时，入渗水总成本的曲线函数，箭头表示入渗水的最大化利润

当 *TB* 和 *TC* 的距离最大时，就产生了最大化利润。图 15.1 中已用箭头表示。从图 15.1 所示的信息中我们得出了一些结论：入渗水的最大利润取决于作物水分生产函数的形状和灌溉水成本。提高灌溉水的均匀性可降低入渗水的经济效率。同样，提高水的成本也会降低渗水的经济效率。比起更均匀的灌溉系统，不均匀系统提高水的成本对降低渗水的经济效率有更大的影响。当然，对大多数的均匀系统来说，提高水的成本对改变入渗水的经济效率的影响相对较小。（单词个数：145）

为了提高灌溉的均匀性转而到灌溉技术和管理上，这将带来负担。增加的成本不会因为增加的效益和提高的灌溉均匀性而抵消。这个因素在具体案例分析中必须被考虑在内。（单词

个数：47）

两种灌溉不均匀性系统的入渗水平均收益函数如图 15.1 所示。平均收益等于总收益除以入渗水。平均收益指每公顷入渗水的平均收益。如图所示，入渗值越低平均收益越大。即使平均收益降低，每增加一定量的灌溉水获得的收益大于付出的成本。当边际收益等于边际成本时，利润才最大。边际收益指的是每增加一件产品获得的收益，同样边际成本指的是每增加一件产品增加的成本。如图 15.1 所示，*TB* 曲线指边际收益，*TC* 曲线指边际成本。正如图 15.1 箭头所示那样，*TB* 曲线等同于 *TC* 曲线才是最佳灌溉条件。（单词个数：186）

图 15.1 中 *AB* 曲线的形状与产量（*Y*）与入渗水（*IW*）之比的形状相同。该比率（*Y*/*IW*）是水利用效率的通用定义。注意，通过该定义所实现的最大水利用效率的 *IW* 并不是最经济有效的。通过灌溉达到最大的水利用效率将导致产量的显著降低。通过这种传统定义最大化水使用效率会导致结果不是利益最大化的。因此，如上所述，提高用水效率可能不是积极的目标。（单词个数：107）

在特定的入渗值情况下，生产功能从相对小的均匀灌溉系统转移到相对大的均匀灌溉系统确实提高了水分利用率。因此，改变灌溉管理增加了水分利用数值，这使生产函数产生积极的改变。然而，这未必经济。生产功能从相对小的均匀灌溉系统转移到相对大的均匀灌溉系统的效率并不明显。因此最主要的结论是不能一概而论，必须从产量和经济具体分析每种情况。（单词个数：99）

第三节 水分利用监测

灌溉制度指灌溉日期、灌溉持续时间和灌溉定额。虽然作物水分灌溉函数图已经为优化灌溉提供了一个科学经济的基础，但是农民并没有完整的详细信息来指导他们的灌溉管理。然而，一般原则仍适用。因为灌溉的目的是补充根部储蓄区散失的水分，所以知道自上次灌溉后的水分蒸腾量是很重要的。或者，农民可以通过检测土壤的含水量随时间的变化来决定土壤何时需要灌溉水。（单词个数：107）

气候条件也会导致蒸腾，因此利用蒸发器或其他的气候资料可以监测潜在的蒸腾。加州已经建立了一些气象站形成了加州灌溉管理信息系统。农民通过计算机系统可以从农田附近的气象站获取每日的信息。（单词个数：55）

气候条件决定了作物潜在蒸发量。作物蒸发量并不常等于潜在蒸发量。例如，每年种植庄稼的早期作物很小，作物的蒸腾量少于潜在蒸发量。随着作物的成长覆盖度增加，作物的蒸发量达到潜在蒸发量。因此，估计作物蒸发量和潜在蒸发量要被乘以作物系数。作物系数随着作物生长过程而变化。研究结果确定了加州集中作物的作物系数。这些研究成果可以被用来指导灌溉管理。然而，研究结果并不是绝对的，农民必须在农田灌溉过程中做出判断并观察，以确保灌溉的合理性。（单词个数：138）

监测土壤的含水量随着时间的变化需要监测仪器。中子水分探测仪可以被用来监测土壤剖面的水分含量，但是这种方法是劳动密集型的。这需要安装中子水分探测仪管并在既定的时间进行监测。还可以在田间的各种深度层安装其他的设备，如土壤湿度计，并按时去查看。有些设备有电子信号可以连接记录仪，以最小的劳动投入获得连续的监测。土壤湿度监测需要资金的投入和某种程度上的运营费用。（单词个数：101）

Chapter 16 Irrigation Culture

Part 1 Irrigation: Industry or Culture?

Agriculture is viewed by the public in two not necessarily consistent ways. The first is that agriculture, including irrigated agriculture, is a business, an industry, albeit an industry essential to human existence. Competing with this pragmatic view is one that sees irrigated agriculture as a complex system, one that has spawned an individual culture. If society takes the position that irrigated agriculture is an industry, it would logically move toward a situation where the user bears all the costs of production.[1]

Since the 1930s, the United States has favored the agriculture-as-culture model, but the trend in recent years is changing. More and more, agriculture, including irrigated agriculture, is seen as a business that must compete in a global economy. This new emphasis stresses the entrepreneurial side of farming, but it may diminish the cultural assets associated with the irrigated agriculture community.

The United States is not alone in this dilemma—it is playing out throughout the world. How does this question of irrigation as industry or culture affect policy making? If the industry view of irrigated agriculture is pursued, one option is to make an even bigger push for markets in water so as to subject the industry to full market discipline. There could be more price pressure on the industry, fewer subsidies, and full-cost pricing of federally supplied water. In return, irrigators might be granted transferrable water rights and limitations on the acreage eligible for federal water might be removed.

On the other hand, if irrigation is viewed more as a culture, policy decisions would tend to insulate agriculture from direct market forces. The prevailing view—irrigation as industry or culture—varies from region to region and person to person. Rather than imply that one view is more right or wrong, it is the committee's intention to say simply that both views exist and will continue to exist as irrigation evolves into the future.[2]

Part 2 The Cultures of Irrigation

The obvious dimensions of irrigation are tangible—how much water is used, what acreage of land is irrigated, what crops are grown, what forces of change and responses are seen. But to really understand irrigation and how it might evolve in the future, we must consider the more intangible, subjective dimensions of irrigation—in a sense, the context in which change must occur. We call this the culture of irrigation.

At one level, irrigation is simply the application of water to grow plants. At another level, it is the basis for an economy and a way of life. In a very real sense, irrigation made possible the highly intensive settlement of a landscape that otherwise would not readily support large numbers of people. Irrigation has transformed that landscape, literally and figuratively. The bands of green fields sometimes spreading out to considerable distances from the banks of the rivers of the western United States, the circles of green covering the Great Plains, the urban oases filled with trees, flowers, and lawns—these are the products of irrigation. Culture, as used in this chapter, refers to the "ideas, customs, skills, arts, etc. of a given people in a given period."

Irrigation, as it has been practiced in agriculture, is a distinctive activity. It is sufficiently distinctive that it has its own history, its own governmental policies, its own institutions, its own practices, and, historically at least, its own communities. Modern irrigation in the United States probably began with the Mormons whose existence as a community in the Great Salt Basin depended on its practice (Arrington, 1975). It grew in places such as California and Colorado, first in support of mining, and then to support settlement itself. Later in the nineteenth century, it outgrew its utilitarian origins and took on the aura of a movement, becoming for some the basis for building utopian communities (Boyd, 1897), for making the desert bloom (Maas and Anderson, 1978), and for civilizing the Great American Desert of the West (Smythe, 1905). Congress created a federal agency—now called the Bureau of Reclamation—dedicated solely to the task of expanding irrigation in the West (Pisani, 1992).

Part 3 Understanding the Culture of Irrigation

Is there a culture of irrigation in the United States? There are many distinctive regional patterns and processes of irrigation. The massive agricultural projects and businesses of California, for example, stand in sharp contrast with smaller operations of the Rocky Mountain states. The center pivot systems of the Great Plains have little in common with sugarcane irrigation of the Gulf Coast. There is an enormous diversity of irrigation cultures.

At the same time, several irrigation patterns and movements assumed national significance in the nineteenth and twentieth centuries. Collectively, the perspectives and projects associated with irrigation established a "culture of irrigation," some aspects of which persist, while others face fundamental challenges. Three aspects of this culture seem particularly relevant for the future of irrigation.

1. The Reclamation Ethic

To this day, many irrigators maintain strong views about the inherent value of "reclamation." Whether draining the bottomlands of the lower Mississippi valley or irrigating the deserts of the West, the historical transformation of "waste into wealth" is a source of enormous satisfaction for irrigators. Many irrigation communities seek to maintain or revive the original values associated with reclamation. They disagree with views that see reclamation as environmentally harmful. Indeed, in some respects the irrigated agricultural community now is paying a price for not responding sooner to early criticism about the harms of reclamation by popular critics such as Reisner (1986).

Although many share contemporary concerns for such things as fish, wildlife, and environmental quality, the community overall was badly served by those who initially dismissed the critics. Reclamation agencies adapted slowly and awkwardly to the changing cultural context.

2. Attitudes Toward Water

In the western states, water often is described as the "lifeblood" of the region. Many in the West still believe that land without water has little value, which is literally true for irrigated cropland. This fundamental dependence on water gave rise to several deeply rooted concepts that guide agricultural water use and have profoundly influenced western water law.

At base, irrigators view water as an essential means to an end—that is, an input needed to grow crops. This highly instrumental view of water promotes the importance of clarity respecting relative rights to use water as well as the value of certainty in those rights. Thus, the principle of priority—"first in time, first in right"—holds great importance for irrigators. Not only does priority help to sort out competing claims to water, it also serves to protect the substantial claims of irrigated agriculture to water since much of this use was established early enough in the settlement of the West to give agricultural users seniority over most other water uses.[3] One consequence of a priority rule is to emphasize time as the most important factor in determining rights to use water rather than, for example, place, value, or purpose of use (Bates et al., 1993).

Related to this strong desire for certainty is the importance of stability and protection against change. Dependent as they are on the availability of water, irrigators understandably fear the diminishment of their water supply. With a historical record of generally increasing land areas coming under irrigation until the past decade or so, irrigators have jealously guarded their claims. Changes of the use of irrigation water rights, particularly for non-agricultural uses outside the original place of use, have been resisted (MacDonnell and Rice, 1994). The legal concept of no injury has emerged to protect the water rights of existing users against change.

Given the increasing competition for water use that often pits irrigation agriculture against urban, tribal, and environmental interests, it is perhaps important to understand these cultural views of water. They help to explain the fervor with which irrigation users sometimes defend their traditional water use prerogatives. They shed light on the resistance of many irrigators to the increased efforts to market water as a means of changing its use from agriculture to cities. They help explain how irrigators may see water as a collective good in relation to the needs of the irrigation community but resist the notion of water as a public, instream resource.[4]

Part 4 Cultural Conflict and Cooperation

Some of the most inspiring, and painful, lessons of irrigation date to the middle and late nineteenth century, when large-scale population movements displaced indigenous cultures and reworked the water resources of their settlement frontier for mining, farming, and ranching purposes. These processes involved remarkable instances of cooperation, and also bitter conflicts. In addition to conflicts with American Indians over land and water, irrigators had uneven relations with other economic groups. Irrigators established themselves in some areas to serve small populations of miners, travelers, trappers, forts, ranchers, and traders. Where mixed activities flourished, they sometimes came into conflict—as when hydraulic mining destroyed water quality, stream channels, and downstream irrigated lands. Federal reclamation policy sought, in part, to "settle" the western territories, that is, to populate them and to substitute a sedentary stable agrarian economy and society for more volatile and transient activities. Later, when flourishing irrigation economies contributed to population growth and commercial expansion, it sometimes led to competition for limited water supplies and environmental conflict.[5]

Cooperation and conflict are perennial in irrigated areas. Prehistoric irrigation systems involved high levels of community cooperation for construction, maintenance, cultivation, and settlement. The archeological record of economic competition and political conflict is limited, but ethno historical evidence suggests that prehistoric irrigators faced a variety of internal and external conflicts that affected the sustainability of their irrigation systems.[6]

Before turning to the legacy of conflict, it is important to review, again, the extraordinary diversity of nineteenth century irrigation practices. A wide variety of farmers and communities adapted practices from the eastern states for the arid West. African-Americans had transferred rice irrigation practices from Gambia to South Carolina (Carney, 1993). Chinese immigrants played a central role in the reclamation of the Sacramento River floodplains and delta (Chan,1986). Indians from the Punjab irrigated lands in the Imperial and Central valleys of California, regions very similar to, and influenced by irrigation practices in, colonial India (Jensen, 1988; Leonard, 1992; Wescoat, 1994). American water engineers and lawyers drew practical lessons from Italy and France, as well as India and Egypt (Davidson, 1875; Hilgard, 1886; Jackson et al., 1990; Kinney, 1912; Wilson, 1890-1891). It is little wonder that new types of conflict, and conflict resolution, arose in this rapidly changing heterogeneous environment.

In future efforts to reform irrigation institutions, three points seem important. First, irrigation is a cultural as well as economic and political system. The cultural character of irrigation institutions accounts in part for their resistance to change; an understanding of this cultural dimension might facilitate constructive change and conflict resolution. Second, it is important not to lose sight of the enduring value and efficacy of modern examples of cooperation and conflict resolution in irrigation. Otherwise, future generations may find themselves trying to recover what was lost from the middle and late twentieth century as well as from earlier periods. Third, it is important to focus on inventing new forms of cooperation that transcend the costly and historically entrenched patterns of conflict that involve water and related land resources, such as alternative methods of dispute resolution and mediation.[7]

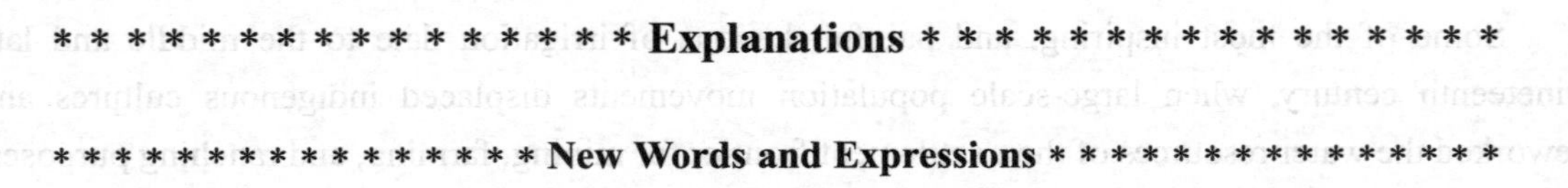

****************** Explanations ******************

****************** New Words and Expressions ******************

1. albeit [ɔːl'biːɪt] conj. 尽管，固然，即使

2. pragmatic [præg'mætɪk] adj. 实际的，实用主义的，国事的

3. dilemma [dɪ'lemə; daɪ-] n. 困境，进退两难，两刀论法

4. federally ['fɛdərəli] adv. 联邦地，联邦政府地，同盟地

5. prevailing [prɪ'veɪlɪŋ] adj. 流行的，一般的，最普通的，占优势的，盛行很广的

v. 盛行，流行（prevail 的现在分词形式），获胜

6. intangible [ɪn'tæn(d)ʒɪb(ə)l] adj. 无形的，触摸不到的，难以理解的

7. distinctive [dɪ'stɪŋ(k)tɪv] adj. 有特色的，与众不同的

8. utilitarian [jʊˌtɪlɪ'teərɪən] adj. 功利的，功利主义的，实利的

n. 功利主义者

9. pivot ['pɪvət] n. 枢轴，中心点，旋转运动

vt. 以……为中心旋转，把……置于枢轴上

adj. 枢轴的，关键的

n. (Pivot)人名（德）皮福特（法）皮沃

vi. 在枢轴上转动，随…转移

10. contemporary [kən'tɛmpərɛri] n. 同时代的人，同时期的东西

adj. 当代的，同时代的，属于同一时期的

11. instrumental [ˌɪnstrə'mɛntl] adj. 乐器的，有帮助的，仪器的，器械的

n. 器乐曲，工具字，工具格

12. seniority [ˌsinɪ'ɔrəti] n. 长辈，老资格，前任者的特权

13. tribal ['traɪbl] adj. 部落的，种族的

n. (Tribal)人名，(法)特里巴尔

14. flourish ['flɜːrɪʃ] vt. 夸耀，挥舞

n. 兴旺，茂盛，挥舞，炫耀，华饰

vi. 繁荣，兴旺，茂盛，活跃，处于旺盛时期

15. substitute ['sʌbstɪtut] n. 代用品，代替者

v. 替代

16. Prehistoric [prihɪ'stɔrɪk] adj. 史前的，陈旧的

17. transcend [træn'sɛnd] vt. 胜过，超越

18. dispute ['dɪs'pjʊt] n. 辩论，争吵

vt. 辩论，怀疑，阻止，抗拒

********************** Complicated Sentences **********************

1. If society takes the position that irrigated agriculture is an industry, it would logically move toward a situation where the user bears all the costs of production.

【译文】如果社会认为灌溉农业是一个产业，那么将会在逻辑上走向一个用户承担所有生产成本的状况。

【说明】that irrigated agriculture is an industry 为定语从句，修饰 position；where the user bears all the costs of production 是对 situation 的展开说明。Move toward a situation 意为“迈向一个局面”。

2. Rather than imply that one view is more right or wrong, it is the committee's intention to say simply that both views exist and will continue to exist as irrigation evolves into the future.

【译文】与指出哪一种看法的对错相比，委员会只是想简单地说，这两种观点都存在，而且随着灌溉发展到未来，二者将继续存在。

【说明】Rather than 引导的从句表示轻微否定，可理解为“与其……不如……”；as 引导的从句表示“随着……的到来”。

3. Not only does priority help to sort out competing claims to water, it also serves to protect

the substantial claims of irrigated agriculture to water since much of this use was established early enough in the settlement of the West to give agricultural users seniority over most other water uses.

【译文】优先原则不仅有助于整顿竞争性的水资源，而且还要能保障灌溉农业用水的大量需求，因为大部分的用水都是在西部地区早早建立起来的，要让农业用水比其他用水更有优先权。

【说明】Not only 与 also 为固定搭配，表示“不但……而且……”；Not only 放在句首时，后跟从句需用倒装结构，此处动词 does 放在主语 priority 之前。

4. They help explain how irrigators may see water as a collective good in relation to the needs of the irrigation community but resist the notion of water as a public, instream resource.

【译文】它们有助于解释灌溉者如何将水看作是与灌溉社区的需求相一致的集体利益，但却拒绝将水作为一种公共的河流资源。

【说明】居中 as 引导的从句到 community 截止，in relation to 从句修饰的是 collective good; but 引导的从句到句尾截止，表示“而不是……”。

5. Later, when flourishing irrigation economies contributed to population growth and commercial expansion, it sometimes led to competition for limited water supplies and environmental conflict.

【译文】此后，当繁荣的灌溉经济促进人口增长和商业扩张时，它有时会导致有限水资源供应的竞争和环境冲突。

【说明】contribute to 与 lead to 含义类似，均为“导致”的意思。Competition for limited water supplies 与 environmental conflict 为并列短语。

6. The archeological record of economic competition and political conflict is limited, but ethno historical evidence suggests that prehistoric irrigators faced a variety of internal and external conflicts that affected the sustainability of their irrigation systems.

【译文】经济竞争和政治冲突的考古记录是有限的，但是民族史学证据表明：史前灌溉者面临着各种各样的内部和外部冲突，这些冲突影响了他们灌溉系统的可持续性。

【说明】suggest that 此处意为“表明……”，that 引导从句。internal 和 external 为一对反义词，意为内部和外部；sustainability 为可持续性的意思。

7. Third, it is important to focus on inventing new forms of cooperation that transcend the costly and historically entrenched patterns of conflict that involve water and related land resources, such as alternative methods of dispute resolution and mediation.

【译文】第三，创造新的合作形式很重要，例如调解纠纷和仲裁的替代方式，这些合作方式要超越在水资源和相关土地资源时代价高昂，且根深蒂固的冲突模式。

【说明】文中第一个 that 引导的从句到句尾截止，是对 cooperation 的补充；第二个 that 引导的从句是大从句的内嵌从句，such as 分句与第二个 that 从句并列。

******************* **Summary of Glossary** *******************

1. global economy　　全球经济
2. policy making　　政策制定

3. market forces 市场力量
4. fundamental challenges 根本的挑战
5. Mississippi valley 密西西比河谷
6. American Indians 美洲印第安人
7. agrarian economy 农业经济
8. archeological record 考古记录
9. conflict resolution 冲突解决
10. lose sight of 忽略
11. substantial claim 实质性要求

* * * * * * * * * * * * * * * * * * Abbreviations (Abbr.) * * * * * * * * * * * * * * * * * *

USBR United States Bureau of Reclamation 美国垦务局

* Exercises *

(1) Irrigation was regarded as culture in the _______ but now more and more regarded as a _______ that competes in a global economy.

(2) The priority rule for western America water use can be explained as _______, _______, which can sort out the _______ claims to water.

(3) In the nineteenth century irrigation practices, _______ immigrants played a central role in the reclamation of the _______ River floodplains and delta.

(4) In future efforts to reform irrigation institutions, irrigation is a combination of _______, _______ and _______ system.

* * * * * * * * * * * * * * Word Building (16) bio-;-ful * * * * * * * * * * * * * *

1. bio- [词根] 表示：生物，生命

logical	adj.	biological	adj.	生物学的，生物的
logy	adj.	biology	n.	生物，生物学
mass	adj.	biomass	n.	生物质
energy	n.	bioenergy	n.	生物能

2. -ful [形容词词尾]，表示：可……的，有……的

power	n./vt.	powerful	adj.	有力的
doubt	n./v.	doubtful	adj.	可疑的
hope	n./v.	hopeful	adj.	有希望的
fear	n./v.	fearful	adj.	可怕的

* * * * * * * * * * * * * * * * **Text Translation** * * * * * * * * * * * * * * * *

第十六章 灌 溉 文 化

第一节 灌溉：产业还是文化?

公众以两种并不一致的方式来看待农业。第一，农业，包括灌溉农业，是一种商业，是一个产业，是人类生存不可或缺的产业。与这种实用主义角度相对应，另一种观点是将灌溉农业视为一个复杂的体系，一种孕育出个人文化的农业。如果社会认为灌溉农业是一个产业，那么将会在逻辑上走向一个用户承担所有生产成本的状况。（单词个数：81）

自 20 世纪 30 年代以来，美国一直赞成农业文化模式，但近些年的这种趋势正在发生改变。越来越多的农业，包括灌溉农业，被视为必须在全球经济中进行竞争的企业。这些新的重点更加强调农业的企业化，但同时，也将减少其与灌溉农业群体相关的文化价值。（单词个数：60）

美国并不是唯一面临这个难题的国家——世界各地都在发生。灌溉作为产业还是文化的这个问题如何影响政策制定？如果追求灌溉农业的产业观点，其中的一个选择就是去进一步推动水资源市场的发展，使产业受到充分的市场纪律约束。那么行业的价格压力将可能更大，补贴水平更低，联邦供水供应全额定价。作为回报，灌溉者可能被授予可转让水权，并且有可能取消联邦水资源合格面积上的限制。（单词个数：100）

另一方面，如果灌溉更多的被视为一种文化，政策决定往往趋向于将农业与直接的市场力量隔绝开来。普遍的看法是，灌溉作为产业还是文化是因地区因人而异的。与指出哪一种看法相比，委员会只想简单地说，这两种观点都存在，而且随着灌溉发展到未来，二者将继续存在。（单词个数：74）

第二节 灌 溉 文 化

灌溉实实在在的内容是灌溉用水量、灌溉面积、种植作物的种类、变化力和产生的影响等。但是要真正了解灌溉以及今后如何发展，我们必须考虑更多无形的、主观的方面——从某种意义上说，即这种变化必然发生的环境。我们称之为灌溉文化。（单词个数：70）

在某一层面上，灌溉仅仅是应用水种植作物；在另一层面上，它是经济和生活方式的基础。在实际意义上，灌溉使得高度集约化的景观成为可能，否则难以维持大量人口的生存。灌溉已经从表面上和含义上改变了景观。从美国西部河岸间或蔓延很长距离的绿色田野，覆盖大平原的绿色圈子，充满树木、花卉和草坪的城市绿洲——这些都是灌溉的产物。本章所说的文化，是指某一特定时期的某一特定人群的思想、风俗、技能、艺术等。（单词个数：128）

已经在农业中实践的灌溉，是一个独特的活动。它有着足够的独特性，它有自己的历史、自己的政府政策、自己的机构、自己的做法，至少在历史上它有自己的群体。美国的现代灌溉大概是从摩门教徒开始，他们在大盐盆地形成社区，生存必须依赖于灌溉。灌溉在诸如加利福尼亚和科罗拉多州等地区进一步壮大，首先服务于采矿业，然后服务于居住地自身。此后在19世纪，它超越了功利主义的起源，并采取了运动的形式，成为建立乌托邦社区的基础，使沙漠绽放了生命之花，并让美国西部大沙漠逐渐开发利用起来。国会创建了一个联邦机构——现在称为垦务局，专门负责扩大西部灌溉的任务。（单词个数：160）

第三节 灌溉文化的理解

在美国是否有灌溉文化？灌溉有很多独特的区域模式和过程。例如，加州的大规模农业项目和企业与落基山脉各州的较小的经营形成鲜明对比。大平原的中心枢纽系统与墨西哥湾沿岸的甘蔗灌溉几乎没有任何共同点。灌溉文化有着巨大的差异。（单词个数：69）

与此同时，在19～20世纪，几种灌溉模式和运动都具有国家意义。总体来说，与灌溉相关的观点和项目构成了“灌溉文化”，其中一些方面依然存留，而另一些则面临着根本的挑战。这种文化的三两个方面似乎与灌溉的未来息息相关。（单词个数：54）

1. 开垦道德

至今，许多灌溉者对“开垦”的内在价值持有强烈的看法。不论是密西西比河下游河谷洼地的排水，还是西部荒漠的灌溉，“变废为宝”的历史性转变是灌溉者极大满足感的根源。许多灌溉社区追求维持或恢复与开垦相关的原始价值观念。他们不认同开垦对环境有害的观点。事实上，在某些方面，因为对诸如Reisner等流行的批判人士早期的开垦危害批评并未作出回应，灌溉农业社区现在正在付出代价。（单词个数：101）

虽然许多人为鱼类、野生动物和环境质量等方面的问题担心，但整个社会被最初驳回那些批评的人严重摧毁了。开垦机构对文化内容的变迁适应性缓慢而笨拙。（单词个数：38）

2. 对待水的态度

在西方国家，水经常被描述为该地区的“命脉”。西方许多人仍然认为没有水的土地没有任何价值，对于灌溉农田来说，这是正确真实的。这种对水的根本依赖导致了几个根深蒂固的概念，它们指导着农业用水并深刻影响西方水法。（单词个数：58）

基本上，灌溉者把水看作是达到目的的一种重要手段。也就是说，水是种植作物所需要的投入。水的高度工具性的观点，促进了明确尊重使用水的相对权利的重要性，以及这些权利的确定性价值。因此，优先的原则——“先到先得”——对灌溉者来说是非常重要的。优先原则不仅有助于整顿竞争性的水资源，而且还要能保障灌溉农业用水的大量需求，因为大部分的用水都是在西部地区早早建立起来的，要让农业用水比其他用水更有优先权。优先原则的一个结果是强调时间是决定用水权的最重要因素，而不是诸如地点、价值或使用目的等其他因素。（单词个数：148）

与这个强烈的确定性愿望相关的是稳定和防止变革的重要性。由于对水的依赖，灌溉者们担心水供应的减少，这是可以理解的。在过去10年左右的时间里，根据史料记载农业土地灌溉面积一直在不断地增加，而灌溉者们却一直谨慎地维护他们的主张。灌溉用水权的改变，特别是在原使用地点以外的非农业用途的变化，一直受到抵制。为了保护既有使用者的水权，无损害的法律概念应运而生。（单词个数：101）

由于水资源的竞争日益激烈，灌溉农业常常与城市、种族和环境利益冲突，因此了解这些水的文化观点可能是很重要的。它们帮助解释了灌溉用户有时为他们传统的用水特权辩护的热情。它们揭示了许多灌溉者对增加市场用水的阻力，这是将水资源从农业转变为城市的一种手段。它们有助于解释灌溉者如何将水看作是与灌溉社区的需求相一致的集体利益，但却拒绝将水作为一种公共的河道资源。（单词个数：105）

第四节 文化冲突与合作

一些最令人鼓舞和痛苦的灌溉教训，要追溯到十九世纪中叶和末期，当时大规模的人口迁移取代了当地的土著文化，重新改造了他们定居地的水资源，用于采矿、农业和牧场用途。这些过程既涉及令人瞩目的合作实例，也有激烈的矛盾冲突。除了与美洲印第安人在陆地和水域上的冲突之外，灌溉与其他经济群体的关系也不平衡。在一些地区，灌溉工人建立起自己的服务，为小矿工、旅行者、捕猎者、军事要塞、牧场主和贸易商提供服务。在各种活动蓬勃发展的地区，有时也会出现冲突，因为水力采矿毁坏了水质、河流渠道和下游灌溉区域。在某种程度上，联邦开垦政策寻求的是"解决"西部地区的问题，就是说向西部地区移民，并取代原有稳定的农业经济和社会，进行更不稳定和短暂的活动。后来，当繁荣的灌溉经济促进人口增长和商业扩张时，它有时会导致有限水资源供应的竞争和环境冲突。（单词个数：166）

在灌溉区，合作和冲突是长期共存的。史前灌溉系统涉及建设、维护、种植和安置方面的高水平社区合作。经济竞争和政治冲突的考古记录是有限的，但是民族史学证据表明史前灌溉者面临着各种各样的内部和外部冲突，这些冲突影响了灌溉系统的可持续性。（单词个数：58）

在讨论冲突的遗留问题之前，回顾一下 19 世纪灌溉实践的非比寻常的多样性是很重要的。各种各样的农民和社区适应了东部各州干旱西部的做法。非洲裔美国人将水稻灌溉方式从冈比亚转移到南卡罗来纳州。中国移民在萨克拉门托河泛滥平原和三角洲的开垦中发挥了核心作用。旁遮普邦的印度人在加利福尼亚帝王谷和中央谷灌溉土地，这些地区与殖民地印度非常相似并受到其灌溉方式的影响。美国的水利工程师和律师从意大利、法国、印度和埃及等地吸取了实践经验。毫无疑问，在这个快速改变的多样化环境中，出现了新的冲突类型和冲突的解决方式。（单词个数：149）

在未来改革灌溉体系的工作中，有三点显得很重要：第一，灌溉是一种文化、经济和政治体系。灌溉体系的文化特征在某种程度上归因于它对变革的抵制；这种对于文化维度的理解可能会促进建设性的变革和冲突的解决。第二，不要忽略现代农业合作实例和冲突解决方案，它在灌溉方面经久不衰的价值和功效很重要。否则，后代可能会发现自己正试图从20世纪中期和更早的时期中，找回丢失的东西。第三，创造新的合作形式很重要，例如调解纠纷和仲裁的替代方式，这些合作方式要超越在水资源和相关土地资源时代价高昂且根深蒂固的冲突模式。（单词个数：137）

Chapter 17 Irrigation Methods

（Link: https://en.wikipedia.org/wiki/Irrigation#Controllers.2C_zones.2C_and_valves）

Part 1 Typical Methods

There are several methods of irrigation. They vary in how the water is supplied to the plants. The goal is to apply the water to the plants as uniformly as possible, so that each plant has the amount of water it needs, neither too much nor too little.

1. Surface irrigation

Surface irrigation is the oldest form of irrigation and has been in use for thousands of years. In surface (furrow, flood, or level basin) irrigation systems, water moves across the surface of an agricultural lands, in an order to wet it and infiltrate into the soil. Surface irrigation can be subdivided into furrow, border strip or basin irrigation. It is often called flood irrigation when the irrigation results in flooding or near flooding of the cultivated land. Historically, this has been the most common method of irrigating agricultural land and still used in most parts of the world.

2. Micro-irrigation

Micro-irrigation, sometimes called localized irrigation, low volume irrigation, or trickle irrigation is a system where water is distributed under low pressure through a piped network, in a pre-determined pattern, and applied as a small discharge to each plant or adjacent to it. [1]Traditional drip irrigation using individual emitters, Subsurface Drip Irrigation (SDI), micro-spray or micro-sprinkler irrigation, and mini-bubbler irrigation all belong to this category of irrigation methods.

3. Drip irrigation

Drip (or micro) irrigation, also known as trickle irrigation, functions as its name suggests. In this system water falls drop by drop just at the position of roots. Water is delivered at or near the root zone of plants, drop by drop. This method can be the most water-efficient method of irrigation, if managed properly, since evaporation and runoff are minimized. The field water efficiency of drip irrigation is typically in the range of 80 to 90 percent when managed correctly.In modern agriculture, drip irrigation is often combined with plastic mulch, further reducing evaporation, and is also the means of delivery of fertilizer. The process is known as fertigation.

Deep percolation, where water moves below the root zone, can occur if a drip system is operated for too long or if the delivery rate is too high. Drip irrigation methods range from very high-tech and computerized to low-tech and labor-intensive. Lower water pressures are usually needed than for most other types of systems, with the exception of low energy center pivot systems and surface irrigation systems, and the system can be designed for uniformity throughout a field or for precise water delivery to individual plants in a landscape containing a mix of plant species.[2] Although it is difficult to regulate pressure on steep slopes, pressure compensating emitters are available, so the field does not have to be level. High-tech solutions involve precisely calibrated emitters located along lines of tubing that extend from a computerized set of valves.

4. Sprinkler irrigation

In sprinkler or overhead irrigation, water is piped to one or more central locations within the field and distributed by overhead high-pressure sprinklers or guns. A system utilizing sprinklers, sprays, or guns mounted overhead on permanently installed risers is often referred to as a solid-set irrigation system. Higher pressure sprinklers that rotate are called rotors are driven by a ball drive, gear drive, or impact mechanism. Rotors can be designed to rotate in a full or partial circle. Guns are similar to rotors, except that they generally operate at very high pressures of 40 to 130 lbf/in² (275 to 900 kPa) and flows of 50 to 1200 US gal/min (3 to 76 L/s), usually with nozzle diameters in the range of 0.5 to 1.9 inches (10 to 50 mm). Guns are used not only for irrigation, but also for industrial applications such as dust suppression and logging.

Sprinklers can also be mounted on moving platforms connected to the water source by a hose. Automatically moving wheeled systems known as traveling sprinklers may irrigate areas such as small farms, sports fields, parks, pastures, and cemeteries unattended. Most of these utilize a length of polyethylene tubing wound on a steel drum. As the tubing is wound on the drum powered by the irrigation water or a small gas engine, the sprinkler is pulled across the field. When the sprinkler arrives back at the reel the system shuts off. This type of system is known to most people as a "water-reel" traveling irrigation sprinkler and they are used extensively for dust suppression, irrigation, and land application of waste water.

Other travelers use a flat rubber hose that is dragged along behind while the sprinkler platform is pulled by a cable.

Part 2 Large Scaled Irrigation

1. Center pivot

Center pivot irrigation is a form of sprinkler irrigation consisting of several segments of pipe (usually galvanized steel or aluminum) joined together and supported by trusses, mounted on wheeled towers with sprinklers positioned along its length. The system moves in a circular pattern and is fed with water from the pivot point at the center of the arc. These systems are found and used in all parts of the world and allow irrigation of all types of terrain.

Most center pivot systems now have drops hanging from a u-shaped pipe attached at the top of the pipe with sprinkler head that are positioned a few feet (at most) above the crop, thus limiting evaporative losses. Drops can also be used with drag hoses or bubblers that deposit the water directly on the ground between crops. Crops are often planted in a circle to conform to the center pivot. This type of system is known as LEPA (Low Energy Precision Application). Originally, most center pivots were water powered. These were replaced by hydraulic systems (T-L Irrigation) and electric motor driven systems (Reinke, Valley, Zimmatic). Many modern pivots feature GPS devices.

2. Irrigation by Lateral move (side roll, wheel line, wheel-move)

A series of pipes, each with a wheel of about 1.5 m diameter permanently affixed to its midpoint, and sprinklers along its length, are coupled together. Water is supplied at one end using a large hose. After sufficient irrigation has been applied to one strip of the field, the hose is removed, the water drained from the system, and assembly rolled either by hand or with a purpose-built mechanism, so that the sprinklers are moved to a different position across the field. [3]The hose is reconnected. The process is repeated in a pattern until the whole field has been irrigated.

This system is less expensive to install than a center pivot, but much more labor-intensive to operate – it does not travel automatically across the field: it applies water in a stationary strip, must be drained, and then rolled to a new strip. Most systems use 4 or 5-inch (130 mm) diameter aluminum pipe. The pipe doubles both as water transport and as an axle for rotating all the wheels. A drive system (often found near the center of the wheel line) rotates the clamped-together pipe sections as a single axle, rolling the whole wheel line. Manual adjustment of individual wheel positions may be necessary if the system becomes misaligned.

Wheel line systems are limited in the amount of water they can carry, and limited in the height of crops that can be irrigated. One useful feature of a lateral move system is that it consists of sections that can be easily disconnected, adapting to field shape as the line is moved. They are most often used for small, rectilinear, or oddly-shaped fields, hilly or mountainous regions, or in regions where labor is inexpensive.

3. Lawn sprinkler systems

A lawn sprinkler system is permanently installed, as opposed to a hose-end sprinkler, which is portable. Sprinkler systems are installed in residential lawns, in commercial landscapes, for churches and schools, in public parks and cemeteries, and on golf courses. Most of the components of these irrigation systems are hidden under ground, since aesthetics are important in a landscape. A typical lawn sprinkler system will consist of one or more zones, limited in size by the capacity of the water source. Each zone will cover a designated portion of the landscape. Sections of the landscape will usually be divided by microclimate, type of plant material, and type of irrigation equipment. A landscape irrigation system may also include zones containing drip irrigation, bubblers, or other types of equipment besides sprinklers.

Although manual systems are still used, most lawn sprinkler systems may be operated automatically using an irrigation controller, sometimes called a clock or timer. Most automatic systems employ electric solenoid valves. Each zone has one or more of these valves that are wired to the controller. When the controller sends power to the valve, the valve opens, allowing water to flow to the sprinklers in that zone.

4. Hose-end sprinklers

There are many types of hose-end sprinklers. Many of them are smaller versions of larger agricultural and landscape sprinklers, sized to work with a typical garden hose. Some have a spiked base allowing them to be temporarily stuck in the ground, while others have a sled base designed to be dragged while attached to the hose.

Part 3 Subirrigation methods

1. Subirrigation

Subirrigation has been used for many years in field crops in areas with high water tables. It is a method of artificially raising the water table to allow the soil to be moistened from below the plants' root zone. Often those systems are located on permanent grasslands in lowlands or river valleys and combined with drainage infrastructure. A system of pumping stations, canals, weirs and gates allows it to increase or decrease the water level in a network of ditches and thereby control the water table.

Subirrigation is also used in commercial greenhouse production, usually for potted plants. Water is delivered from below, absorbed upwards, and the excess collected for recycling. Typically, a solution of water and nutrients floods a container or flows through a trough for a short period of time, 10–20 minutes, and is then pumped back into a holding tank for reuse. Subirrigation in greenhouses requires fairly sophisticated, expensive equipment and management. Advantages are water and nutrient conservation, and labor savings through reduced system maintenance and automation. It is similar in principle and action to subsurface basin irrigation.

Another type of subirrigation is the self-watering container, also known as a subirrigated planter. This consists of a planter suspended over a reservoir with some type of wicking material such as a polyester rope. The water is drawn up the wick through capillary action.

2. Subsurface textile irrigation

Subsurface Textile Irrigation (SSTI) is a technology designed specifically for subirrigation in all soil textures from desert sands to heavy clays. A typical subsurface textile irrigation system has an impermeable base layer (usually polyethylene or polypropylene), a drip line running along that base, a layer of geotextile on top of the drip line and, finally, a narrow impermeable layer on top of the geotextile (see diagram). Unlike standard drip irrigation, the spacing of emitters in the drip pipe is not critical as the geotextile moves the water along the fabric up to 2 m from the dripper. The impermeable layer effectively creates an artificial water table.

* * * * * * * * * * * * * * * * Explanations * * * * * * * * * * * * * * * *

*** * * * * * * * * * * * * * New Words and Expressions * * * * * * * * * * * * * ***

1. border strip [ˈbɔrdɚstrɪp] n. 分区
2. adjacent [əˈdʒesnt] adj. 邻近的，毗连的
3. mulch [mʌltʃ] n. 覆盖物
4. fertigation [ˌfəːtiˈgeiʃən] n. 水肥灌溉
5. percolation [ˌpɚkəˈleʃən] n. 渗透，过滤
6. steep [stip] adj. 陡峭的，不合理的；夸大的；急剧升降的
   n. 峭壁；浸渍
   vt. 泡；浸；使……充满
   vi. 泡；沉浸
7. spray [spreɪ] n. 喷雾，喷雾器；水沫
   vt. 喷射

vi. 喷

8. mounte[ˈmaʊntɚ] v. 安装

9. pasture [ˈpæstʃɚ] n. 牧场

vt. 放牧；吃草

10. cemeteries [ˈsemitəris] n. 墓园

11. lateral [ˈlætərəl] n. 侧面

adj. 侧面的，横向的

12. misaligned [ˌmɪsəˈlaɪnd] adj. 不对齐的，方向偏离的；不重合的

13. rectilinear[ˈrɛktəˈlɪnɪɚ] adj. 直线的

14. aesthetics [ɛsˈθɛtɪks] n. 美学

15. subirrigation[sʌbˈirigeiʃən] n. 地下灌溉

16. impermeable [ɪmˈpɝmɪəbl] adj. 不透水的，不能渗透的

* * * * * * * * * * * * * * * * * Complicated Sentences * * * * * * * * * * * * * * * * *

1. Micro-irrigation, sometimes called localized irrigation, low volume irrigation, or trickle irrigation is a system where water is distributed under low pressure through a piped network, in a pre-determined pattern, and applied as a small discharge to each plant or adjacent to it.

【译文】微灌，有时称为局部灌溉、低容量灌溉或滴灌，是一种水以预定模式通过分布管网低压输送，并以小流量输送到每珠植物，以及跟它邻近的植物的系统。

【说明】sometimes called localized irrigation, low volume irrigation 为主语 Micro-irrigation 的补充语。under low pressure 指低压输水；adjacent to 意为连接。

2. Lower water pressures are usually needed than for most other types of systems, with the exception of low energy center pivot systems and surface irrigation systems, and the system can be designed for uniformity throughout a field or for precise water delivery to individual plants in a landscape containing a mix of plant species.

【译文】除了低能耗中心支轴式喷灌系统和地面灌溉系统之外，大多数其他类型的系统都通常需要低水压，系统可以通过设计来保证灌溉均匀度或者是控制水流精确的流向包括不同农作物的独立田块。

【说明】with the exception of 表示除了……以外。center pivot systems 中心支轴式喷灌系统。landscape containing a mix of plant species 表示有不同种类植物的景观。后半句意为可以通过对单独植物进行精确输水来保证灌溉均匀度。

3. After sufficient irrigation has been applied to one strip of the field, the hose is removed, the water drained from the system, and assembly rolled either by hand or with a purpose-built mechanism, so that the sprinklers are moved to a different position across the field.

【译文】在对一片田地进行充分的灌溉之后，将软管从系统中移走，将系统中的水排干，并将组件用手或特制机械滚动，使喷头穿过田地移动到不同位置。

【说明】has been applied 和 is removed 分别为现在完成时态和现在时态，表示灌溉结束后，再将软管移走。Either by…or with…表示“要不……要不……”，purpose-built mechanism 表示专门为此安装的装置。

* * * * * * * * * * * * * * * * * Summary of Glossary * * * * * * * * * * * * * * * * *

1. flood irrigation 漫灌

2. cultivated land 耕地
3. trickle irrigation 滴灌
4. mini-bubbler irrigation 微涌灌
5. sprinkler irrigation 喷灌
6. conform to 符合
7. labor-intensive 劳动密集型的
8. oddly-shaped 奇形怪状的
9. lawn sprinkler systems 草坪喷灌系统
10. water table 地下水位
11. capillary action 毛细现象

******************* **Abbreviations (Abbr.)** *******************

1. LEPA Low Energy Precision Application 法律实施计划局
2. SSTI Subsurface Textile Irrigation 地下织物灌溉

******************** **Exercises** ********************

(1) ________, which is also called ________, has been the most common method of irrigating agricultural land and still used in most parts of the world.

(2) In drip irrigation, water is delivered ________ to crop root, which can be the most water-efficient method of irrigation, since ________ and runoff are minimized. The field water efficiency of drip irrigation is typically in the range of 80 to ________ percent when managed correctly.

(3) The spacing of emitters in the drip pipe of SSTI is not critical as the ________ moves the water along the fabric up to 2 m from the dripper. The impermeable layer effectively creates an ________ water table.

(4) Irrigation by Lateral move system is less expensive to install than a center pivot, but much more ________ to operate - it does not travel ________ across the field.

************* **Word Building (17) semi-;super-** *************

1. semi- [前缀]，表示：半；部分；不完全

semiautomatic adj. 半自动的
semicircle n. 半圆形
semiconductor n. 半导体
semidiameter n. 半径

2. super- [前缀]，表示：超；高级；在……之上

superhigh adj. 超高的

superimpose	vt.	添加；重叠，叠合
superterrene	adj.	地上的
superaqueous	adj.	水上的
supersaturate	adj.	过饱和的
supermarket	n.	超级市场

*** * * * * * * * * * * * * * * Text Translation * * * * * * * * * * * * * * ***

第十七章 灌 溉 方 法

（链接: https://en.wikipedia.org/wiki/Irrigation#Controllers.2C_zones.2C_and_valves）

第一节 典 型 方 法

有几种典型的灌溉方法，它们在如何给植物供应水的方面各不相同，但目标都是尽可能均匀地将水应用到植物上，使每个植物得到所需的水量，既不太多也不太少。（单词个数：48）

1. 地面淹灌

地面灌溉是最古老的灌溉形式，已经应用了数千年。在地表（沟渠、洪泛区或平坦流域）灌溉系统中，水流过农田的表面，以便将其润湿并渗入土壤。地面灌溉可以细分为沟灌、畦灌，流域灌溉。当灌溉导致耕地形成地面漫流或接近漫流时，通常称为漫灌。从历史上看，这是最常用的灌溉农田的方法，并且仍然在世界大部分地区使用。（单词个数：98）

2. 微灌

微灌，有时称为局部灌溉、低容量灌溉或滴灌，是一种水以预定模式分布，通过管网低压输送，并以小流量输送到每珠植物，以及跟它邻近的植物的系统。传统的滴灌使用单个滴头，地下滴灌、微喷灌、微涌灌都属于这一灌溉类型。（单词个数：67）

3. 滴灌

滴灌（或微灌）也称滴流灌溉，其功能正如它的名字。在这个系统里，水仅仅一滴滴落在植物根部位置。水是在植物的根区或附近一滴滴输送的。如果妥善管理，使蒸发损失和径流最小化，这种方法可以说是最节水的灌溉方法。当正确管理时，滴灌的田间水利用率通常在80%～90%的范围内。在现代农业中，滴灌通常与塑料地膜相结合，进一步减少蒸发，同时也用作施肥的手段。这个过程称为灌溉施肥。（单词个数：109）

如果滴灌系统运行时间过长或输送速度过高，可能会发生深层渗漏，水会流到根系以下。滴灌包括从高科技化、计算机化到低科技和劳动密集型的多种类型。除了低能耗中心支轴式喷灌系统和地面灌溉系统之外，大多数其他类型的系统都通常需要低水压，系统可以通过设计来保证灌溉均匀度或者是控制水流精确的流向不同农作物的独立田块。虽然很难调整陡坡上的压力，但压力补偿发射器是可用的，所以田地不一定要水平。高科技的解决方案，包含

在数控阀门延伸出的管道线路上沿途分布的精准刻度滴头。（单词个数：138）

4. 喷灌

在喷灌或高架灌溉系统中，水通过管道输送到一个或多个中心位置，并通过高压洒水器或洒水枪分配。使用在永久安装立管上架设的喷头、喷雾器或喷枪的系统通常称为固定式灌溉系统。旋转的高压喷头称为转子，由滚珠驱动，齿轮传动或冲击机构驱动。转子可以设计成以全部或部分圆周旋转。 喷枪类似于转子，除了通常在 40～130lbf/in^2（275～900kPa）的高压和 50～1200US 加仑/min（3～76L/s）的流量下运转。喷嘴直径通常在 0.5～1.9 英寸（10～50mm）的范围内。喷枪不仅用于灌溉，还用于工业应用，如抑尘和伐木。（单词个数：147）

喷头也可以安装在通过软管连接到水源的移动平台上。 自如移动的轮式系统称为移动喷灌装置，可以灌溉小农场、运动场、公园、牧场和无人值守的墓地等地区。这些系统中，很多都是使用了一段卷在滚筒上的聚乙烯管。当管子被灌溉水或小型气体发动机驱动从卷轴上拉下时，喷头就会被拉过田地。当喷头回到卷轴上时，系统就会自动关闭。这就是大多数人熟知的“水卷”式移动喷灌机，广泛用于除尘，灌溉和废水的土地利用。（单词个数：119）

其他移动喷灌器使用扁平橡胶软管，沿着被电缆拉着的喷洒平台后面拖动。（单词个数：21）

第二节 大 规 模 灌 溉

1. 中心枢轴灌溉

中心枢轴灌溉是一种喷灌的形式，由连接在一起并由桁架支撑的多个管段（通常为镀锌钢管或铝管）组成， 安装在沿管长方向分布喷头的塔车上。系统以圆形图案移动，并从圆弧中心的枢转点进给水。在世界各地都可以见到这些系统的使用，并适用于灌溉各种类型的地形。（单词个数：78）

现在大多数中心枢轴系统都有连接在支管顶部悬垂下来的装有喷头的 U 形管，喷头位于作物上方最多几英尺处，因此限制了蒸发损失。滴下的水也可以在作物之间的地面阻力软管或喷水头中存放使用。作物通常种植在一个圆圈里，以符合中心枢轴系统。这种类型的系统称为 LEPA（低能耗精密应用系统）。最初，大多数中枢枢纽都是水力供能的。后来这些由液压系统（T-L 灌溉）和电动机驱动系统（Reinke，Valley，Zimmatic）取代。许多现代枢纽甚至配有 GPS 设备。（单词个数：111）

2. 移动支渠灌溉（侧滚、轮线、轮动）

一系列在中点固定安装直径约 1.5m 的轮子并沿途分布喷头的管道连接在一起。一端用大软管供水。在对一片田地进行充分的灌溉之后，将软管从系统中移走，将系统中的水排干，并将组件用手或特制机械滚动，使喷头穿过田地移动到不同位置，再将软管重新连接。该过程可以模式重复，直到整个场被灌溉。（单词个数：100）

该系统的安装成本比中心枢轴系统低，但操作起来更加劳动密集，它不能在田地自动运行：供水是在固定渠道上，必须排干，然后再滚动到新的渠道上。大多数系统使用直径 4 英寸或 5 英寸（130 mm）的铝管。管道既可以用作水运，也可以用作旋转所有车轮的轴。驱动系统（经常建在轮线的中心附近）将夹紧的管段一起作单轴旋转，滚动整个车轮线。如果系统没对齐，可能需要手动调整单个车轮位置。（单词个数：109）

轮线系统的水量受到限制，可以灌溉的作物高度也受到限制。横向移动系统的一个有用的特征是它由可以容易分离的部分组成，适用于形状可线性移动的田地。它们最常用于小型、直线型或奇怪形状的田地、丘陵或山区，或劳动力廉价地区。（单词个数：74）

3. 草坪喷灌系统

草坪喷灌系统是固定安装的，而不是一个便携式的软管端喷灌系统。喷灌系统安装在住宅草坪、商业景观、教堂、学校、公园、墓地以及高尔夫球场上。因为美观在景观中的重要性，这些灌溉系统的大部分组成部分隐藏在地下。典型的草坪洒水系统将由一个或多个由水源给水能力限制大小的区域组成。每个区域将覆盖景观的指定部分。景观通常按微气候、植物种类和灌溉设备类型分开。景观灌溉系统可能还包括部分区域载有滴灌、涌灌或者除喷灌外的其他灌溉设备。（单词个数：128）

虽然仍然使用手动系统，但是大多数草坪喷洒系统可以使用灌溉控制器（有时称为计时器或定时器）自动操作。大多数自动化系统采用电磁阀。每个区域都有一个或多个连接到控制器的阀门。当控制器向阀门供电时，阀门打开，允许水流入该区域的喷头。（单词个数：67）

4. 软管喷灌

软管喷灌系统有很多类型。它们中的许多都是大型农业和景观喷灌器的小型版本，它们的大小与典型的花园软管一样。一些喷灌设备有一个带尖基座，允许软管暂时插在地面上，而另一些有一个连接软管可拖动的滑车基座。（单词个数：56）

第三节 地下灌溉方法

1. 地下灌溉

在地下水位高的地区，常年使用地下灌溉法灌溉农作物。这是一种人工抬高地下水位使土壤从植物的根区下面湿润的方法。这些系统通常位于低地或河谷的永久性草地上，并与排水基础设施相结合。一个由泵站、运河、堰和闸门组成的系统在沟渠网络中抬高或降低水位，从而控制地下水位。（单词个数：86）

地下灌溉也用于商业温室生产，通常用于盆栽植物。水从下方输送，向上吸收，多余的水分可以收集再循环利用。比较典型的做法是，水和营养素会在短时间内淹没或流经一个水槽，约 10~20min，然后将其泵回到储存罐中以便重新使用。温室中的地下灌溉需要相当复杂、昂贵的设备和管理。其优点是通过减少系统维护和自动化来保存水和养分、节省人力，大体上和地下流域灌溉相似。（单词个数：96）

另一种类型的地下灌溉是自给水容器，也称为再灌溉播种机。这个装置由悬挂在蓄水池上的种植机和像聚酯绳一样的毛细材料组成。水通过毛细管作用被吸收。（单词个数：44）

2. 地下织物灌溉

地下织物灌溉（SSTI）是专门为从沙漠到重黏土的所有土壤质地进行亚灌溉设计的技术。典型的地下织物灌溉系统含有不透水的基层（通常是聚乙烯或聚丙烯）、沿该基层延伸的滴灌线、滴灌线顶部的土工织物层，以及最后在土工织物上的狭窄不透水层。与标准滴灌不同的是，要将水沿着土工织物移动到离滴灌器 2 m 处的位置，因而滴灌管中滴头的间距并不重要。不透水层有效地创造了人造水位。（单词个数：106）

Chapter 18 Irrigation Performance Evaluation

Part 1 Assessment of Irrigation Performance

Assessment of irrigation performance is a prerequisite for improving water use in the agricultural sector to respond to perceived water scarcity. Between 1996 and 2000, we conducted a comprehensive assessment of the performance of the Genil-Cabra Irrigation Scheme (GCIS) located in Andalusia, Southern Spain. The area has about 7000 ha of irrigated lands distributed in 843 parcels and devoted to a diverse crop mix, with cereals, sunflower, cotton, garlic and olive trees as principal crops. Irrigation is on demand from a pressurized system and hand-moved sprinkler irrigation is the most popular application method. Six performance indicators were used to assess the physical and economic performance of irrigation water use and management in the GCIS, using parcel water-use records and a simulation model. The model simulates the water-balance processes on every field and computes an optimal irrigation schedule, which is then checked against actual schedules.[1] Among the performance indicators, the average irrigation water supply: demand ratio (the ratio of measured irrigation supply to the simulated optimum demand) varied among years from 0.4 to 0.64, indicating that the area is under deficit irrigation. When rainfall was included, the supply: demand ratio increased up to 0.87 in one year, although it was only 0.72 in the driest year, showing that farmers did not fully compensate for the low rainfall with sufficient irrigation water.

The availability of water for irrigation will probably decrease in the future due to increased demands from other sectors, such as municipal, tourism, recreation and the environment. In Spain, fresh- water demand is estimated as $35\times10^5 m^3$/year with about 70% devoted to irrigation and the rest to other uses. Additionally, the government anticipates that irrigation demand in southern Spain will increase by about 17% in the next 10 years.

Improvement in water management and the modernization/rehabilitation of the Spanish irrigation schemes are important objectives to achieve more efficient use of water. Only 27% of the irrigated area in Spain (approximately 915 000ha) is less than 20 years old, whereas 37% is more than 90 years old. In recent years, the water administration emphasized system modernization and rehabilitation but comparatively little attention was paid to the improvement of irrigation management.

The improvement of water management in an irrigation scheme requires the assessment of irrigation performance as a point of departure. Computer simulation using hydrologic models has been useful for this task. Many models have been used to simulate parts of the hydrologic cycle in irrigated agriculture, from empirical or functional to mechanistic. Additionally, to facilitate data acquisition and carry out spatial analyses, recently developed tools, such as remote sensing and geographic information systems, have been combined with hydrologic models to assess the behavior of irrigation schemes.

Several authors have defined sets of indicators that characterize irrigation system performance, intending to evaluate current practices and recommend improvements in irrigation efficiency and water productivity. These performance indicators are also used to quantify the system ability and to achieve the objectives established for an irrigation area or to assess the current performance of the system relative to its potential.

The different types of performance indicators are related to: 1) the water balance; 2) economic, environmental and social objectives; 3) system maintenance. Several authors have used these indicators for: ①assessing trends in performance; ②comparing performance among irrigation schemes; ③resource optimization; ④determining a compromise solution between equity and efficiency within an irrigation area. The complexity of models used for calculating the water balance-based performance indicators varies from one-dimensional, physically based, hydrologic model to very simplified models, such as those based on the FAO (Food and Agriculture Organization) methodology. At the scheme level, input information to compute performance indicators is normally obtained from total water delivery records and from water consumption estimates derived from the cropped areas. Such information often has substantial uncertainty and does not allow for in-depth analysis at levels below the scheme. Nevertheless, scheme-level assessments are needed for comparative purposes and are the only approach feasible when there is no access to information at sub-scheme levels.

The objective of this work was to conduct a comprehensive assessment of the irrigation performance of an area using on farm water-use information and a simulation model. The area selected was the Genil-Gabra irrigation scheme (GCIS) located in Andalusia, southern Spain. This area was chosen because it was possible to obtain accurate information on water use and on the cropping patterns of individual parcels during four irrigation seasons.

Part 2 Efficient Irrigation Systems: The Flow of the Future

Irrigation has been the technology underlying many of the world's greatest civilizations. The ancient irrigation techniques depended mostly on the terrain, water supply, and the engineering skills of the civilization.

Many different irrigation systems were developed very early in agriculture history. Some of the first irrigation systems were developed in Egypt and Mesopotamia. The prosperity of the civilizations of the Nile River Valley has been recorded throughout history and their success depended on the efficiency with which the government organized the best use of the river water. Abundant crops could be stored for years and irrigation techniques could be improved.

At first, the Egyptians based their agriculture on growing winter crops after the annual floods were gone. The Egyptian irrigation systems depended on several factors. There was only one source of water, the Nile, which was too powerful to control. [2] Therefore, irrigation systems had to be simple in terms of construction, and built high along the riverbank in order to deal with only the peak of the flood. Since the river valley is never more than 15 miles wide, irrigation systems could not carry water a great distance from the river (Ancient, 1999).

The Egyptians built large, flat basins for growing crops along the riverbanks. They developed simple sluices, or water channels with gates, that diverted water into the basins at the peak of the flood. It was simple in terms of both engineering and labor to arrange for good water flow through several basins controlled by simple gates. Water was allowed to sit in the fields for 40 to 60 days. It was then drained using ditches and canals, at the right time during the growing cycle, downstream back into the river. Salts never built up in the soil because there was plenty of water and the flow of water in the canals and ditches was strong enough to avoid silting.

Irrigation systems were completely different in dry, hilly areas. Rivers and springs don't have the flow of the major rivers, so water storage systems needed to be developed. One of the earliest successful techniques was a diversion dam. One of the most impressive dams used for ancient irrigation is located near Marib, the ancient capital of Yemen. It was built around 600 BC. The dam is over 1600 feet long and is built of carefully crafted masonry, strengthened by copper fastenings.

As irrigation developed, civilizations learned that there are some things needed to have a successful irrigation system. A region that can be irrigated on a long-term basis has to have an abundant supply of good water, well-drained soil, good regional drainage, and a supply of fertilizer for the soil.

Irrigation can only be maintained on a long-term basis if certain conditions apply. Water has to be applied in a way that salt is not allowed to build up in the soil. This usually means that a lot of good-quality water is applied, and the drainage system is rapid and efficient. Soils also need a large supply of fertilizer to balance the flushing that is required to keep them free of salt.[3]

It took several hundred years and much trial and error before researchers figured out the principals of irrigation and engineers developed mechanical water control systems that drastically changed irrigation techniques, and irrigation into a form of science.

By 1995, due to the advancements in irrigation technology, the total area of irrigated land in the world was just over 555 750 000 acres.

* * * * * * * * * * * * * * * * * * * Explanations * * * * * * * * * * * * * * * * * *

* * * * * * * * * * * * * * * * * New Words and Expressions * * * * * * * * * * * * * * *

1. prerequisite ['pri:'rekwizit] n. 先决条件
2. parcel ['pa:sl] n. 包裹；v. 打包
3. cereal ['siəriəl] n. 谷类食物
4. garlic ['gɑ:lik] n. 大蒜
5. olive ['ɔliv] n. 橄榄
6. pressurized ['preʃəraizd] adj. 加压的，受压的
7. municipal [mju:'nisipəl] adj. 市政的，市立的，地方性的，地方自治的
8. rehabilitation [ri: hə,bili'teiʃən] n. 复原，康复
9. empirical [em'pirikəl] adj. 完全根据经验的，经验主义的；[化]实验式

10. data acquisition 数据采集
11. spatial analysis 空间分析
12. terrain ['terein] n. 地形
13. sluice [slu: s] n. 水闸，蓄水
v. 开闸放水，流出，冲洗，奔泻
14. silting ['siltiŋ] n. 淤积；淤塞；充填
15. masonry ['meisnri] n. 石工术，石匠职业
16. drastically ['dræstikəli] adv. 激烈地，彻底地
17. evolution [i:və'lu:ʃən] n. 进展，发展，演变，进化
18. corrugate ['kɔrugeit] v. 弄皱，起皱，（使）成波状
19. versatile ['və: sətail] adj. 通用的，万能的，多才多艺的，多面手的
20. consistency [kən'sistənsi] n. 连结，结合，坚固性，浓度，密度，一致性，连贯性
21. pivot ['pivət] n. 枢轴，支点，（讨论的）中心点，重点
adj. 枢轴的
vi. 在枢轴上转动

****************** **Complicated Sentences** ******************

1. The model simulates the water-balance processes on every field and computes an optimal irrigation schedule, which is then checked against actual schedules.

【译文】模型模拟了不同地块的水量平衡过程，计算了最优灌溉方案，并与实际灌溉方案对比分析。

【说明】and 连接两个动词 simulates 和 computes；which 引导定语从句修饰前面的 schedule.

2. There was only one source of water, the Nile, which was too powerful to control.

【译文】只有一个水源——尼罗河，其水势很大，难于控制。

【说明】句中 which was too powerful to control 是非限制性定语从句，修饰 the Nile。

3. Soils also need a large supply of fertilizer to balance the flushing that is required to keep them free of salt.

【译文】土壤也需要有大量的肥料供应以保持冲淤平衡并去除盐分。

【说明】to balance 为目的状语；that 引导定语从句修饰 the flushing。

****************** **Summary of Glossary** ******************

1. performance evaluation 性能评估
2. terrain 地形
3. water channel 水槽，水道
4. growing cycle 生长周期
5. set system 固定系统
6. mobile system 流动系统
7. portable system 轻便系统
8. water-balance 水量平衡
9. optimal irrigation schedule 最优灌溉方案
10. simulation model 仿真模型

* * * * * * * * * * * * * * * * * * * Abbreviations (Abbr.) * * * * * * * * * * * * * * * * * * *

1. GCIS. Genil-Cabra Irrigation Scheme 赫尼尔-卡布拉灌区灌溉计划
2. FAO Food and Agriculture Organization of the United Nations 联合国粮农组织

* * * * * * * * * * * * * * * * * * * Exercises * * * * * * * * * * * * * * * * * * *

(1) The area has about 7000 ha of irrigated lands distributed in 843 parcels and devoted to a ______ crop mix, with ______, sunflower, cotton, ______ and olive trees as principal crops.

(2) Irrigation is on demand from a ____ system and hand-moved ______irrigation is the most popular ______ method.

(3) The ancient irrigation ________ depended mostly on the ________, water supply, and the ________ skills of the civilization.

(4) Therefore, irrigation systems had to be simple in terms of _______, and built high along the ______ in order to deal with only the ______ of the flood.

* * * * * * * * * * * * * * Word Building (18) agr-; auto- * * * * * * * * * * * * * *

1. agr- [词根]，表示：田、地

agrichemicals	n.	农用化学品
agrimotor	n.	农用机动车
agrology	n.	农业土壤学
agrotechnique	n.	农业技术

2. auto- [前缀]，表示：自动

autoregulation	n.	自动调整
autonomous	adj.	自治的，自主的
automatic	adj.	自动的，无意识的
autoflow	n.	自动流程图
autofocus	n.	自动对焦

* * * * * * * * * * * * * * * * * Text Translation * * * * * * * * * * * * * * * * *

第十八章 灌溉运行评价

第一节 灌溉运行评估

灌溉运行评价是提高农业用水效率、缓解缺水矛盾的必要条件。1996～2000年间，我们对位于西班牙南部Andalusia的Genil-Cabra灌区（GCIS）运行进行了系统性评价。该灌区灌溉面积7000公顷，分为843个单元，种植作物种类繁多，以向日葵、棉花、大蒜、橄榄树为主。灌区灌溉由承压系统供水，手动喷灌是该灌区最流行的喷灌方法。本项研究以单元用水资料和模拟模型为基础，分别用6个运行指标来评价GCIS灌区的灌溉用水和管理经济运行状态。模型模拟了不同地块的水量平衡过程，计算了优化灌溉方案，并与实际灌溉方案对比分析。在运行指标中，平均灌溉供需比从0.45～0.64（实际灌溉供水与模拟优化需水之比）呈年际变化，说明该地区为缺水灌溉。如果考虑降雨因素，一年的供需比增加到0.87，但是极端干旱年份为0.72，说明干旱年份没有得到充足的灌溉水量补充。（单词个数：220）

由于其他因素（如城市供水、旅游、娱乐和环境需水）的影响，未来灌溉可利用水量有可能减少。西班牙的淡水资源水量估计为$35\times10^5m^3$/年，其中有70%用于灌溉，其余用水占30%。另外，政府部门预计在未来10年内，西班牙南部的灌溉需水量将增加17%。（单词个数：68）

进一步提高西班牙灌区的水利管理和现代化，以及工程改造是实现高效用水的重要目标。西班牙只有27%的灌区（总面积约915000公顷），建成年限少于20年，而有37%的灌区年限超过了90年。近年来，水利管理组织重视了灌区系统现代化建设和灌区工程改造，但是对于提高灌溉管理重视不够。（单词个数：69）

改善灌区水利管理需要以灌溉运行评价为起点。以水文模型为基础，计算机模拟是灌溉运行评价的有用工具。灌溉农业水文循环模拟模型很多，有些是经验型的，有些是函数型的，也有的是过程描述型的。另外，为了便于资料收集和空间分析，近年来开发了一些工具如遥感、地理信息系统，这些工具与水文模型相结合，可以满足灌溉计划效果评价的需要。（单词个数：86）

许多作者提出了不同的评价指标，用来反映灌溉系统运行特性、评价运行效果、提出改进灌溉效率和水分生产率的建议。也可以用这些指标对系统能力进行定量分析，为实现灌区目标提供依据，或者以灌溉系统潜力为标准，用这些指标评价灌溉系统现状运行情况。（单词个数：60）

不同类型的运行指标有：①水量平衡；②经济、环境和社会目标；③系统维护。许多作者将这些指标用于：①评价运行趋势；②不同灌溉计划方案的对比分析；③资源优化；④确定一个灌区的整体平衡与局部高效之间的协调方案。用来计算水量平衡运行指标的模型有一维模型，物理模型、水文模型以及非常简化的模型，如联合国粮农组织（Food and Agriculture Organization of the United Nations-FAO）提出的FAO方法。当编制灌区规划时，计算运行指标的输入信息，通常是用总供水资料和由作物种植面积，计算用水资料而得到的。这些信息往往带有很大的不确定性，不能用来进行总体计划的深度分析。所以，需要针对不同的目的和灌溉计划分层评价，当缺乏计划层以下的资料时，这是唯一的可行性方法。（单词个数：159）

本章的目的是利用田间用水资料和模拟模型，对灌区的灌溉运行开展系统性评价。所选灌区位于西班牙南部 Andalusia 的 Genil-Cabra 灌溉系统。选择该灌区的原因是在 4 次灌溉时期内，它具有精确的用水资料和各个单元的作物布局资料。（单词个数：68）

第二节 有效的灌溉系统：未来的趋势

灌溉是许多世界上最伟大的文明中所包含的技术。古代灌溉技术大都依赖地势、供水和文明中所蕴含的工程技术。（单词个数：30）

农业史记载，很早就开发了许多不同的灌溉系统。有些早期的灌溉系统产生于埃及和底格里斯与幼发拉底两河之间的美索不达米亚地区。尼罗河流域文明的繁荣载满了历史，其成功取决于政府对河水的有效利用。充足的收成可储存好几年，灌溉技术也得到改进。（单词个数：70）

起初，埃及人的农业主要依靠洪水过后种植一些冬季作物。其灌溉系统的运行也受好几个因素影响。水源只有一个，即尼罗河，水势很大，而且难于控制。因此，灌溉系统只好建得很简单，而且建在河岸高处，以对付洪水高峰。且该河流宽度一般不超过 15 英里，灌溉系统很难将河水送到较远的地方。（单词个数：93）

埃及人沿着河岸开辟了巨大而平坦的盆地来种植作物。他们修建了简单的水闸或带有闸门的水渠，在洪水高峰期时将水引到盆地。从工程和劳力角度来说，用简单的水闸把水引到数个盆地都可实现。田里的水可保留 40～60 天。之后，在生长周期过程中，适时地通过沟渠排放到河里。盐分从不会停留在土壤里，因为水量充足，沟渠里的水流势很大，避免了淤泥堵塞。（单词个数：119）

干旱地区、山区的灌溉系统完全不同。河水和泉水流量不如干流的多，所以有必要开发储水系统。早期最成功的技术之一就是大坝。古代灌溉最令人瞩目的大坝之一就位于也门古都 Marib 附近。该坝建造于公元前 600 年，坝长 1600 多英尺，用经过细致加工的砖石建成，以铜制成的固定物加固。（单词个数：81）

随着灌溉技术的发展，人们意识到需要有成功的灌溉系统。一个地区要得到长期灌溉，就必须有足够优质的水量，排水很好的土壤，良好的地区排水系统和土壤肥料的供应。供水要既安全也可靠。（单词个数：49）

只有在一定条件合适时，灌溉才能持续较长的时间。输水时，必须用采用一种使盐分无法进入土壤的方法。这通常是指利用优质水，并且排放系统快而有效。土壤也需要有大量的肥料供应，以平衡使之不带盐分的冲刷。（单词个数：72）

研究人员经过了很多尝试和失败，花了几百年的时间，才发现了灌溉原理。后期，工程师开发出机械水控制系统，使灌溉技术发生了根本的变化，并将灌溉发展成为科学。（单词个数：37）

到 1995 年，由于灌溉技术的发展，世界灌溉土地总面积达 555 750 000 英亩。（单词个数：23）

Chapter 19 River Closure and Diversion in Dam Construction

Part 1 River Closure

1. Introduction

Closures on the natural river bed may encounter problems in rivers where the flow remains significant even during the dry season. [1] This closure is made at the beginning of the work, when the dam is built in one stage and the river diverted in temporary tunnels (or sometimes in canal); in wider valleys where the dam is built in two stages the closure is made at the end of the first stage and water is diverted in temporary or permanent openings left in the first stage structures.

Although the direct cost of closure is generally a small percentage of the total cost of the dam, it has to be studied carefully as a failure or delay may cause a serious and costly delay to the whole scheme and because the design of the temporary diversion relates to the maximum water head chosen for the closure.

It is consequently of interest to define safe and realistic conditions for closures and to estimate the relevant costs according to maximum water head in order to reach overall optimization of the whole scheme (closure, diversion, programme, etc.) or to give more flexibility for programming closure beyond the low flow season if suitable.

A number of theoretical approaches have been made for closure analysis and model tests may represent hydraulic conditions quite well (such as water levels and average speeds) but they seem to define with much less precision the dimension and characteristics of required materials. It is consequently of interest to compare and classify the large number of actual closures effected around the world (including most of the largest rivers), to analyze the difficulties encountered and try to define some simple rules which may tentatively be used as a guide for a preliminary study of future schemes.

2. Type of river closure

Except for some special cases which will be considered later in this report (instantaneous closure by blasting in steep sided valleys and sand hydraulic filling in flat plains) nearly all important closures have been achieved by one of the following methods. [2]

End dumping (or point tipping or vertical closure) method when advancing the embankment above water from one bank or from both until the gap is closed.

Frontal dumping (or horizontal tipping) method when materials are placed uniformly across the whole width of the gap in the river channel (generally requiring a special structure such a bridge or cable crane for transport of materials).

Part 2 River Diversion

The idea of diverting the river for the construction of the dam clearly has important influence on the time required for completion and is also of great economic importance to the cost of project, so therefore it is a subject well worth investigating in detail. It is not easy to lay down definite rules on this problem, since the diversion of the river is always subject to basic data which one cannot easily fit into a standard pattern, such as the flow of the river and the characteristics of the terrain. Nevertheless, we shall attempt to put forward the conclusions we have reached after having taken part in the diversion of a large number of rivers, and it is our hope that these conclusions may be of some use for similar cases, i.e. for the diversion of rivers medium size. In our opinion a doctrine could be built on the following principles:

(1) The diversion flow should depend on the problems that would arise should there occur an overflow over the dam itself during the construction period.[3] These problems can arise from the erosionability of the materials of the dam (earth dam) or of the foundations (weak foundations). Subject to neither of these problems being serious, it will then be cheaper to envisage not being able to work on the dam foundations during a series of days, rather than forcing the flow of the diversion.

(2) For rivers of medium flow, the advantages of building a diversion above the level of the river bed are normally greater than the higher value of the cofferdams. In the case of diversions at a higher level than that of the river bed, one should not underrate the advantage of being able to use the bottom outlets, by simple cofferdamming of the mouth of the diversion, for the closing operation and, better still, as a suitable place for the location of the outlet elements of rockfill dams.

(3) Diversion tunnels are a very safe solution, but not always the cheapest. We feel they should be used when required by the topographical conditions of the dam or in cases where the levels of the foundations are difficult to determine during the designing stage.[4] For all other cases a diversion by enclosures, for the large flows, or by external ducts or channels, for medium flows, comes out more economical.

(4) The possibility of using diversion tunnels as a place for the bottom outlet is a clearly feasible solution in the case of embankment dams, but not so much as for concrete dams.

(5) The possibility of using the diversion tunnels as location of the dam spillways, implies in the case of dams of medium or large height, certain hydraulic problems of erosion of linings which may be most difficult to solve, especially if it is provided that the tunnel may empty when under load.

(6) In rockfill dams of medium or small height, the solution of providing an incorporated bulk of concrete housing spillways and outlets, renders diversion operations much easier.

In accordance with these principle, we shall set out our subject under the following headings:

(1) Diversion flow.

(2) Suitable layout in elevation of the diversion conduct and of the bottom outlets for the diversion of rivers of medium flow.

(3) Possible alternative solutions to the diversion tunnel.

(4) Using the diversion tunnel as site for the bottom outlet.

(5) Using the diversion tunnel as spillway.

(6) Incorporating concrete bulks for the housing of outlets in the case of embankment dams, for diverting the river.

* * * * * * * * * * * * * * * * * Explanations * * * * * * * * * * * * * * * * *

* * * * * * * * * * * * * * * * * New Words and Expressions * * * * * * * * * * * * * * * *

1. closure [ˈkləuʒə] n. 截流，关闭，终止
2. temporary [ˈtempərəri] adj. 暂时的，临时的
3. realistic [riəˈlistik] adj. 现实的；实际可行的；现实主义的
4. relevant [ˈreləvənt] adj. 有关的，切题的
5. optimization [ˈɔptimaiziʒən] n. 最优化
6. beyond [biˈjɔnd] prep. 远于；迟于；越出

   adj. 在更远处
7. theoretical [ˌθiəˈretikəl] adj. 理论（上）的
8. precision [priˈsiʒən] n. 精确（性），精密（度）
9. classify [ˈklæsifai] vt. 把…分类，把…分级
10. tentatively [ˈtentətivli] adv. 试探（性）地，试验（性）地
11. preliminary [priˈliminəri] n. [pl.] 初步做法

    adj. 预备的，初步的
12. end dumping 立堵法截流
13. frontal dumping 平堵法截流
14. crane [krein] n. 起重机；鹤

    vt. 伸长（脖子等）
15. investigate [inˈvestigeit] v.调查，调查研究
16. doctrine [ˈdɔktrin] n. 教义，教条，主义
17. series [ˈsiəri: z] n. 一系列，连续；丛书，连续剧
18. cofferdam [ˈkɔfədæm] n. 围堰
19. underrate [ˌʌndəˈreit] v. 低估
20. rockfill [ˈrɔkfil] adj. 用石头填充的
21. topographical [ˌtɔpəˈgræfikl] adj. 地形的，地形的
22. tunnel [ˈtʌnl] v. 挖（地道），开（隧道）

    n. 隧道，地道

* * * * * * * * * * * * * * * * * Complicated Sentences * * * * * * * * * * * * * * * * *

1. Closures on the natural river bed may encounter problems in rivers where the flow remains significant even during the dry season.

【译文】对天然河床的截流可能会遇到河流流量很大的问题，甚至是在干旱季节也会遇到。

【说明】where 引导定语从句修饰前面的 rivers。

2. Except for some special cases which will be considered later in this report (instantaneous closure by blasting in steep sided valleys and sand hydraulic filling in flat plains) nearly all important closures have been achieved by one of the following methods.

【译文】除了在本节稍后考虑的一些特殊情况外（如在陡峭的山谷爆破合龙和在平原喷砂填充沙流），几乎所有重要的截流都通过以下方法实现。

【说明】原文中 which 引导定语从句，其中 which 指代前面的 some special cases。

3. The diversion flow should depend on the problems that would arise should there occur an overflow over the dam itself during the construction period.

【译文】引水流量的大小应取决于在施工期间可能出现的水流溢顶的问题。

【说明】句中，should there occur an overflow…为倒装句形式的虚拟语态，含义相当于 if 引导的条件句，正常的语序是 if there should occur an overflow … construction period。

4. We feel they should be used when required by the topograthical conditions of the dam or in cases where the levels of the foundations are difficult to determine during the designing stage.

【译文】只有在大坝的地质状况允许，或者是在设计期间基础的安全等级很难确定的情况下才采用隧洞导流。

【说明】when 和 in case 分别表达了 2 种不同的条件状语。

* **Summary of Glossary** *

1. river closure 截流
2. river diversion 导流
3. overflow 溢出，泛滥
4. diversion tunnel 导流隧洞
5. spillway 溢洪道，泄洪道

* **Exercises** *

(1) It is consequently of interest to define safe and ________ conditions for closures and to estimate the ________ costs according to maximum water head in order to reach overall ________ of the whole scheme (closure, diversion, programme, etc...) or to give more flexibility for programming closure ________ the low flow season if suitable.

(2) The possibility of using diversion tunnels as a place for the bottom outlet is a clearly feasible solution in the case of ________ dams, but not so much as for concrete dams.

(3) In rockfill dams of medium or small height, the solution of providing an incorporated ________ of concrete housing spillways and ________, renders diversion operations much easier.

* * * * * * **Word Building (19) electr-(electri-, electro-); inter-** * * * * * *

1. electr-（electri-，electro-） [前缀]，表示：电，电的

electric adj. 电的，导电的，电动的，电气

electrical adj. 电的，有关电的

electrician	n.			电工，电气技师
electronic	adj.			电子的
electromechanical	adj.			机电的，电动机械的
electromagnetic	adj.			电磁的

2. inter- [前缀]，表示：互，互相

change	n./v.	interchange	v.	相互交换
connect	v.	interconnect	vt.	（使）互相连接
action	n./vt.	interaction	n.	相互作用，交感
national	adj.	international	adj.	国际的，世界的
dependent	adj.	interdependent	adj.	相互依赖的，互助的
view	n./vt.	interview	vt.	会见，接见
marry	vt.	intermarry	vi.	近亲结婚

*** * * * * * * * * * * * * * * Text Translation * * * * * * * * * * * * * * ***

第十九章 大坝施工的截流与导流

第一节 截 流

1. 概述

对天然河床的截流，可能会遇到河流流量很大的问题，甚至是在干旱的季节也会遇到。截流一般是大坝不分期建设或河流暂时改道（有时是疏通开始前就进行）；在河床比较宽的地方，大坝要分两期建设，第一期末要完成合龙，水流从第一期建设的暂时或永久的开口处泄流。（单词个数：88）

虽然截流的费用一般只占一个大坝的总成本的很小比例，但必须对其加以研究，因为如果失败或延迟可能会导致严重的后果，并付出昂贵的代价--使整个计划延迟，因为暂时导流的设计与合龙最高水头的选择息息相关。（单词个数：58）

因此，为确定截流的安全和实际条件，并根据最大水头估计相关成本，就成为实现全流域最优的整个计划，以及在大流量季节灵活规划截流的关键。（单词个数：54）

大量用于截流合龙分析的理论方法和测试模型，会比较好的代表水文条件（例如水位和平均流速），但它们似乎不能界定更精确的尺寸和所需材料的特点。对世界各地合龙工程的比较和分类（包括世界大多数的最大型河流），分析遇到的困难，对初步研究和未来计划作为实验性指导的规则就成为关键。（单词个数：95）

2. 截流类型

除了在本节稍后考虑的一些特殊情况外（如在陡峭的山谷爆破截流合龙和在平原喷砂填充沙流），几乎所有重要的截流都通过以下两种方法得以实现。（单词个数：42）

立堵法截流（垂直封闭）从岸边一侧或两侧增高大坝直到完成合龙（单词个数：26）

平堵法截流（水平）横跨河道间隙全部宽度，将材料均匀的填充于全部河道范围（通常需要特殊的结构例如运输材料的桥式或缆式吊车）。（单词个数：37）

第二节 导 流

河流导流对大坝建设在时间上有重要的影响，而且也对整个工程的造价有重要的影响，因此弄清每个细节是十分必要的。对这个问题制定明确的规则并不是容易的事，因为河流导流，总是以实际中得到的数据为标准，不能轻易地概括为一个标准模式，如：河流流量，地质地形的特点等。不过，我们试图将我们得到的大型河流的导流结论，在中等大小的河流中得以直接引用。在我们看来，可以按照以下原则来实现：（单词个数：154）

（1）引水流量的大小应取决于在施工期间可能出现的水流漫溢情况。可能会有侵蚀大坝（土坝）或基础、地基（薄弱地基）的问题。对于这些都不是严重的问题，在连续一段时期内期望它们不发生，相对于强迫河流改道要经济的多。（单词个数：73）

（2）对一些中等河流，在河床上建导流设施通常比高造价的围堰更经济。水位大大高出河床的情况下，要侧重使用导流底孔，因为导流口围堰截流，闭气更简单，也适于堆石坝合理泄流。（单词个数：87）

（3）导流隧洞是一个非常安全的解决方案，但并不是最经济的。只有在大坝的地质状况和要求，或者是在设计期间基础的安全等级很难决定的情况时才使用它。对于其他情况下，大流量，或由外部管道或渠道，使用中型导流会来得更经济。（单词个数：69）

（4）如果是土石坝，利用作为底孔的导流隧洞可能是比较可行的解决方案，而对于混土大坝，效果并不理想。（单词个数：32）

（5）利用施工期的大坝泄洪导流隧洞作为它的溢洪道，意味着这个大坝可能是中型坝或高坝，特别是在一定水压下隧洞空载时，出现的一些水力侵蚀问题可能会很难解决。（单词个数：51）

（6）对于中等坝或低坝，若提供了包括混凝土溢洪道和泄洪建筑等设施问题的具体解决措施，这对于导流的操作会更加的容易。（单词个数：26）

根据这些原则，我们提出如下建议：（单词个数：15）

（1）导流流量。（单词个数：2）

（2）合理布置导流的实施高程和河流底孔。（单词个数：22）

（3）导流隧洞的可能替代解决方案。（单词个数：8）

（4）使用导流隧洞作为底孔泄水隧洞。（单词个数：11）

（5）使用导流隧洞作为泄洪隧洞。（单词个数：7）

（6）把土坝河流导流孔作为泄水孔。（单词个数：19）

Chapter 20 Embankment and Fills

Part 1 Introduction of Embankment

1. Type of embankment

In general, there are two types of embankment dams: earth (earthfill dam) and rockfill (rock fill dam). The selection is dependent upon the usable materials from the required excavation and available borrow. It should be noted that rockfills can shade into soil fills depending upon the physical character of the rock and that no hard and fast system of classification can be made. Rocks which are soft and will easily break down under the action of excavation and placement can be classified with earthfills. Rocks which are hard and will not break down significantly are treated as rockfills.

The selection and the design of an earth embankment are based upon the judgement and experience of the designer and is to a large extent of an empirical nature. The various methods of stability and seepage analysis are used mainly to confirm the engineer's judgement.

2. Freeboard

All earth dams must have sufficient extra height known as freeboard to prevent overtopping by the pool. The freeboard must be of such height that wave action, wind setup, and earthquake effects will not result in overtopping of the dam. [1] In addition to freeboard, an allowance must be made for settlement of the embankment and the foundation which will occur upon completion of the embankment.

3. Top width

The width of the earth dam top is generally controlled by the required width of fill for ease of construction using conventional equipment. In general, the top width should not be less than 30 ft. if a danger exists of an overtopping wave caused either by massive landslides in the pool or by seismic block tipping, then extra top width of erosion resistive fill will be required.

4. Alignment

The alignment of an earthfill dam should be such as to minimize construction costs but such alignment should not be such as to encourage sliding or cracking of the embankment. Normally the shortest straight line across the valley will be satisfactory, but local topographic and foundation conditions may dictate otherwise. Dams located in narrow valleys often are given an alignment which is arched upstream so that deflections of the embankment under pool load will put the embankment in compression thus minimizing transverse cracking. [2]

5. Abutments

Three problems are generally associated with the abutments of earth dams: ①seepage; ②instability; ③transverse cracking of the embankment. If the abutment consists of deposits of pervious soils it may be necessary to construct an upstream impervious blanket and downstream drainage measures to minimize and control abutment seepage.

Where steep abutments exist, especially with sudden changes of slopes or with steep bluff, there exists a danger of transverse cracking of the embankment fills. This can be treated by excavation of the abutment to reduce the slope, especially in the impervious and transition zones. The transition zones, especially the upstream, should be constructed of fills which have little or no cohesion and a well-distributed gradation of soils which will promote self-healing should transverse cracking occur. [3]

6. Stage construction

It is often possible, and in some cases necessary, to construct the dam embankment in stages. Factors dictating such a procedure are: ①a wide valley permitting the construction of the diversion or outlet works and part of the embankment at the same time; ②a weak foundation requiring that the embankment not be built <u>too</u> rapidly <u>to</u> prevent overstressing the foundation soils ; ③a wet borrow area which requires a slow construction to permit an increase in shear strength through consolidation of the fill. In some cases it may be necessary to provide additional drainage of the foundation or fill by means of sand drain wells or by means of horizontal pervious drainage blanket. [4]

7. Embankment soils

Most soils are suitable for use for embankment construction, however, there are physical and chemical limitations, soils which contain excessive salts or other soluble materials should not be used.[5] Substantial organic content should not exist in the soils. Lignite sufficiently scattered through the fill to prevent the danger of spontaneous combustion, is not objectionable. Fat clays with high liquid limits may prove difficult to work and should be avoided.

8. Compaction requirements

The strength of the impervious and semi-impervious soils depends upon the compacted densities. These depend in turn upon the water content and weight of the compacting equipment. The design of the embankment is thus influenced by the water content of the borrow soils and by the practicable alternations to the water content either prior to placement of the fill or after placement but prior to rolling. [6] If the natural water content is too high, then it may be reduced in borrow area by drainage, or by harrowing. If the soil is too dry it should be moistened in the borrow area either by sprinkling or by ponding and then permitted to stabilize the moisture content before use. The range of placement water content is generally between 2 percent dry to 2 or 3 percent wet of the standard Proter optimum water content. Pervious soils should be compacted to at least 80 percent of relative density.

If necessary, test fills should be constructed with variations in placement water content, lift thickness, number of roller passes and type of rollers. For cases of steep abutment, the fill must be placed in thin lifts and compacted by mechanical hand tampers. All overhangs should either be removed or filled with lean concrete prior to fill placement.

9. Types of instruments

The type of instrumentation depends upon the size and complexity of the project. The devices in common use are: ①piezometers; ②surface movement monuments; ③settlement gages; ④inclinometers; ⑤internal movement and strain indicators; ⑥pressure cells; ⑦seismic acceleration meter; ⑧movement indicators at conduit joints and other concrete structures.

Part 2 Placing and Protecting Fill

Fill shall be placed so that no part of the final foundation surface remains exposed for more than 72 hours.

Fill shall be placed in such methods as will prevent segregation of the material. Where the contract requires the placing of different types of fill in separate zones, the contractor shall carry out the work so as to prevent mixing of different types of fill, such mixed materials shall be removed to a spoil tip and replaced with fresh fill.

Any undesirable material accumulated on the fill surface shall be removed before placing the next layer of fill. No fill material shall be placed on a previous layer of fill that has dried out, become saturated or in any way deteriorated by exposure or by spilling of other material or disturbance by mechanical transport or by deposition of wind blown particles or by any other means. Before fresh fill material is placed, all such deteriorated fill or foreign material shall be removed to a depth at which material of an acceptable standard is exposed. The surface of each layer is to be approved by the engineer before the next layer is placed.[7]

Moreover, any fill shall be placed in uniform layers not greater than the approved thickness as specified hereafter and in an orderly sequence approximately horizontal along the centerline of the embankments. Except where specified or directed otherwise, no portion of any embankment shall be stepped more than 3 feet higher than any immediately adjacent portion except where permitted by the engineer and the slope formed by such steps shall be not exceed 1V: 3H and not less than 1V: 4H from one level to another.[8] Except as shown on the drawings or as otherwise directed, all fill placement surfaces shall be slope at right angles to the centerline of the embankment in both the upstream and downstream direction from the downstream edge of the core so as to allow run-off and prevent the accumulation of water. The drainage slope on the temporary surface of any zone shall not exceed 1 on 30 and the highest point shall be the downstream edge of the core.

Construction of any one embankment shall be carried out over the maximum possible length, no less than 1500 feet, of that embankment in such a manner that no temporary construction slope crosses the axis of the embankment except as approved by the engineer. Where a temporary construction slope crossing the axis of the embankment is permitted by the Engineer it shall be formed at a gradient of 1V : 5H. When subsequently placing material against this slope it shall be cut back in steps equal to the layer thickness to avoid feather edges.[9] The contractor shall complete each layer of fill fully up to the abutment contacts and structures and against sloping foundations and ensure that the fill is compacted as specified throughout.

At the end of each working day, or if it starts to rain, the surface of the fill shall be made smooth and compacted with a smooth drum roller with a drainage slope to induce runoff from the filled areas and leave no areas that can retain water. Where necessary, grips, drainage ditches and the like shall be formed to assist drainage and to prevent runoff from damaging placed material.[10] Runoff from heavy rain shall be controlled to prevent gully erosion of the placed fill. Any gully erosion shall be repaired with material compacted <u>in accordance with</u> the Specification, and eroded surfaces shall be restored and graded to ensure a proper bond with new fill placed on them. Any eroded material other than gravel and any contaminated material shall be removed from the embankment and placed in designated spoil tips.

Any part of the fill that becomes saturated or attains excessive moisture content or that is rendered unsuitable due to poor surface drainage, uncontrolled traffic, or for any other reason, shall be excavated and removed to a spoil tip and replaced by the fresh fill. [11]

Temporary access ramps shall be removed when work in that area is completed. Any ramps or other areas within the limits of an embankment which, in the opinion of the engineer have been over-compacted or damaged by the concentrated use by construction equipment, shall be reworked and re-compacted or, if the engineer requires, shall be excavated, removed to spoil tip and replaced by the fresh fill. [12]

* * * * * * * * * * * * * * * * * * Explanations * * * * * * * * * * * * * * * * * *

* * * * * * * * * * * * * * * * * New Words and Expressions * * * * * * * * * * * * * * * *

1. embankment [em'bæŋkmənt] n. 堤防，筑堤
2. earth fill n. 填土，[地] 泥流
3. rockfill ['rɔkfil] n. 填石
4. empirical nature 实证性
5. seepage analysis 渗流分析
6. freeboard ['fri: bɔ: d] n. 超高，干舷，吃水线以上的船身
7. overtopping ['əuvə'tɔping] n. 漫顶；漫溢
8. foundation [faun'deiʃən] n. 地基，房基；建立，设立，创办，创建；基础，基本原理，根据；基金（会）；粉底霜
9. seismic ['saizmik] adj. 地震的，由地震引起的；震撼世界的
10. resistive [ri'zistiv] adj. 抗（耐、防）…的；电阻的；抵抗的，有抵抗力的，有耐力的
11. alignment [ə'lainmənt] n. 定线；（国家、团体间的）结盟；排成直线
12. topographic [ˌtɔpə'græfik] adj. 地志的，地形学的
13. transverse ['trænzvə: s] n. 横向物，横轴，横断面；[数] 横截轴；[解] 横肌 adj. 横向的，横断的，横切的；[数] 横截的
14. bluff [blʌf] n. 断崖，绝壁，诈骗 adj. 绝壁的，直率的 v. 诈骗
15. transition zone 移行区，过渡区
16. overstressing ['əuvə'stresiŋ] n. 大应力，过度（超限）应力，超负载
17. shear strength 抗剪强度
18. consolidation [kənˌsɔli'deiʃən] n. 固结作用
19. soluble ['sɔljubl] adj. 可溶的
20. organic content 有机质成分
21. lignite ['lignait] n. 褐煤
22. scatter ['skætə] v. 分散，散开，驱散
23. spontaneous [spɔn'teiniəs] adj. 自动的，自发的

24. clay [klei] n. 黏土，泥土，肉体，人体，似黏土的东西，陶土制的烟斗

25. semi-impervious 半透水的

26. density [ˈdensiti] n. 密集，稠密；密度

27. standard Proter optimum water content 标准普氏最优含水量

28. lift thickness 铺层厚度

29. piezometer [ˌpaiəˈzɔmitəs] n. 压力计，压强计，测压计

30. movement monument 位移标志

31. settlement gage 沉陷量测仪（计）

32. inclinometer [ˌɪnkliˈnɔmitə] n. 测斜仪

33. strain indicator 应变指示仪

34. conduit joint 管道接头

35. segregation [ˌsegriˈgeiʃən] n. 分离，隔离

36. disturbance [disˈtə: bəns] n. 扰动，骚扰

37. feather edge 削边，薄边，羽毛边

38. gully [ˈgʌli] n. 冲沟，溪谷，集水沟，雨水口，檐槽

39. render [ˈrendə] vt. 造就，使得 vi. 给予补偿

*** * * * * * * * * * * * * * * * * * Complicated Sentences * * * * * * * * * * * * * * * * * ***

1. The freeboard must be of such height that wave action, wind setup, and earthquake effects will not result in overtopping of the dam.

【译文】超高的高度必须足以在波浪作用、风浪壅高和地震影响下，不会导致坝的漫顶。

【说明】此句为 such … that 结构的复合句。

2. Dams located in narrow valleys often are given an alignment which is arched upstream so that deflections of the embankment under pool load will put the embankment in compression thus minimizing transverse cracking.

【译文】对于峡谷中的坝，常采用向上游拱出的坝轴线，以便在坝体受库水压力作用而发生变形时，能使坝体压紧，从而尽量减少其横向开裂。

【说明】此句的主体结构就是“主语+谓语+宾语”简单句的被动语态。主语中过去分词短语 located in narrow valleys 作定语，修饰整个句子的主语 Dams；which 引导的定语从句修饰 alignment；词组 so that 表示“以使……，以便……”，引导目的状语从句。而 the embankment under pool load 修饰 deflections，will put the embankment in compression 及 minimizing transverse cracking 的主语都是 deflections。

为了便于理解，这个句子可以改写为：

Dams, located in narrow valleys, often are given an alignment, which is arched upstream so that deflections of the embankment under pool load will put the embankment in compression and will minimize transverse cracking.

3. The transition zones, especially the upstream, should be constructed of fills which have little or no cohesion and a well-distributed gradation of soils which will promote self-healing should transverse cracking occur.

【译文】过渡区，尤其是在上游侧的过渡区，必须用黏着力很小或无黏着力且颗粒级配良好的土料来填筑，这种土料发生横向裂缝时能自行愈合。

【说明】该句主体结构为一简单句，但其中包含了两个由 which 引导的定语，第一个 which

引导的定语修饰 fills；第二个 which 引导的定语修饰 soils，该从句中还有一个省略的 if 状语从句的倒装结构。

4. In some cases it may be necessary to provide additional drainage of the foundation or fill by means of sand drain wells or by means of horizontal pervious drainage blanket.

【译文】在某些情况下，需要增设基础排水设施或填筑排水沙井，或采用水平透水的排水铺盖。

【说明】此句中的 it 为形式主语，真正的主语是 to provide additional drainage 及 fill，fill 又由 or 引导了两个 by means of 的并列形式，即 sand drain wells 和 horizontal pervious drainage blanket。

5. Most soils are suitable for use for embankment construction, however, there are physical and chemical limitations, soils which contain excessive salts or other soluble materials should not be used.

【译文】大多数土料适合于坝体填筑，然而，在物理和化学性质上也有一定的局限，含有过多盐或其他可溶性物质的土料不可以使用。

【说明】此句含有 however 引导的转折结构，which 引导的定语从句修饰 soils。

6. The design of the embankment is thus influenced by the water content of the borrow soils and by the practicable alternations to the water content either prior to placement of the fill or after placement but prior to rolling.

【译文】料场土料的含水量及其在填筑前或堆筑后而未碾压前的填土实际含水量的变化，都会影响坝体的设计。

【说明】此句的主体结构是被动语态。但真正的主语（被动语态中 by 引导）是由 and 连接的并列结构，即 by the water content of the borrow soils 和 by the practicable alternations to the water content，而后种情况下又由 either…or…（或者……或者……）连接了两种并列情况。

7. Before fresh fill material is placed, all such deteriorated fill or foreign material shall be removed to a depth at which material of an acceptable standard is exposed. The surface of each layer is to be approved by the engineer before the next layer is placed.

【译文】在填筑新填料前，所有受到破坏的物质或者杂质都应该清除，清除的深度达到合格的标准材料露出时为止。在填筑新一层的填料前，每一层填料都必须经工程师认可。

【说明】本段由两个被动语态的句子组成的。前一句中，before 引导一个时间状语从句，主句中 depth 由 at which 引导的定语从句修饰；后一句也同样含有 before 引导的时间状语从句。

8. Except where specified or directed otherwise, no portion of any embankment shall be stepped more than 3 feet higher than any immediately adjacent portion except where permitted by the engineer and the slope formed by such steps shall be not exceed 1V∶3H and not less than 1V∶4H from one level to another.

【译文】除非另有规定或者另有指示，任何填筑部分都不应该比相邻层的填料高出 3 英尺以上，除非工程师许可。由于此种填筑而形成的从一个高程到另外一个高程阶梯坡度不能比 1∶3（V∶H）还陡，也不能比 1∶4（V∶H）还缓。

【说明】此句含有两个 except 引导的条件状语从句，分别表明了例外的具体条件（规定、指示或工程师的许可）。内容上包含填筑的高度要求和坡度要求两部分，高度方面，要求相邻高度差小于 3 英尺；坡度方面要求不能比 1 : 3（V : H）还陡，也不能比 1 : 4（V : H）还缓，分别用了 not exceed（不超出……）和 not less than（不少于……）。

9. When subsequently placing material against this slope, it shall be cut back in steps equal to the layer thickness to avoid feather edges.

【译文】后来再在此坡面填料时，应将此坡面削成台阶式的，以避免出现羽毛边，每一阶级的厚度与铺设层厚度一样。

【说明】此句包含一个由 when 引导的时间状语从句，主句中 it 是形式主语，谓语为 be cut back，equal to 前省略 and，后面的不定式短语 to avoid feather edges 做目的状语。

10. Where necessary, grips, drainage ditches and the like shall be formed to assist drainage and to prevent runoff from damaging placed material.

【译文】必要时，挖一些沟壑等帮助排水，防止流水损坏填筑材料。

【说明】此句是虚拟语气，主语包括 grips, drainage ditches and the like，其中 and the like 是“如此等等”的意思，由 and 连接两个动词不定式做目的状语。

11. Any part of the fill that becomes saturated or attains excessive moisture content or that is rendered unsuitable due to poor surface drainage, uncontrolled traffic, or for any other reason, shall be excavated and removed to a spoil tip and replaced by the fresh fill.

【译文】如果填筑的任何一部分已经饱和、水分过多，或因表面排水不好、车辆过多的碾压及其他原因造成填筑不合适，所有这些不合适的填筑应该挖掉，运到弃料场，并用新材料代替填筑到原处。

【说明】此句是一简单句，主语是 any part，谓语由 excavated、removed 和 replaced 组成。主语 any part 后是一定语从句，该从句并列了几种情况（saturated、attains excessive moisture content、rendered unsuitable due to poor surface drainage、uncontrolled traffic、for any other reason）。分清了句子成分，就比较容易理解了。

12. Any ramps or other areas within the limits of an embankment which, in the opinion of the engineer have been over-compacted or damaged by the concentrated use by construction equipment, shall be reworked and re-compacted or, if the engineer requires, shall be excavated, removed to spoil tip and replaced by the fresh fill.

【译文】对于大坝填筑范围内的任何坡面便道或其他施工区域，如果工程师认为已经被施工设备的频繁使用而碾压过多或者毁坏了，承包商应该重新作业、重新压实；如果工程师要求，应将其挖除并运到弃料场，且要求用新的替代材料填筑完好。

【说明】此句结构与上句类似，也是简单句，主语 any ramps or other areas，谓语是 shall be reworked and re-compacted、be excavated、removed 及 replaced。主语中含有介词短语作定语，且 be excavated、removed 及 replaced 需要依照工程师的意见执行，即有一条件状语从句 if the engineer requires。

******************** **Summary of Glossary** ********************

1. earthfill　　填土坝，土坝
2. excavation　　挖掘

3. borrow 取土，取料
4. earth embankment 土坝
5. freeboard 超高
6. allowance 余地，裕量
7. abutment 坝肩，桥礅
8. pervious soil 透水性土料
9. liquid limits 液限
10. overtopping 漫顶，溢出
11. roller 压路机，滚筒
12. relative density 相对密度
13. seismic acceleration meter 地震加速仪
14. stage construction 分期施工
15. filter zone 过滤层
16. initial set 初凝
17. moisture content 含水量，水分
18. concrete structure 混凝土结构

* Exercises *

(1) In general, there are two types of embankment dams: earth ________ dam and rock ________ dam.

(2) The alignment of an earthfill dam should be ________ such as to ________ construction costs but such alignment should not be such as to encourage ________ or ________ of the embankment.

(3) If necessary, test fills should be constructed with variations in placement water ________, lift thickness, number of ________ passes and ________ of rollers.

(4) Before fresh fill material is ________, all such deteriorated fill or foreign material shall be removed to a ________ at which material of an acceptable standard is ________.

* * * * * * * * * * * * * * * * Word Building (20) -fill; -ic * * * * * * * * * * * * * * * *

1. -fill [后缀]，表示：……填充的

earth	n./v.	eathfill	adj.	土筑的
rock	n./v.	rockfill	adj.	石筑的

2. -ic [后缀]，表示：学，术，师

logy	adj.	logic	n.	逻辑学
arith	n.	arithmetic	n.	算术
rhetor	n.	rhetoric	n.	修辞学
mech-		mechanic	n.	机械师

* * * * * * * * * * * * * * * * Text Translation * * * * * * * * * * * * * * * *

第二十章 土石坝及其填筑

第一节 土石坝概述

1. 土石坝坝型

一般来说，土石坝有两种类型：土坝和堆石坝。坝型的选择取决于能从开挖的地点和可用的料场（stock ground：料场）（borrow area：取土面积；采料场）处取得合用材料的情况。应当指出的是，根据岩石的物理特性，堆石可以逐渐变化为填土，因而不能对土石料作出严格而固定的分类。那些软弱的和在开挖填筑时容易破碎的岩石可被归入填土类。而坚硬和不会大量破碎的岩石，则列为堆石类。（单词个数：103）

一座土坝的选定和设计都依赖于设计人员的判断和经验，而且在很大程度上是属于经验性的。各种稳定和渗透分析方法，主要是作为证实工程师的判断而使用。（单词个数：45）

2. 超高

所有的土坝都必须有一个足够的额外高度，称为超高，以防止水库漫顶。超高的高度必须足以在波浪作用、风浪壅高和地震影响下，不会导致坝的漫顶。除了超高外，在大坝即将竣工时，大坝和地基的高度还要有余量，以满足坝体沉降的要求。（单词个数：65）

3. 坝顶宽度

土坝的坝顶宽度一般由便于施工的填筑宽度来控制，即能满足施工时一般的常规设备要求。通常，坝顶宽度应不小于30英尺。如果存在大规模塌方进入水库，或者有因地震使岩块倒落而引起波浪漫顶的危险，则需要采用抗冲刷的材料填筑更宽的坝顶高度。（单词个数：67）

4. 定线

土坝的坝轴线选定应尽量使建设费用降到最少，但是也不能因此引起坝体发生滑动或开裂。一般来说，一条横跨河谷的最短直线可能满足要求。但是，当地的地形和地基条件，也可能要求采用另外的方案。对于峡谷中的坝，常采用向上游拱出的坝轴线，以便在坝体受库水压力作用而发生变形时，能使坝体压紧，从而尽量减少其横向开裂。（单词个数：83）

5. 两岸坝肩

一般有三个问题与土坝坝座有关：①渗透；②不稳定；③坝体的横向开裂。如果坝座是由透水的沉积土构成，就可能需要建造一道上游不透水的铺盖和下游排水设施，可尽量减少和控制坝座内的渗透。（单词个数：49）

在坝肩岸坡很陡的地方，特别在边坡突变或有陡壁处，那里的坝体填土会产生横向裂缝的危险。这个问题可以用开挖坝座放缓边坡来处理，这样的处理在不透水区和过渡区特别需要。过渡区，尤其是在上游侧的过渡区，必须用黏着力很小或无黏着力，且颗粒级配良好的土料来填筑，这种土料如发生横向裂缝时能自行愈合。（单词个数：76）

6. 分期施工

土坝的分期施工往往是可能的，而且在一些情况下是必须的。要求这样施工程序的因素有：①河谷宽阔，可以允许导流或泄水工程与一部分坝体同时施工；②地基软弱，要求坝体不能过快填筑，以防止地基中产生过大的应力；③料场潮湿，要求放慢施工，以使土料能通过固结作用来增加抗剪强度。在某些情况下，可能需要增设基础排水设施或填筑排水沙井，或采用水平透水的排水铺盖。（单词个数：113）

7. 坝体的土料

大多数土料都适合于坝体填筑，然而，在物理和化学性质上也有一定的限制，含有过多盐分和可溶性物质的土料，不可以使用。土壤里不应存在大量的有机质成分。褐煤若能通过填筑而充分分散，不会导致自燃，就不妨碍使用。而具有高度液限的肥黏土，多半难以施工，应避免使用。（单词个数：69）

8. 压实的要求

不透水和半透水的土料的强度取决于压实的密度。压实密度又取决于土料的含水量和压实设备的重量。因此，料场土料的含水量和在填筑前或堆筑后而未碾压前的填土实际含水量的变化，都会影响坝体的填筑。如果天然含水量太高，可以在料场用排水或将土料耙松的办法来减低。如果土料太干燥，则需在料场用洒水和泡水的办法把土料润湿，然后再让土料在使用以前保持稳定的含水量。填筑时的含水量范围一般介于比标准普氏最优含水量低2%到高2%～3%之间。透水性土料至少压实到相对密度的80%。（单词个数：155）

如果需要，应该变换填筑层的含水量、铺层厚度、碾压遍数和碾压机的型式等，进行填筑试验。对于坡度很陡的坝肩部位，必须用薄层填筑，并用手扶打夯机夯实。所有外悬突出部位，均应在填土堆筑以前挖除或用贫混凝土（少灰混凝土）填平。（单词个数：57）

9. 观测仪器的类型

观测仪器的类型取决于工程的规模和复杂性。通用的装置是：①测压计；②表面位移量测仪；③沉陷量测仪（计）；④测斜仪；⑤内部位移和应变指示仪；⑥压力盒（压力传感器或压应力计）；⑦地震加速度仪；⑧在管道接头和其他混凝土结构上的位移标记。（单词个数：52）

第二节　填料的填筑和保护

大坝（坝堤）填筑的方式应做到：无论哪一部分的填料填筑后，最后的基础面暴露的时间均不超过72h。（单词个数：20）

填筑填料的方法还应该防止填筑材料的分离。如果合同要求在不同的地段填筑不同的填料，承包商在施工时，应防止不同类型的填料相互混淆。万一工程师认为过多不同类型填料相互混合了，这些被混合的材料应当清除，并运到废料场去，且用新的填料取代。（单词个数：60）

任何沉积在填筑面上的劣质材料，在填筑下一层填料前都应该清除掉。如果先一层填料已经变干、饱和、因暴露在外受到破坏、有其他物质散落进来、因机械运输遭到破坏、有风吹的外来物沉积或者其他方式的破坏，都不得在其上填筑新一层填料。在填筑新填料前，所有受到破坏的物质或者杂质都应该清除，清除的深度至合格的标准材料露出时为止。在填筑新

一层的填料前，每一层填料都必须经工程师认可。（单词个数：112）

另外，所有填料填筑应该按照统一层厚进行填筑，这个厚度不能超出本节后面的条款规定的许可厚度，并且按照循序渐进的方式沿着堤坝中心线近似水平地向前推进。除非另有规定或者另有指示，任何填筑部分都不应该比（近在咫尺的）相邻层的填料高出 3 英尺以上，除非工程师许可。由于此种填筑而形成的从一个高程到另外一个高程阶梯坡度不能比 1：3（V：H）还陡，也不能比 1：4（V：H）还缓。除非图纸有专门的注明或者有其他的规定，所有的填筑面，无论是在心墙上游还是下游，都应该按照与大坝中心线形成一定角度的坡面形式填筑，以心墙下游边为坡面分界线，便于排水，以防积水。任何临时填筑面的排水斜坡不应超出 1：30，最高位置应该是心墙的下游边。（单词个数：163）

大坝填筑的方式要尽可能最大长度地向前铺设，至少 1500 英尺；同时还应保证临时施工坡面不能穿越大坝的轴线，除非工程师许可那样。如果工程师许可临时施工坡面穿越大坝轴线，那么坡比应为 1：5（V：H）。后来再在此坡面填料时，应将此坡面削成台阶式的，以避免出现薄边，每一阶级的厚度与铺设层厚度一样。承包商应将每一层填料一直铺完至坝肩相连处或其他临界结构物处，避免倾斜的基础面，并保证按照规定将各处的填料都压实。（单词个数：121）

在每个工作日结束或遇下雨时，填筑面应该用平轮压路机将填筑面碾压光滑形成一个排水坡面，便于将填筑区域的水引走，不留下任何积水带。必要时，挖一些沟壑帮助排水，防止流水损坏填筑材料。要控制大雨形成的径流，以防止排水沟对填筑面/材料的侵蚀。因排水沟而造成的任何侵蚀都要用合适的材料按照规范要求碾压、修补。被侵蚀的表面应该恢复成原状，形成的坡面应保证合适的黏结度，使得填于其上的材料能与之很好地结合。除砾石外，任何侵蚀的材料和污染的材料都应该从堤坝清除掉，并运到指定的弃料场。（单词个数：140）

如果填筑的任何一部分已经饱和、水分过多，或因表面排水不好、车辆过多的碾压及其他原因造成填筑不合适，所有这样不合适填筑应该挖掉，运到弃料场，并用新材料代替填筑到原处。如果得到工程师许可，可将这样的填筑翻松并重新压实。（单词个数：45）

某一区域的施工完成以后，通向两个工作区的临时坡路应当拆除。对于大坝填筑范围内的任何坡面便道或其他施工区域，如果工程师认为已经被施工设备的频繁使用而碾压过多或者毁坏了，承包商应该重新作业，重新压实；如果工程师要求，应将其挖除并运到弃料场，且要求用新的替代材料填筑完好。（单词个数：66）

Chapter 21 Main Equipment in Hydropower Plants

Part 1 Various Hydraulic Turbines

Hydraulic turbine is a machine whose runner is powered by water; it transmits mechanical energy to the rotor of a generator to make it turn.[1]

There are two main types of hydro turbines: impulse and reaction. The type of hydropower turbine selected for a project is based on the height of standing water—referred to as "head"—and the flow, or volume of water, at the site. [2] Other deciding factors include how deep the turbine must be set, efficiency, and cost.

1. Impulse turbine

The impulse turbine generally uses the velocity of the water to move the runner and discharges to atmospheric pressure. The water stream hits each bucket on the runner. There is no suction on the down side of the turbine, and the water flows out the bottom of the turbine housing after hitting the runner. An impulse turbine is generally suitable for high head, low flow applications.

(1) Pelton Turbine. A Pelton turbine (as shown in Fig.21.1) has one or more free jets discharging water into an aerated space and impinging on the buckets of a runner. [3] Draft tubes are not required for impulse turbine since the runner must be located above the maximum tailwater to permit operation at atmospheric pressure.

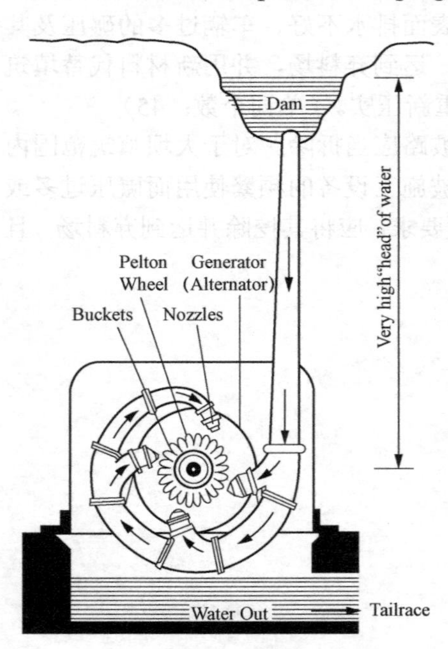

Fig.21.1 Pelton Turbine

(http://www.wpcorp.com.au)

The Pelton turbine resembles the waterwheels used at water mills in the past. The Pelton turbine has small "buckets" all around its rim. Water from the dam is fed through nozzles at very high speed hitting the buckets, pushing the wheel around.

(2) Cross-Flow Turbine. A cross-flow turbine is drum-shaped and uses an elongated, rectangular section nozzle directed against curved vanes on a cylindrically shaped runner.[4] It resembles a "squirrel cage" blower. The cross-flow turbine allows the water to flow through the blades twice. The first pass is when the water flows from the outside of the blades to the inside; the second pass is from the inside back out. A guide vane at the entrance to the turbine directs the flow to a limited portion of the runner. The cross-flow was developed to accommodate

larger water flows and lower heads than the Pelton.

2. Reaction turbine

A reaction turbine develops power from the combined action of pressure and moving water. The runner is placed directly in the water stream flowing over the blades rather than striking each individually. Reaction turbines are generally used for sites with lower head and higher flows than compared with the impulse turbines.

(1) Propeller turbine. A propeller turbine generally has a runner with three to six blades in which the water contacts all of the blades constantly. Picture a boat propeller running in a pipe. Through the pipe, the pressure is constant; if it isn't, the runner would be out of balance. The pitch of the blades may be fixed or adjustable. The major components besides the runner are a scroll case, wicket gates, and a draft tube. There are several different types of propeller turbines:

Kaplan turbines, as shown in Fig.21.2, resemble ship's propellers. However, within the Kaplan turbines, the angle (or pitch) of the blades can be altered to suit the water flow. Actually, both the blades and the wicket gates are adjustable, allowing for a wider range of operation.

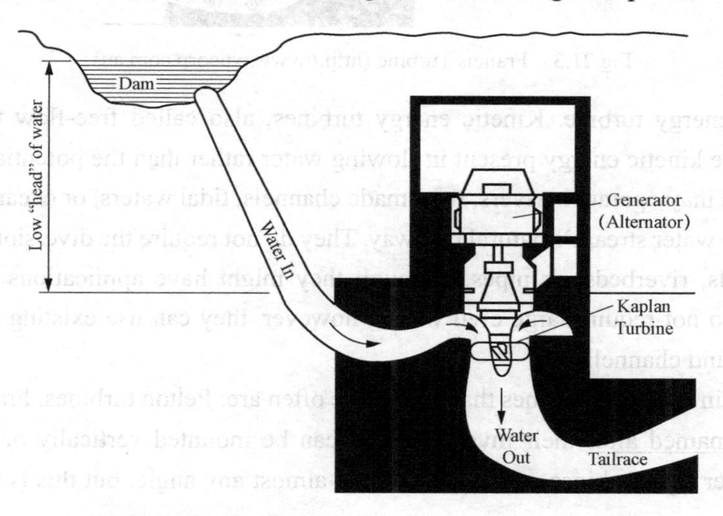

Fig.21.2 Kaplan Turbine (http://www.wpcorp.com.au)

Bulb turbine and generator can make a sealed unit placed directly in the water stream. Straflo turbine is a hydropower turbine that the generator can be attached directly to its perimeter. For a tube turbine, the penstock bends just before or after the runner, allowing a straight line connection to the generator.

(2) Francis turbine. A Francis turbine, as shown in Fig.21.3, has a runner with fixed buckets (vanes), usually nine or more. Water is introduced just above the runner and all around it and then falls through, causing it to spin.[5] Besides the runner, the other major components are the scroll case, wicket gates, and draft tube.

The Francis turbine is also similar to a waterwheel, as it looks like a spinning wheel with fixed blades in between two rims. This wheel is called a 'runner'. A circle of guide vanes surround the runner and control the amount of water driving it. Water is fed to the runner from all sides by these vanes causing it to spin.

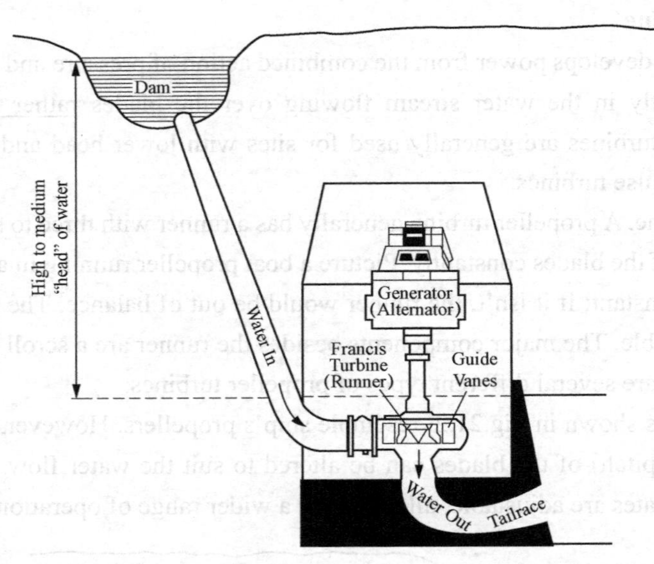

Fig.21.3 Francis Turbine (http://www.wpcorp.com.au)

(3) Kinetic energy turbine. Kinetic energy turbines, also called free-flow turbines, generate electricity from the kinetic energy present in flowing water rather than the potential energy from the head. The systems may operate in rivers, man-made channels, tidal waters, or ocean currents. Kinetic systems utilize the water stream's natural pathway. They do not require the diversion of water through manmade channels, riverbeds, or pipes, although they might have applications in such conduits. Kinetic systems do not require large civil works; however, they can use existing structures such as bridges, tailraces and channels.

The three main types of burbines that used more often are: Pelton turbines, Francis turbines, and Kaplan turbines (named after their inventors). All can be mounted vertically or horizontally. The Kaplan or propeller type turbines can be mounted at almost any angle, but this is usually vertical or horizontal.

The Pelton wheel is used where a small flow of water is available with a 'large head'. The Francis turbine is used where a large flow and a high or medium head of water is involved. Kaplan or propeller type turbines are designed to operate where a small head of water is involved.

Part 2 Generator Unit

Generator unit is the device set with a turbine that transmits the water's mechanical energy to the generator's rotor to make it turn to produce electricity. [6] Fig.21.4 shows the structure of a hydraulic generator unit, parts of which are described as followed.

1. Turbine

Turbine headcover is the structure that covers the upper part of the turbine's runner.

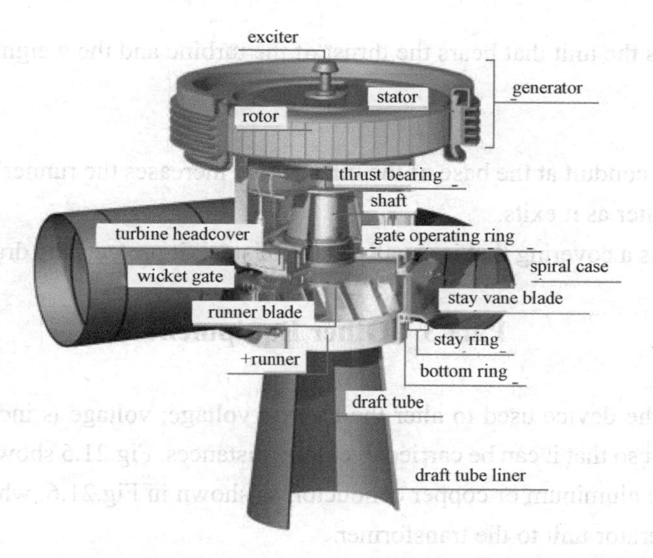

Fig.21.4 Hydraulic Generator Unit (http://visual.merriam-webster.com)

Runner is the movable part of the turbine that transmits the movement of the water to the shaft to which it is attached to turn the rotor.

Runner blade is the stationary curved plate on the turbine's runner; it receives the thrust of the water to turn the runner.

2. Generator

Generator is the machine that consists of a rotor and a stator; it produces an electric current.

Stator is the stationary part of the generator that consists of a coil of copper conductors, which collects the electric current produced by the rotor.

Rotor is the movable part of the generator that is made up of electromagnets; its rotation induces an electric current in the stator.

Exciter is the device that supplies electric current to the rotor's electromagnets.

3. Control parts

Spiral case is a duct shaped like a spiral staircase that is used to distribute water uniformly around the turbine to make it turn smoothly.

Stay ring is the set of two rings linked together by the stay vanes.

Stay vane blade is the fixed panel that receives pressurized water from the spiral case and directs it over the wicket gates.

Wicket gates are movable panels that regulate the flow of water entering the turbine to ensure a constant rotational speed of the runner.

Gate operating ring is a movable device that controls the opening and closing of the wicket gates.

Bottom ring is the circular part under the wicket gates that holds them in place.

4. Driving and supporting parts

Shaft is cylindrical part that communicates the movement of the turbine's runner to the generator's rotor.

Thrust bearing is the unit that bears the thrust of the turbine and the weight of the rotating parts of the generator unit.

5. Other parts

Draft tube is the conduit at the base of the turbine that increases the runner's output by reducing the pressure of the water as it exits.

Draft tube liner is a covering that is usually made of steel; it protects the draft tube from erosion.

Part 3 Other Equipment

Transformer is the device used to alter the electric voltage; voltage is increased as the current leaves the power plant so that it can be carried over long distances. Fig.21.5 shows a large transformer.

Busbar is a large aluminum or copper conductor, as shown in Fig.21.6, which transmits electric current from the generator unit to the transformer.

Fig.21.5 Transformer (http://www.directindustry.com)

Fig.21.6 Busbar (http://www.sidehillcopper.com)

Circuit breaker is a useful mechanism automatically cutting off the power supply in the event of overload so that the devices nearby can be protected from too much current. One type of circuit breaker is shown in Fig.21.7.

Lightning arrester, as shown in Fig.21.8, is the device that protects the electric facilities from power surges caused by lightning.

Fig.21.7 Circuit Breaker
(http://zh.wikipedia.org)

Fig.21.8 Lightning Arrester
(http://www.directindustry.com)

Bushing, as shown in Fig.21.9, is the device that allows the conductor to pass through the wall and separates it from the latter.

Fig.21.9 Bushing (http://www.reuelnc.com)

* * * * * * * * * * * * * * * * * Explanations * * * * * * * * * * * * * * * * *

* * * * * * * * * * * * * * * * New Words and Expressions * * * * * * * * * * * * * *

1. accommodate [ə'kɔmədeit] v. 适应
2. aerated ['eiəreitid] adj. 充气（通气，通风，鼓风）的
3. blade [bleid] n. 叶片，刀片
4. blower ['bləuə] n. 送风机，吹风机
5. bucket ['bʌkit] n. 桶状物，铲斗，叶片
6. civil work 土建工程
7. cross-flow 交叉流动，横向流动
8. elongated ['i:lɔŋgeitid] adj. 加长的，拉长的，伸长的
9. impinge [im'pindʒ] v. 碰撞，撞击
10. impulse ['impʌls] n. 推动
11. jet [dʒet] n. 喷嘴，喷射器，喷射流体
12. mount [maunt] vt. 装上，设置，安放
13. nozzle ['nɔzl] n. 管口，喷嘴
14. pitch [pitʃ] n. 斜度，螺距
15. propeller [prə'pelə] n. 螺旋桨；推进器
16. rectangular section 矩形截面
17. rim [rim] n. 边，轮缘
18. runner ['rʌnə] n. 叶轮，转子
19. scroll case 蜗壳
20. spin [spin] v. 旋转
21. suction ['sʌkʃən] n. 吸入，吸力，抽气，抽气机，抽水泵
22. tailwater [teil'wɔ: tə] n. 尾水
23. vane [vein] n.（风车、螺旋桨等的）翼，叶片
24. wicket ['wikit] n. 导叶，小门
25. coil [kɔil] n. 线圈
26. electric current 电流
27. electromagnet [i,lektrəu'mægnit] n. 电磁石，电磁铁
28. exciter [ik'saitə] n. 励磁机；刺激物，兴奋剂
29. induce [in'dju: s] vt. 感应，引起
30. liner ['lainə] n. 衬垫，衬里
31. rotor ['rəutə] n. 转子，回转轴，转动体
32. stator ['steitə] n. 定子，固定片
33. aluminum [,ælju:'minjəm] n. 铝
34. arrester [ə'restə] n. 避雷器
35. bushing ['buʃiŋ] n. [机]轴衬，[电工]

套管

36. circuit breaker 断路器

37. conductor [kən'dʌktə] n. 导体；导线

38. power supply 电源，电力供应，供电

39. surge [sə:dʒ] n. 电流急冲，电涌；突然增大

*** * * * * * * * * * * * * * * * * Complicated Sentences * * * * * * * * * * * * * * * * ***

1. Hydraulic turbine is a machine whose runner is powered by water; it transmits mechanical energy to the rotor of a generator to make it turn.

【译文】水轮机是一种由水为其叶轮提供动力的机械，它把机械能传递到发电机的转子，使其旋转。

【说明】句中 whose 引导的定语从句修饰前面的 machine。power 在句中用作动词，表示 provide power(动力) for，即“为……提供动力”。

2. The type of hydropower turbine selected for a project is based on the height of standing water—referred to as "head"—and the flow, or volume of water, at the site.

【译文】工程中选用的水轮机类型取决于当地蓄水的高度（称为“水头”）和流动情况或者水量。

【说明】based on 意思是“基于，以……为基础”；referred to as 表示“作为……被说明，称为”；at the site 表示“在现场，当地”。

3. A Pelton turbine has one or more free jets discharging water into an aerated space and impinging on the buckets of a runner.

【译文】Pelton 轮机具有一个或多个自由喷嘴，用于向通气的空间内注水并冲击叶轮的斗状叶片。

【说明】discharging 和 impinging 两个分词短语做定语修饰前面的 jets。

4. A cross-flow turbine is drum-shaped and uses an elongated, rectangular-section nozzle directed against curved vanes on a cylindrically shaped runner.

【译文】双击式水轮机是鼓状的，用一根伸长的矩形截面喷嘴直接对着柱状成型叶轮上的弯曲叶片。

【说明】词句为主+系表+动宾结构，宾语中又含有过去分词(directed)短语做定语，修饰 nozzle。

5. Water is introduced just above the runner and all around it and then falls through, causing it to spin.

【译文】水从叶轮的正上方引入，进入其周围，继而下落穿过，使其旋转。

【说明】此句描写了水流的 3 个连续动作及与叶轮之间的互动关系，简明扼要。

6. Generator unit is the device set with a turbine that transmits the water's mechanical energy to the generator's rotor to make it turn to produce electricity.

【译文】发电机组是成套设备，带有一个轮机，把水的机械能传递给发电机的转子，使其旋转产生电能。

【说明】句子的复杂之处在于，device set 的定语 with a turbine 后面又带了一个由 that 引导

的定语从句，而定语从句中又是三个动词不定式连用。翻译时需要注意意思准确和表达通顺。

* * * * * * * * * * * * * * * * * * * Summary of Glossary * * * * * * * * * * * * * * * * * * *

1. equipment	设备，装备（不可数名词）
2. device	装置，设备（可数名词）
3. hydraulic turbine	水轮机
4. hydro turbine	水轮机
5. hydropower turbine	水轮机
6. runner	叶轮，转子
7. rotor	（发电机）转子
8. stator	（发电机）定子
9. blade	（轮机）叶片
10. vane	（风车、螺旋桨等的）翼，叶片
11. generator	发电机
12. transformer	变压器
13. conductor	导体
14. circuit breaker	断路器
15. arrester	避雷器

* Exercises *

(1) A Pelton turbine has one or more ________ discharging water into an aerated space and impinging ________.

(2) The Francis turbine is also ________ a ________, as it looks like a spinning wheel with fixed blades ________ two rims.

(3) Circuit breaker is a useful ________ automatically cutting off the ________ in the event of overload ________ the devices nearby can be ________ from too much current.

* * * * * * * * * * * * * Word Building (21) -ance; over- * * * * * * * * * * * * * *

1. -ance [后缀]，表示：名词

allow	vt.	allowance	n.	津贴，允许
enter	n./v.	entrance	n.	入口，进入
assist	v.	assistance	n.	协助
acquaint	vt.	acquaintance	n.	熟人
perform	v.	performance	n.	表演
import	n./v.	importance	n.	重要性
endure	v.	endurance	n.	强度，抗磨度，耐疲劳度

2. over- [前缀]，表示：额外的……，越过……

load	n./v.	overload	n.	超载
brim	n./v.	overbrim	v.	溢出，满出
build	n./v.	overbuild	vt.	建造过多
burden	n./v.	overburden	vt.	装载过多
busy	adj.	overbusy	adj.	太忙的
buy	n./v.	overbuy	vt.	买得过多

*** * * * * * * * * * * * * * * * * Text Translation * * * * * * * * * * * * * * * * ***

第二十一章　水电站主要设备

第一节　不同的水轮机

水轮机是一种由水为其叶轮提供动力的机械。它把机械能传递到发电机的转子，使其旋转。（单词个数：25）

水轮机主要有两种类型：冲击式和反击式。工程中选用的水轮机类型，取决于当地存水的高度（称为“水头”）和水流情况或者水量。其他有影响的因素包括水轮机的放置深度、效率以及费用大小。（单词个数：57）

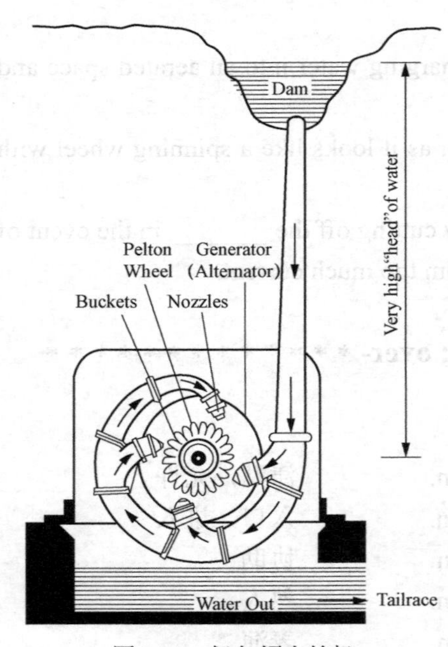

图 21.1　佩尔顿水轮机

1. 冲击式水轮机

冲击式水轮机通常是利用水的速度带动叶轮，然后在大气压力下排出水。水流冲击叶轮的每扇斗状叶片。这种水轮机的底部没有抽水泵，水在冲击叶轮后从水轮机的托架底部流出。冲击式水轮机一般适合于高水头、流量小的场合。（单词个数：66）

（1）佩尔顿水轮机。佩尔顿水轮机具有一个或多个自由喷嘴，用于向通气的空间内注水并冲击叶轮的斗状叶片，如图 21.1 所示。这种冲击式水轮机不需要尾水管，因为是在大气压力下运行，叶轮必须安装在最大尾水高度的上方。（单词个数：52）

佩尔顿轮机，过去是用在水磨坊里。佩尔顿轮机在其边缘周围有小型的水斗。来自大坝的水流经喷口以很高的速度冲击水斗，推动齿轮转动。（单词个数：42）

（2）双击式水轮机。双击式水轮机是鼓状的，用一根伸长的矩形截面喷嘴直接对着柱状成型叶轮上的弯曲叶片。它像一个鼠笼型的吹风机。双击式水轮机让水两次流过叶片。第一次是水从叶片外面流向里面；第二次是从里面流出。在水轮机的入口处有一个导流叶片，引导水流流向叶轮上的指定区域。与水斗式相比，双击式水轮机发展得更适合于大流量和低水头的场合。（单词个数：98）

2. 反击式水轮机

反击式水轮机综合利用水压和水速进行发电。叶轮直接放置在经过叶片的水流中，而不是单独冲击每一个叶片。与冲击式水轮机相比，反击式水轮机一般用于低水头大流量的位置。（单词个数：51）

（1）螺旋桨式水轮机。螺旋桨式水轮机的叶轮一般有3～6扇叶片，水流持续不断地接触它们。仿佛是一根管道里小船的螺旋桨。在这个管道里，压力是恒定的；否则的话，叶轮就会失去平衡。叶片的斜度是固定的或可调的。除了叶轮外的主要部件有蜗壳、导叶和抽水泵。螺旋桨式水轮机包括几种不同的类型：（单词个数：81）

卡普兰涡轮机如图21.2所示，像轮船的螺旋桨。在卡普兰涡轮机内部，叶片的斜度可以根据水速进行调整。事实上，它的叶片和导叶都是可调的，因此有很大的运行范围。（单词个数：46）

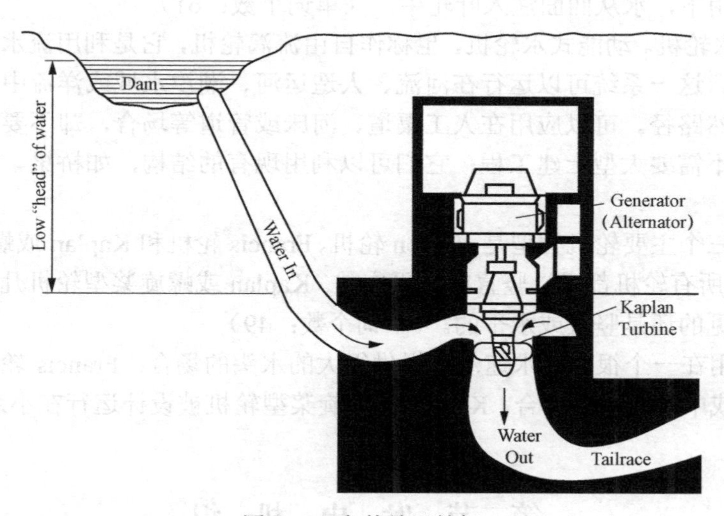

图21.2　卡普兰涡轮机

灯泡式水轮机和发电机可以做成直接安置在水流里的密封单元里。贯流式水轮机是一种可以将发电机直接连在其周围的水轮机。混流式水轮机的压力水管是在叶轮的前面或后面弯曲的，这样可以实现发电机与水轮机的直线连接。（单词个数：52）

（2）Francis 水轮机。如图21.3所示，Francis 水轮机的叶轮一般有9个或更多的固定叶片。水从叶轮的正上方引入，进入其周围，继而下落穿过，使其旋转。除了叶轮外的主要部件有蜗壳、导叶和抽水泵。（单词个数：49）

Francis 轮机也类似水车，因为它看上去像个在两个轮缘之间具有固定叶片的转轮。这样的转轮就叫做叶轮。在叶轮的周围有一圈导流叶片，用来控制驱动叶轮的水量。在使叶轮旋

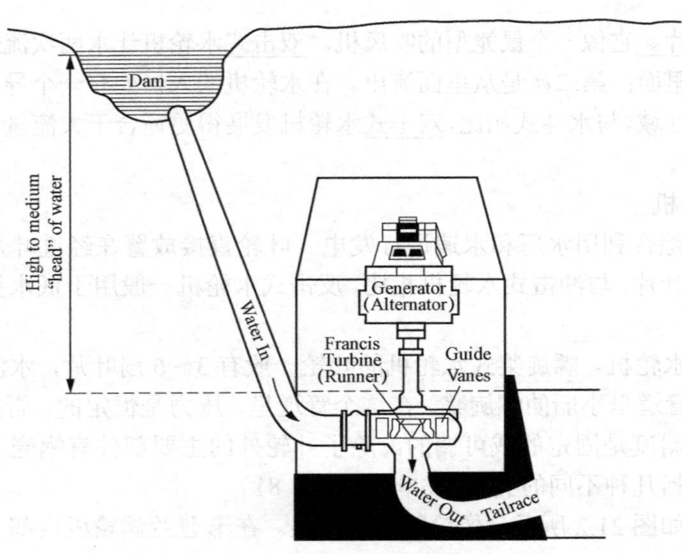

图 21.3 Francis 水轮机

转的导流叶片作用下，水从四面注入叶轮中。（单词个数：61）

（3）动能式水轮机。动能式水轮机，也称作自由流涡轮机，它是利用流水的动能而非上游水的势能来发电。这一系统可以运行在河流、人造运河、潮汐水域或洋流中。动能式系统利用的是水源的自然路径。可以应用在人工渠道、河床或管道等场合，却不要求通过它们来分流。动能式系统不需要大型土建工程；它们可以利用现有的结构，如桥梁、泄水道或水渠。（单词个数：88）

用得较多的三个主要轮机类型是：Pelton 轮机、Francis 轮机和 Kaplan 或螺旋桨型轮机（以其发明者命名）。所有轮机都可以竖直或水平安放。Kaplan 或螺旋桨型轮机几乎可以按任意角度安放，但最常见的还是竖直或水平的。（单词个数：49）

Pelton 轮机用在一个很小的水速却能提供很大的水头的场合；Francis 轮机用在很大的水速，且水头很高或中等高度的场合。Kaplan 或螺旋桨型轮机被设计运行在小水头的场合。（单词个数：53）

第二节 发 电 机 组

发电机组是成套设备，带有一个轮机，把水的机械能传递给发电机的转子，使其旋转产生电能。图 21.4 所示为一套水力发电机组，下文描述了它的组件。（单词个数：44）

1. 水轮机

水轮机帽是覆盖在水轮机叶轮上部的结构。（单词个数：14）

叶轮是水轮机的转动部分，它将水的运动转化到能转动转子的轴上。（单词个数：27）

叶轮叶片是叶轮上的固定的曲面板，它接受水的冲击从而转动叶轮。（单词个数：22）

2. 发电机

发电机是由转子和定子组成的机器，它能产生电能。（单词个数：17）

定子是铜导体线圈组成的发电机的固定部分，铜导体吸收了转子产生的电能。（单词个数：25）

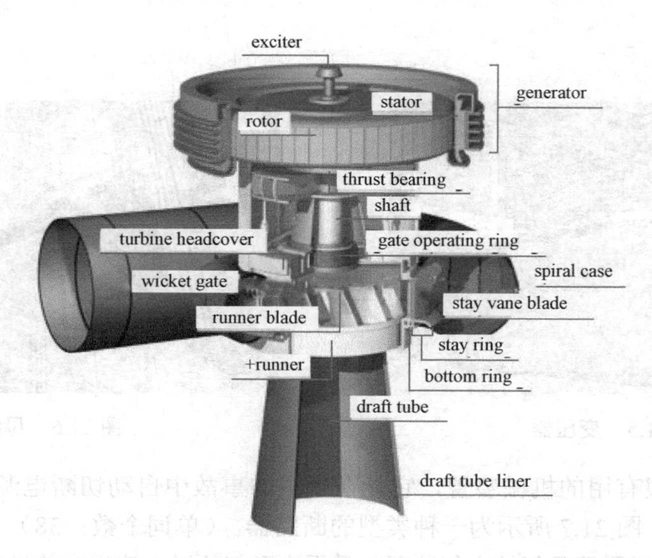

图 21.4 水力发电机组

转子是由电磁铁组成的发电机的转动部分；它的转动在定子中感应出电流。（单词个数：23）

励磁机是向转子电磁铁提供电流的装置。（单词个数：12）

3. 控制部分

蜗壳是一个形如螺旋楼梯的管子，用来使水轮机周围的水均匀分布，以实现其平稳旋转。（单词个数：25）

座环是由固定导叶连接在一起的两个环。（单词个数：14）

固定导叶是承接从蜗壳过来的压水并将其引到导叶的固定板。（单词个数：22）

活动导叶是调节进入水轮机的水流量使其叶轮匀速转动的活动板。（单词个数：23）

导叶操作环是控制活动导叶开启和关闭的活动装置。（单词个数：17）

底环是位于导叶下方使其处于正确位置的圆形部件。（单词个数：15）

4. 驱动及支持部分

轴是连接水轮机叶轮和发电机转子的圆柱形部分。（单词个数：16）

推力轴承是承受水轮机的推力和发电机组转动部分重力的元件。（单词个数：23）

5. 其他部分

尾水管是水轮机底部的管道，通过减小叶轮排出水的压力，提高了叶轮的输出能力。（单词个数：26）

尾水管衬里通常是由钢制造的遮盖物，保护尾水管，使其免受腐蚀。（单词个数：19）

第三节 其 他 设 备

变压器是用来改变电压等级的设备；电流离开发电厂之后要增加其电压使其能够被远距离传输。图 21.5 所示为一台变压器。（单词个数：35）

母线是将电流从发电机组传输到变压器的大型铝制或铜质导体，如图 21.6 所示。（单词个

数：23）

图 21.5 变压器

图 21.6 母线

断路器是一种很有用的机械装置，它能在超载的事故中自动切断电源，以保护周围的设备免受过电流危害。图 21.7 所示为一种类型的断路器。（单词个数：38）

如图 21.8 所示，避雷器是保护电气装置免受雷击引起的过电流损害的设备。（单词个数：20）

图 21.7 断路器

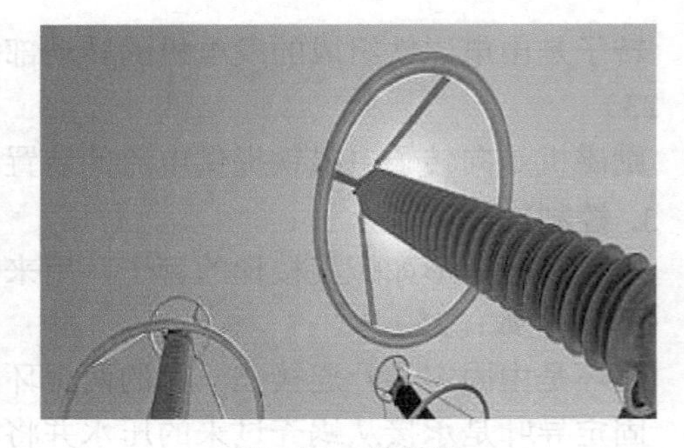

图 21.8 避雷器

如图 21.9 所示，套管是允许导体穿过墙体并与之隔离的设备。（单词个数：23）

图 21.9 套管

Chapter 22 Hydroelectricity and Its Characteristics

Part 1 General Introduction of Hydroelectricity

Hydroelectricity is electricity generated by hydropower, i.e., the production of power through use of the gravitational force of falling or flowing water.[1]

When you look at rushing waterfalls and rivers, you may not immediately think of electricity. But hydroelectric (water-powered) power plants are responsible for lighting many of our homes and neighborhoods.

Most hydroelectric power comes from the potential energy of dammed water. In this case the energy extracted from the water depends on the volume and on the difference in height between the source and the water's outflow. [2] This height difference is called the head. The amount of potential energy in water is proportional to the head. To obtain very high head, water for using may be run through a large pipe called a penstock.

Less common types of hydro schemes use water's kinetic energy or undammed sources. The truth is that any steady current of flowing water from a river or other waterway can be converted to power.

The movement of water as it flows downstream creates kinetic energy. A hydroelectric power plant converts this energy into electricity by forcing water, often held at a dam, through a hydraulic turbine that is connected to a generator. [3] The water exits the turbine and is returned to a stream or riverbed below the dam.

Generally, hydropower is mostly dependent upon precipitation and elevation changes; high precipitation levels and large elevation changes are necessary to generate significant quantities of electricity.

Hydropower is considered a renewable energy resource because it uses the Earth's water cycle to generate electricity. Water evaporates from the Earth's surface, forms clouds, precipitates back to earth, and flows toward the ocean. It is the most widely used form of renewable energy. Once a hydroelectric complex is constructed, the project produces no direct waste, and has a considerably lower output level of the greenhouse gas carbon dioxide (CO_2) than fossil fuel powered energy plants. [4]

Part 2 History of Hydropower

Humans have been harnessing water to perform work for thousands of years. The Greeks used waterwheels for grinding wheat into flour more than 2000 years ago. Besides grinding flour, the power of the water was used to saw wood and power textile mills and manufacturing plants.

Fig.22.1 The Dam Across Fox River in Appleton, Wisconsin, the Site of the First Hydroelectric Power Plant in the World

For more than a century, the technology for using falling water to create hydroelectricity has existed. The evolution of the modern hydropower turbine began in the mid-1700s when a French hydraulic and military engineer, Bernard Forest de Bélidor wrote Architecture Hydraulique. [5] In this four-volume work, he described using a vertical-axis versus a horizontal-axis machine.

During the 1700s and 1800s, water turbine development continued. In 1880, a brush arc light dynamo driven by a water turbine was used to provide theatre and storefront lighting in Grand Rapids, Michigan; and in 1881, a brush dynamo connected to a turbine in a flour mill provided street lighting at Niagara Falls, New York. These two projects used direct-current (DC) technology.

Alternating current (AC) is used today. That breakthrough came when the electric generator was coupled to the turbine, which resulted in the world's first hydroelectric plant located on the Fox River in Appleton, Wisconsin, in 1882, as shown in Fig.22.1.

Part 3 Advantages

The major advantage of hydroelectricity is elimination of the cost of fuel. The cost of operating a hydroelectric plant is nearly immune to increases in the cost of fossil fuels such as oil, natural gas or coal, and no imports are needed.[6]

Hydroelectric plants also tend to have longer economic lives than fuel-fired generation, with some plants now in service which were built 50 to 100 years ago. Operating labor cost is also usually low, as plants are automated and have few personnel on site during normal operation.

Where a dam serves multiple purposes, a hydroelectric plant may be added with relatively low construction cost, providing a useful revenue stream to offset the costs of dam operation.[7] It has been calculated that the sale of electricity from the Three Gorges Dam will cover the construction costs after 5 to 8 years of full generation.

Since hydroelectric dams do not burn fossil fuels, they do not directly produce carbon dioxide. While some carbon dioxide is produced during manufacture and construction of the project, this is a tiny fraction of the operating emissions of equivalent fossil-fuel electricity generation. According to some research projects, hydroelectricity produces the least amount of greenhouse gases and externality of any energy source. The extremely positive greenhouse gas impact of hydroelectricity is found especially in temperate climates.

Reservoirs created by hydroelectric schemes often provide facilities for water sports, and become tourist attractions in themselves. In some countries, aquaculture in reservoirs is common. Multi-use dams installed for irrigation support agriculture with a relatively constant water supply. Large hydro dams can control floods, which would otherwise affect people living downstream of the project.

Part 4 Disadvantages

Dam failures have been some of the largest man-made disasters in history. Also, good design and construction are not an adequate guarantee of safety. Dams are tempting industrial targets for wartime attack, sabotage and terrorism. Also, the creation of a dam in a geologically inappropriate location may cause disasters like the one of the Vajont Dam in Italy, where almost 2000 people died, in 1963. Smaller dams and micro hydro facilities create less risk, but can form continuing hazards even after they have been decommissioned. For example, the Kelly Barnes small hydroelectric dam failed in 1967, causing 39 deaths with the Toccoa Flood, ten years after its power plant was decommissioned in 1957.

Almost all rivers convey silt. Dams on those rivers will retain silt in their catchments, because by slowing the water, and reducing turbulence, the silt will fall to the bottom. [8] Siltation reduces a dam's water storage so that water from a wet season cannot be stored for use in a dry season. Often at or slightly after that point, the dam becomes uneconomic. Near the end of the siltation, the basins of dams fill to the top of the lowest spillway, and even storage from a storm to the end of dry weather will fail. Some especially poor dams can fail from siltation in as little as 20 years. Larger dams are not immune.

Hydroelectric projects can be disruptive to surrounding aquatic ecosystems both upstream and downstream of the plant site. Generation of hydroelectric power changes the downstream river environment. Water exiting a turbine usually contains very little suspended sediment, which can lead to scouring of river beds and loss of riverbanks. Since turbine gates are often opened intermittently, rapid or even daily fluctuations in river flow are observed.

Another disadvantage of hydroelectric dams is the need to relocate the people living where the reservoirs are planned. In February 2008, it was estimated that 40-80 million people worldwide had been physically displaced as a direct result of dam construction. Additionally, historically and culturally important sites can be flooded and lost.

Changes in the amount of river flow will correlate with the amount of energy produced by a dam. The result of diminished river flow can be power shortages in areas that depend heavily on hydroelectric power.

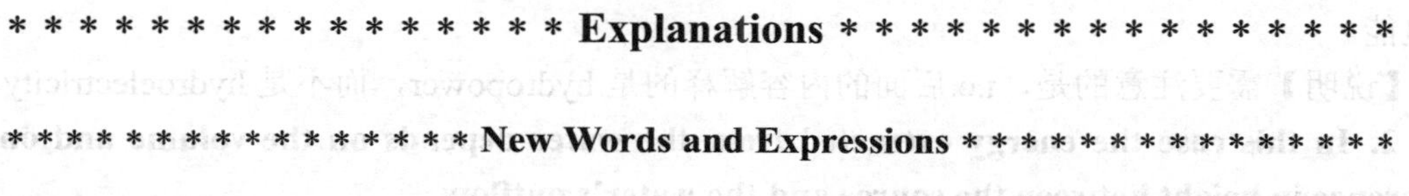

* * * * * * * * * * * * * * * * Explanations * * * * * * * * * * * * * * * *

* * * * * * * * * * * * * * * * New Words and Expressions * * * * * * * * * * * * * * * *

1. carbon dioxide 二氧化碳（CO_2）
2. fossil fuel 化石燃料
3. generator [ˈdʒenəreitə] n. 发电机
4. head [hed] n. 水头，落差

5. hydraulic [haiˈdrɔ: lik] adj.水力的，水压的
6. hydro [ˈhaidrəu] n. 水电厂
adj. 水电的，相当于 hydroelectric
7. hydroelectricity [ˌhaidrəuiˌlekˈtrisiti] n. 水电
8. hydropower [ˈhaidrəupauə] n. 水力发电；水力，水电
9. potential energy 势能
10. power plant 发电厂
11. renewable energy 可再生能源
12. turbine [ˈtə: bain] n. 涡轮，轮机，透平
13. complex [ˈkɔmpleks] n. 综合物；综合性建筑
14. penstock [ˈpenstɔk] n. 压力水管，水道，水渠
15. kinetic energy 动能
16. gravitational force 引力，重力，地心吸力
17. breakthrough [ˈbreikˌθru:] n. 突破
18. couple [ˈkʌpl] vt. 耦合，连接
19. dynamo [ˈdainəməu] n. 发电机
20. harness [ˈha: nis] vt. 利用（河流、瀑布等）产生动力（尤指电力）
21. versus [ˈvə: səs] prep. 与...相对
22. waterwheel [ˈwɔ: təwi: l] n. 水车，吊水机
23. elimination [iˌlimiˈneiʃən] n. 排除，除去
24. immune to 对……免疫，不受……影响
25. Three Gorges 三峡
26. revenue [ˈrevənju:] n. 收入，国家的收入，税收
27. decommission [ˌdi:kəˈmiʃən] vt. 使退役
28. diminished [diˈminiʃt] adv. 减少了的
29. disruptive [disˈrʌptiv] adj. 分裂（性）的；破坏性的
30. fluctuation [ˌflʌktjuˈeiʃən] n. 波动，起伏
31. inappropriate [ˌinəˈprəupriit] adj. 不适当的，不相称的
32. intermittently [intəˈmitəntli] adv. 间歇地
33. sabotage [ˈsæbəta: ʒ] n. 阴谋破坏
34. scour [ˈskauə] vt. 冲刷
35. siltation [sliteiʃən] n. 沉积作用，淤积
36. terrorism [ˈterərizəm] n. 恐怖主义，恐怖行动
37. turbulence [ˈtə: bjuləns] n. 动荡，（液体或气体的）紊乱
38. spillway [ˈspilwei] n. 溢洪道，泄洪道

******************* Complicated Sentences *******************

1. Hydroelectricity is electricity generated by hydropower, i.e., the production of power through use of the gravitational force of falling or flowing water.

【译文】水电是由水力发电（即通过利用落差或者流动的水的重力实现的电力生产）产生的电能。

【说明】需要注意的是，i.e.后面的内容解释的是 hydropower，而不是 hydroelectricity。

2. In this case the energy extracted from the water depends on the volume and on the difference in height between the source and the water's outflow.

【译文】在这种场合，从水中提取的电能取决于水量，也取决于源头与排出水之间的高差。

【说明】extracted from 意思是"从……提取"。两个 depends on 并列，其中第二个 depends

省略了，意思是“取决于”。

3. A hydroelectric power plant converts this energy into electricity by forcing water, often held at a dam, through a hydraulic turbine that is connected to a generator.

【译文】通过强制水（通常由大坝控制）流过与发电机相连的水轮机，水电厂将这种能量转换为电能。

【说明】句子的复杂性在于 by 所引导的状语本身比较复杂，其中 water 带有定语 often held at a dam，turbine 带有一个由 that 引导的定语从句。翻译时需要理清修饰关系，调整表达顺序。

4. Once a hydroelectric complex is constructed, the project produces no direct waste, and has a considerably lower output level of the greenhouse gas carbon dioxide (CO_2) than fossil fuel powered energy plants.

【译文】水电综合利用工程一旦建成，工程不产生直接废物，温室气体（CO_2）的输出水平明显低于以化石燃料为动力的电厂。

【说明】句中 powered 表示“由……驱动的，由……提供动力的”。

5. The evolution of the modern hydropower turbine began in the mid-1700s when a French hydraulic and military engineer, Bernard Forest de Bélidor wrote Architecture Hydraulique.

【译文】现代水轮机的发展始于 18 世纪中叶，当时法国的一位水力和军事工程师 Bernard Forest de Bélidor 写了《建筑水力学》。

【说明】句中 when 引导的状语从句修饰动词 begin，在翻译时可以灵活处理。

6. The cost of operating a hydroelectric plant is nearly immune to increases in the cost of fossil fuels such as oil, natural gas or coal, and no imports are needed.

【译文】运营水电站的成本几乎不受石油、天然气、煤等化石燃料成本增长的影响，而且不需要进口。

【说明】immune to 的意思是“对……免疫，不受……影响”。

7. Where a dam serves multiple purposes, a hydroelectric plant may be added with relatively low construction cost, providing a useful revenue stream to offset the costs of dam operation.

【译文】对于综合利用大坝来说，用相对较低的建设成本即可增建一个水电厂，提供有用的收入来源，可抵消大坝运行的成本。

【说明】分词短语 providing a useful… operation.做目的状语。

8. Dams on those rivers will retain silt in their catchments, because by slowing the water, and reducing turbulence, the silt will fall to the bottom.

【译文】那些河流上的大坝会在其集水处沉积泥沙，这是因为，减缓水流速度，减轻水的湍流，泥沙就会沉到水底。

【说明】句子的结构看上去比较奇怪，可整理如下：

Dams on those rivers will retain silt in their catchments by slowing the water, and reducing turbulence, because the silt will fall to the bottom.

******************** **Summary of Glossary** ********************

1. electricity	电，电能
2. hydroelectricity	水电
3. hydropower	水力发出的电力
4. power plant	发电厂
5. renewable energy	可再生能源
6. hydroelectric complex	水电枢纽，水电综合建筑
7. potential energy	势能
8. kinetic energy	动能
9. head	水头，落差
10. hydraulic turbine	水轮机
11. waterwheel	水车，吊水机
12. vertical-axis	垂直轴
13. horizontal-axis	水平轴
14. catchment	集水；集水处（水库或集水盆地），流域（面积）
15. siltation	沉积作用，淤积

******************** **Abbreviations (Abbr.)** ********************

1. DC	direct current	直流（电）
2. AC	alternating current	交流（电）

******************** **Exercises** ********************

(1) A hydroelectric power plant converts this energy into electricity by forcing water, often held at a ________, through a ________ that is connected to a________. The water exits the turbine and is returned to ______ below the dam.

(2) The major ___________of hydroelectricity is elimination of the cost of fuel. The cost of operating a hydroelectric plant is nearly immune to ___________ in the cost of _________ such as _________ , and no imports are needed.

(3) Dam failures have been some of the largest in history. Also, good design and construction are not an adequate. Dams are tempting industrial targets for _________, sabotage and terrorism.

************** **Word Building (22) pro-, under-** **************

1. pro- [词头]，表示：前，进

Proceed	vi.	向前
Progress	n.	进行

Prospect	n.	展望		
Promote	vt.	增进		

2. under- [前缀]，表示：欠，低，在……下面

ground	n./v.	underground	ad.	地下（的），秘密（的）
water	n./v.	underwater	ad.	在水下(的)，在水中(的)
utilize	v.	underutilized	adj.	未充分利用的

********************* **Text Translation** *********************

第二十二章 水电及其特性

第一节 水电概述

水电是由水力发电（即通过利用落差，或者流动的水的重力实现的电力生产）产生的电能。(单词个数：22)

当看到奔流的瀑布和江河时，你也许并不会立刻联想到电能，但是水力发电厂却承担着照亮我们千家万户的责任。(单词个数：30)

大多数水电的能量来自大坝蓄水的势能，这种情况下，从水中提取的电能取决于水量，也取决于源头与排出水位之间的高差。这个高差称为水头，水中势能大小与水头成正比。为了得到很高的水头，发电用水可能会流过一种称为压力管道的巨型管道。(单词个数：75)

一些特殊的水力发电方案利用的是非大坝蓄水的动能。事实上，任何产生于河流或其他水路的稳定流动的水流，都可以转化成电能。(单词个数：34)

水流向下游的运动产生了动能。通过强制水（通常由大坝控制）流过与发电机相连的水轮机，水电厂将这种能量转换为电能。水从水轮机流出后返回了河流或大坝后面的河床。(单词个数：55)

通常，由水力发出的电能的大小，很大程度上取决于降水量和水位高度的变化；丰富的降水和大的水位差，都是产生大量电力的必备条件。(单词个数：25)

水电可看作一种可再生能源，因为它利用了地球的水循环来产生电能。水从地球表面蒸发，形成了云，又变成降水返回地面，然后又流进了海洋。它是可再生能源最广泛使用的形式。水电综合利用工程一旦建成，不产生直接废物，温室气体 CO_2 的输出水平明显低于以化石燃料为动力的火电厂。(单词个数：77)

第二节 水力发电的历史

人类对水的利用已经有数千年的历史了。希腊人早在两千多年前，就利用水车将小麦研磨成面粉。除了磨面，水力还用于锯木、为纺织品厂、制造厂提供动力。(单词个数：46)

利用流动的水进行发电的技术在一个多世纪前就已经存在了。现代水轮机的发展始于18世纪中叶，当时法国的一位水力和军事工程师 Bernard Forest de Bélidor 写了《建筑水力学》一书。在该部4卷的著作中，他描写了如何应用水平轴及垂直轴机械。（单词个数：56）

图 22.1 坐落在威斯康星州阿普尔顿市福克斯河上的世界第一座水电站大坝

在1700～1800年间，水轮机得到了不断发展。1880年，用一个由水轮机驱动的弧光刷发电机，给密歇根州大瀑布城的剧院和商铺提供照明用电；1881年，一台与面粉厂的水轮机相连的有刷发电机提供了纽约州尼亚加拉瀑布地区的街道照明用电。这两项工程使用的都是直流电技术。（单词个数：62）

我们现在使用的是交流电技术。这一突破产生于将发电机直接耦合在水轮机上，最早实现这一技术的，是1882年在威斯康星州阿普尔顿市的福克斯河上建成的世界上第一座水电站（见图22.1）。（单词个数：36）

第三节 水电的优点

水电的主要优势在于省去了燃料的费用。运营水电站的成本几乎不受石油、天然气、煤等化石燃料成本增长的影响，而且不需要进口。（单词个数：42）

与使用燃料的发电相比，水电厂也往往有更长的经济寿命，有些50～100年前建成的水电厂至今仍在运行。运行劳动成本也通常较低，这是由于电厂是自动运行的，正常运行期间只有少数人员在现场。（单词个数：46）

对于综合利用枢纽来说，用相对较低的建设成本即可增建一个水电厂，从而提供有用的发电收入，抵消大坝运行的成本。据计算，三峡大坝满额发电运行5～8年的发电价值就可以抵偿其建设费用。（单词个数：57）

因为水电大坝不燃烧化石燃料，所以它们不直接产生二氧化碳。大坝和工程建设过程中，会产生一些二氧化碳，这只相当于产生等额发电量的化石燃料排放量的极小部分。一些研究项目显示，在产生等效经济效益的条件下，水电产生的温室气体量在所有能源中是最少的。水电的这一特点在温带气候中体现得尤为突出。（单词个数：75）

水电建设规划中的水库为水上运动的开展提供了条件，而它们本身也成了旅游景点。在一些国家，水库周围的水产养殖也很普遍。为灌溉而建造的综合利用水坝，通过提高相对稳定的供水支持了农业生产。大型水坝还可以控制那些可能影响下游人们生活的洪水。（单词个数：54）

第四节 水电的缺点

一些大坝事故属于历史上最严重的人为事故。而且，良好的设计和建造并不能保证足够的安全，在军事战争、阴谋破坏和恐怖主义活动中，大坝都是引人注意的工业目标。而且，在地质条件不合适的地方建造大坝，也可能引起灾难，比如1963年意大利的维昂特大坝事故导致几乎2000人死亡。小型水坝和微型水电设施带来的风险小一些，但却可能形成持续性的危害，甚至于在它们报废之后都还有危害。例如，凯利巴恩斯的小型水坝事故导致1967年托科阿洪水中39人死亡，而此时距其1957年报废已经过去了10年时间。（单词个数：113）

几乎所有的河流都会携运泥沙。那些河流上的大坝会在其集水处沉积泥沙，这是因为水流到坝前速度减缓，减轻了水的湍流，泥沙就会沉到水底。泥沙淤积减少了大坝的蓄水量，因此，雨季存储的水量就不够旱季的使用。往往是从这时起或稍后一段时间，大坝就变得不经济了。在淤积的后期，大坝的底部会填到最低泄洪道的顶部，导致从暴风雨时节到旱季末期蓄水都变得不可能。一些特别不幸的大坝，在泥沙开始淤积的20年后就报废了，大一点儿的水坝也不能幸免。（单词个数：115）

水电项目对其所在位置的上下游周围水域中的生态系统有破坏性影响，水力发电改变了下游河流的环境。离开轮机的水通常含有很少的悬浮沉淀物，这会导致河床的冲刷和河岸的损失。水轮机的阀门经常间歇性的开启，导致了河流中的快速变化，甚至是每天都出现的起伏波动。（单词个数：65）

水力大坝的另一个缺点，是需要搬迁走在建水库周边生活的人们。据估计，截止到2008年2月，在世界范围内已经有4000万到8000万人因水坝建设而搬迁移居。另外，历史和人文的重要景观也可能因此而被淹没或消失。（单词个数：51）

河水流量的变化会影响到大坝的发电量，河水流量的减少，会造成对水电严重依赖的地区电力不足。（单词个数：36）

Chapter 23 Pumped-Storage Plants

Part 1 General Introduction

Pumped storage hydroelectricity is a type of hydroelectric power generation used by some power plants for load balancing. In other words, pumped storage reservoirs are not really a means of generating electrical power.[1] They are a way of storing energy so that we can release it quickly when we need it.

Demand for electrical power changes throughout the day. Sometimes a sudden peak in demand will appear. If power stations don't generate more power immediately, there will be power cut. The problem is that most of our power is generated by fossil fuel power stations, which take half an hour or so to crank themselves up to full power. Nuclear power stations take much longer. We need something that can go from nothing to full power immediately, and keep us supplied for around half an hour or so until the other power stations catch up. [2] Pumped storage reservoirs are the answer we have chosen.

Pumped storage is the largest-capacity form of grid energy storage now available. The method stores energy in the form of water, pumped from a lower elevation reservoir to a higher elevation. Low-cost off-peak electric power is used to run the pumps.

At times of low electrical demand, excess generation capacity is used to pump water into the higher reservoir. When there is higher demand, water is released back into the lower reservoir through a turbine, generating electricity. Reversible turbine/generator assemblies act as pump and turbine (usually a Francis turbine design). Some facilities use abandoned mines as the lower reservoir, but many use the height difference between two natural bodies of water or artificial reservoirs.

Plants that do not use pumped-storage are referred to as conventional hydroelectric plants; conventional hydroelectric plants that have significant storage capacity may be able to play a similar role in the electrical grid as pumped storage, by deferring output until needed.

Pure pumped-storage plants just shift the water between reservoirs, but combined pump-storage plants also generate their own electricity like conventional hydroelectric plants through natural stream flow.

Part 2 Performance

Taking into account evaporation losses from the exposed water surface and conversion losses, approximately 70% to 85% of the electrical energy used to pump the water into the elevated reservoir can be regained.[3] The technique is currently the most cost-effective means of storing large amounts of electrical energy on an operating basis, but capital costs and the presence of appropriate geography are critical decision factors.

Although the losses of the pumping process makes the plant a net consumer of energy overall, the system increases revenue by selling more electricity during periods of peak demand, when electricity prices are highest.[4]

The relatively low energy density of pumped storage systems requires either a very large body of water or a large variation in height. For example, 1000 kilograms of water (1 cubic meter) at the top of a 100 meter tower has a potential energy of about 0.272 kWh. The only way to store a significant amount of energy is by having a large body of water located on a hill relatively near, but as high as possible above, a second body of water.[5] In some places this occurs naturally, in others one or both bodies of water have been man-made.

This system may be economical because it flattens out load variations on the power grid, permitting thermal power stations such as coal-fired plants and nuclear power plants and renewable energy power plants that provide base-load electricity to continue operating at peak efficiency (base load power plants), while reducing the need for "peaking" power plants that use costly fuels.[6] However, capital costs for purpose-built hydrostorage are high.

Along with energy management, pumped storage systems help control electrical network frequency and provide reserve generation. Pumped storage plants, like other hydroelectric plants, can respond to load changes within seconds.

A new use for pumped storage is to level the fluctuating output of intermittent power sources. The pumped storage absorbs load at times of high output and low demand, while providing additional peak capacity. In certain conditions, electricity prices may be close to zero or occasionally negative, indicating there is more generation than load available to absorb it; although at present this is rarely due to wind alone, increased wind generation may increase the likelihood of such occurrences.[7] It is particularly likely that pumped storage will become especially important as a balance for very large scale photovoltaic generation.

Part 3 History and Development

The first use of pumped storage was in the 1890s in Italy and Switzerland. In the 1930s reversible hydroelectric turbines became available. These turbines could operate as both turbine-generators and in reverse as electric motor driven pumps. The latest in large-scale engineering technology are variable speed machines for greater efficiency. These machines generate in synchronism with the network frequency, but operate asynchronously (independent of the network frequency) as motor-pumps.

Between 1976 and 1982 at Dinorwig, in North Wales, a huge project was built, as shown in Fig.23.1. Dinorwig has the fastest "response time" of any pumped storage plant in the world - it can provide 1320 Megawatts in 12 seconds. There was a big height difference between two existing lakes, so less work was needed to build the station. When there's a sudden demand for power, the "headgates"

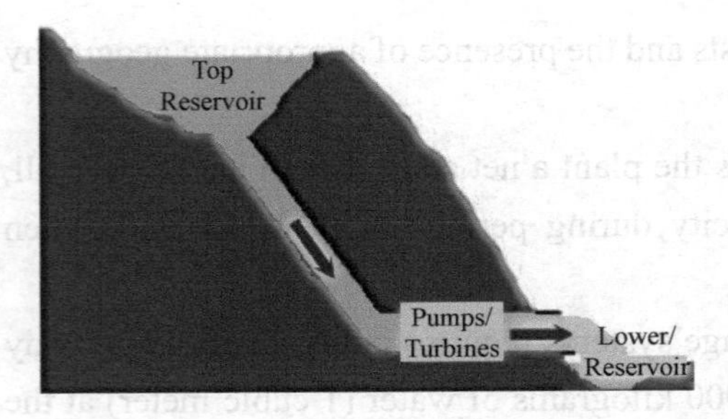

Fig.23.1 Pumped storage plant at Dinorwig

(huge taps) are opened, and water rushes down the tunnels and falls 600 metres on its way to the turbines, which drive the powerful generators. The water then collects in the bottom reservoir, ready to be pumped back up later.

In 2000 the United States had 19.5 GW of pumped storage generating capacity, accounting for 2.5% of baseload generating capacity. Pumped hydroelectricity system generated (net) 5500 GWh of energy because more energy is consumed in pumping than is generated; losses occur due to water evaporation, electric turbine/pump efficiency, and friction.

In 2007 the EU had 38.3 GW net capacity of pumped storage out of a total of 140 GW of hydropower and representing 5% of total net electrical capacity in the EU (Eurostat, consulted August 2009). [8]

The use of underground reservoirs as lower dams has been investigated. Salt mines could be used, although ongoing and unwanted dissolution of salt could be a problem. If they prove affordable, underground systems might greatly expand the number of pumped storage sites. Saturated brine is about 20% denser than fresh water.

* * * * * * * * * * * * * * * * * * * Explanations * * * * * * * * * * * * * * * * * *

* * * * * * * * * * * * * * * * * * New Words and Expressions * * * * * * * * * * * * * * *

1. assembly [ə'sembli] n. 集合，装配
2. crank…up 加快，做好准备
3. grid [grid] n. 电网
4. load balancing 负载平衡，负荷平衡
5. mine [main] n. 矿，矿山，矿井
6. nuclear power 核动力，核电
7. off-peak 非高峰（期）的
8. reversible [ri'vəːsəbl] adj. 可逆的
9. cost-effective ['kɔsti'fektiv] adj. 有成本效益的，划算的
10. cubic ['kjuːbik] adj. 立方体的，立方的
11. decision factor 决定因素
12. flatten ['flætn] vt. 使变平
13. fluctuating ['flʌktjueitiŋ] adj. 变动的，上下摇动的
14. frequency ['friːkwənsi] n. 频率
15. hydrostorage ['haidrəu'stɔːridʒ] n. 水存储
16. net [net] adj. 净余的，净的
17. photovoltaic [ˌfəutəuvɔl'teiik] adj. 光电的，光伏发电的
18. purpose-built ['pəːpəsbilt] adj. 为特定目的建造的
19. reserve [ri'zəːv] n. 储备，保留，备用
20. affordable [ə'fɔːdəbl] adj. 供应得起
21. asynchronous [ei'siskrənəs] adj. 异步的
22. brine [brain] n. 盐水
23. dissolution [ˌdisə'luː ʃən] n. 分解
24. friction ['frikʃən] n. 摩擦，摩擦力
25. headgate [hedgeit] n. 水头阀门，闸门

26. ongoing [ˈɔnˌgəuiŋ] adj. 正在进行的
27. response [riˈspɔns] n. 响应，反应
28. saturated [ˈsætʃəreitid] adj. 饱和的
29. synchronism [ˈsiŋkrənizəm] n. 同步性

* * * * * * * * * * * * * * * * * * **Complicated Sentences** * * * * * * * * * * * * * * * * * *

1. In other words, pumped storage reservoirs are not really a means of generating electrical power.

【译文】换句话说，抽水蓄能水库其实不是产生电力的一种途径。

【说明】句中 means 意思是“途径，手段，方法”，虽然看上去像是复数形式，实际上是可数名词的单数形式。in other words 表示“换句话说”。

2. We need something that can go from nothing to full power immediately, and keep us supplied for around half an hour or so until the other power stations catch up.

【译文】我们需要某些能够立即从零到满功率的方式，使我们在 0.5 小时左右的时间里能够保持供电，直到其他电站赶上来。

【说明】something 意思是“某种事物，某些事物”，nothing 表示“没有，数值为零”。catch up 的意思是“追赶，跟上”。

3. Taking into account evaporation losses from the exposed water surface and conversion losses, approximately 70% to 85% of the electrical energy used to pump the water into the elevated reservoir can be regained.

【译文】考虑到暴露水面的蒸发损失和能量转换损失，用于把水抽到高库的电能中大约能够重新获得 75%~80%的电能。

【说明】句中 taking into account 意思是“考虑到，把……纳入考虑”。

4. Although the losses of the pumping process make the plant a net consumer of energy overall, the system increases revenue by selling more electricity during periods of peak demand, when electricity prices are highest.

【译文】虽然抽水过程的损耗使其在整体上成为一个净耗能者，但该系统通过在电价最高的用电高峰期销售电量，可以增加收入。

【说明】句中 net 表示“净，正反抵消后剩余的”，overall 表示“总体上，大体上”。when 引导的定语从句修饰 periods of peak demand。

5. The only way to store a significant amount of energy is by having a large body of water located on a hill relatively near, but as high as possible above, a second body of water.

【译文】存储巨额能量的唯一途径，是拥有距离另一水体相对较近但位置又尽可能比它高的庞大水体。

【说明】as ... as possible 表示“尽可能地”。

6. This system may be economical because it flattens out load variations on the power grid, permitting thermal power stations such as coal-fired plants and nuclear power plants and renewable energy power plants that provide base-load electricity to continue operating at peak efficiency (base load power plants), while reducing the need for "peaking" power plants that use costly fuels.

【译文】这种系统可能是经济合算的，因为它使电网的负荷变化变平，允许提供基本负荷的燃煤电厂等热电站、核电厂、可再生能源电厂（统称基荷电厂）持续运行在最高效率，而降低对采用昂贵燃料的峰荷电厂的需求。

【说明】句子的主体很短，就是 This system may be economical。后面的内容整体是 because 引导的从句。该从句又是由 while 连接的两个部分，每一部分的结构也都不简单，因而翻译起来比较困难。需要在正确理解的基础上，采用合理的顺序进行表达。

7. In certain conditions, electricity prices may be close to zero or occasionally negative, indicating there is more generation than load available to absorb it; although at present this is rarely due to wind alone, increased wind generation may increase the likelihood of such occurrences.

【译文】在特定情况下，电价可能接近为零或者偶尔为负的，表明发电量大于能够用于吸收它的负荷。虽然目前这很少单独源于风电，但增长的风力发电会增大这种事件发生的可能性。

【说明】close to 意思为“接近”，at present 意思是“现在，目前”，due to 意思是“由于”。

8. In 2007 the EU had 38.3 GW net capacity of pumped storage out of a total of 140 GW of hydropower and representing 5% of total net electrical capacity in the EU.

【译文】2007 年，欧盟的 140GW 水电总量中有 38.3GW 的抽水蓄能净容量，占欧盟净发电总容量的 5%。

【说明】out of a total 表示“在总量中”。

****************** **Summary of Glossary** ******************

1. power plant	电厂，发电厂
2. power station	电站，发电厂
3. stream flow	流量
4. capacity	容量，生产量
5. power grid	电网
6. peak load	峰荷，用电高峰
7. base load	基荷，基本负荷
8. net capacity	净容量
9. load balancing	负荷平衡
10. electricity prices	电价

****************** **Abbreviations (Abbr.)** ******************

1. kW・h	kilowatt-hour	千瓦・时
2. MW	megawatt	兆瓦
3. GW	gigawatt	千兆瓦
4. EU	European Union	欧盟

*** * * * * * * * * * * * * * * * * * * Exercises * * * * * * * * * * * * * * * * * * ***

(1) Taking into ________ evaporation losses from the exposed water surface and conversion losses, approximately 70% to 85% of the electrical energy used to pump the water into the elevated reservoir can be ________.

(2) In certain conditions, electricity prices may be close to zero or occasionally negative, indicating there is ________ generation than load available to absorb it; although at present this is rarely due to wind alone, ________ wind generation may increase the likelihood of such occurrences.

(3) Plants that do not use pumped-storage are referred to as ________ hydroelectric plants; conventional hydroelectric plants that have significant storage capacity may be able to play a ________ role in the electrical grid as pumped storage, by ________output until needed.

*** * * * * * * * * * * * * Word Building (23) less-;meg(a)- * * * * * * * * * * * ***

1. -less [形容词后缀]，表示：无……，没有……的，不……的

care	n./v.	careless	adj.	不关心的；不注意的；粗心的
colour	n./v.	colourless	adj.	无色的，乏味的
use	n./v.	useless	adj.	无用的，无价值的
end	n./v.	endless	adj.	无尽的
price	n./v.	priceless	adj.	无价的

2. meg(a)- [前缀]，表示：大，巨大；meglo-表示：巨大，扩大

megadebt	n.	巨额债务
megacity	n.	（人口超过 100 万的）大城市
megalopolis	n.	巨大都市，人口稠密地带
megastar	n.	超级巨星

*** * * * * * * * * * * * * * * * Text Translation * * * * * * * * * * * * * * * ***

第二十三章 抽水蓄能电站

第一节 概 述

抽水蓄能发电是被一些电站用来负荷调节的一种水力发电形式。换句话说，抽水蓄能水库其实不是产生电力的一种途径，而是一种储能方式，所以我们可以在需要的时候把它迅速泄放。（单词个数：51）

人们对电能的需求量在一天内是随时变化的。有时会出现用电高峰，如果电厂不能迅速保证发电量足够多，将会断电。问题是我们的电力大多由化石燃料电站生产，需要经过 0.5h 左右才能增加到满功率输出。核电站则需要时间更长。我们需要某些能够立即从零到满功率的方式，使我们在 0.5h 左右的时间里能够保持供电，直到其他电站赶上来。抽水蓄能电站正好能够解决这一问题。（单词个数：104）

抽水蓄能是目前能够实现的最大容量的电网储能形式。它把能量以水体存储起来，将低海拔水库的水抽到一个更高海拔的水库。用低成本非高峰期电力来运行水泵。（单词个数：40）

在电力负荷出现低谷时，用剩余电力把水抽到上水库。当电力负荷出现高峰时，将上水库的水泄放下来，通过水轮机发电。可逆式水轮机/发电机组件作为泵和涡轮机（通常是一混流式水轮机设计）。一些设施如废矿等，作为下游水库，但主要是利用上下库两种水体的自然或人工水库的高差。（单词个数：73）

不用抽水蓄能的电厂称为常规水电厂；具有相当数量存储容量的常规水电厂，通过推迟输出（除非需要），也能够在电网中扮演与抽水蓄能类似的功能。（单词个数：41）

单纯的抽水蓄能电站只是在水库之间转移水体，但混合式水电站像常规水电站一样通过天然流量发电。（单词个数：25）

第二节　特　　性

考虑到暴露水面的蒸发损失和能量转换损失，用于把水抽到高库的电能，大约能够重新获得 75%～80%的电能。该技术是目前运行大量存储电能的最划算的方式，但资金成本和地形是否合适，是严格的决定因素。（单词个数：65）

虽然抽水过程的损耗使其在整体上成为一个净耗能者，但该系统通过在电价最高的用电高峰期销售电量，可以增加收入。（单词个数：35）

抽水蓄能系统的能量密度相对较低，这就要求足够多的水量，或很大的高度差。例如，1000kg 的水（$1m^3$）在 100m 高的水塔处具有约 0.272kW 的势能。存储巨额能量的唯一途径是需要大量水体处于相对靠近另一水体的山上，或者尽可能地高于另一水体。一些地方有天然的这种条件，而在没有这种条件的地方，则要人工建造一个或两个水库。（单词个数：101）

这种系统可能是经济合算的，因为它使电网的负荷变化变得平坦，允许燃煤电厂、核电长、可再生能源电厂等提供基本负荷的热电站（基荷电厂）持续在最高效率区运行，从而降低对采用昂贵燃料的峰荷电厂的需求。然而，为这种特定目的而建造水库的成本，也是很高的。（单词个数：67）

配合能量管理，抽水蓄能系统有助于控制电网频率并提供发电备用。抽水蓄能电站能像其他水电站一样在数秒内响应负荷变化。（单词个数：30）

抽水蓄能的一种新的用途是减缓间歇性电源输出波动。抽水蓄能在高（发电）输出、低（用电）需求的时段吸收负荷，而提供额外的峰荷（发电）容量。在特定情况下，电价可能接近为零，或者偶尔为负的，表明发电量大于能够用于吸收它的负荷。虽然目前这很少单独源于风电，但增长的风力发电会增大这种偶然事件的可能性。特别地，抽水蓄能在大规模光伏发电平衡中越来越重要。（单词个数：98）

第三节 发 展 历 史

19 世纪 90 年代在意大利和瑞士首次使用抽水蓄能。20 世纪 30 年代，发明了可逆水轮机，这种轮机既可以作为轮机一发电机运行，也可以反过来作为电动机驱动的水泵。最新的大规模工程技术是能够提高效率的变速机。这些电机发电与电网频率同步，但作为电动机一水泵时异步运行（独立于电网频率）。（单词个数：69）

从 1976 年到 1982 年，一座巨大的工程在威尔士北部的迪诺威克建造（见图 23.1）。在世界上所有抽水蓄能电站中，迪诺威克有最快的“响应时间”——它能在 12s 中达到 1320MW。天然存在的两个湖有着很大的高度差，所以建造电站时不需做很多工作。当出现负荷高峰时，进水闸（巨大龙头）打开，水从水道落下 600m 而到达水轮机，再有水轮机带动发电机发电。然后水量存储在下游水库中以被抽到上游水库。（单词个数：105）

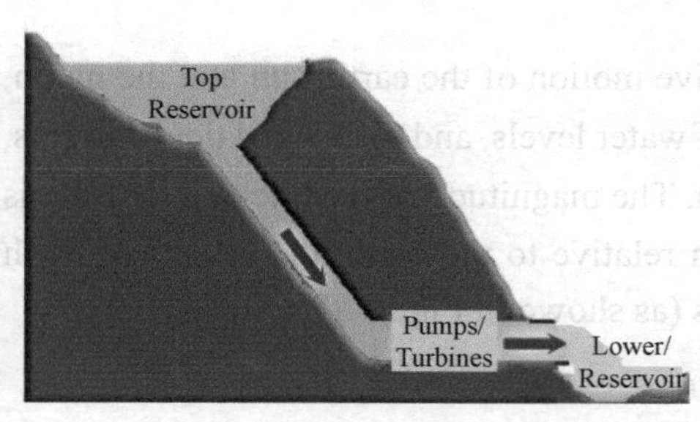

图 23.1 迪诺威克抽水蓄能电站

2000 年，美国有 19.5GW 的抽水蓄能发电容量，占基荷发电容量的 2.5%。抽水蓄能电站水力发电系统产生-5500 亿 W·h 的能量（净值）。因为更多的能量被抽水所损耗。产生损耗主要是因为水汽蒸发、电涡轮/泵的效率、摩擦。（单词个数：50）

2007 年，欧盟的 140GW 水电总量中，有 38.3GW 的抽水蓄能净容量，占欧盟净发电总容量的 5%。（欧盟统计局，2009 年 8 月）（单词个数：36）

人们对地下水库作为低水坝进行了研究，可以使用盐矿，虽然目前多余的盐溶解可能是个问题。如果能够证明这些都能负担得起，地下系统可能大大扩展抽水蓄能电站数量。饱和溶液要比淡水密度大 20%。（单词个数：51）

Chapter 24 Tidal Power Station

Part 1 General Introduction of Tidal Power

Tidal power, sometimes called tidal energy, is the only form of energy which derives directly from the relative motions of the earth–moon system, and to a lesser extent from the earth–sun system.[1] The tidal forces produced by the moon and sun, in combination with earth's rotation, are responsible for the generation of the tides.

In other words, tidal energy is generated by the relative motion of the earth, sun and the moon, which interact via gravitational forces. Periodic changes of water levels, and associated tidal currents, are due to the gravitational attraction by the sun and moon. The magnitude of the tide at a location is the result of the changing positions of the moon and sun relative to the earth, the effects of earth rotation, and the local shape of the sea floor and coastlines (as shown in Fig.24.1).

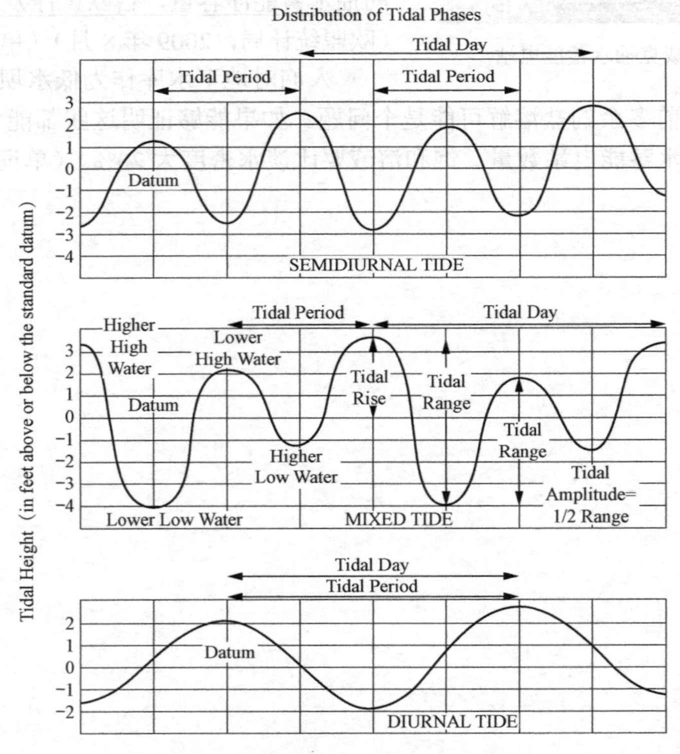

Fig.24.1 Variation of Tides Over a Day

Because the Earth's tides are caused by the tidal forces due to gravitational interaction with the moon and sun, and the earth's rotation, tidal power is practically inexhaustible and classified as a renewable energy source.

A tidal energy generator uses this phenomenon to generate energy. The stronger the tide, either in water level height or tidal current velocities, the greater the potential for tidal energy generation.[2]

Tidal movement causes a continual loss of mechanical energy in the earth–moon system due to pumping of water through the natural restrictions around coastlines, and due to viscous dissipation at the seabed and in turbulence.[3] This loss of energy has caused the rotation of the earth to slow in the 4.5 billion years since formation. During the last 620 million years the period of rotation has increased from 21.9 hours to the 24 hours we see now; in this period the earth has lost 17% of its rotational energy.[4] While tidal power may take additional energy from the system, increasing the rate of slowdown, the effect would be noticeable over millions of years only, thus being negligible.

Tidal power is a form of hydropower that converts the energy of tides into electricity or other useful forms of power.Although not yet widely used, tidal power has potential for future electricity generation. Tides are more predictable than wind energy and solar power. Historically, tide mills have been used, both in Europe and on the Atlantic coast of North America. The earliest occurrences date from the Middle Ages, or even from Roman times.

The incoming and outgoing tides of the sea can be harnessed to produce electrical power. Tides are often very powerful and the sea can move very quickly when the tide is coming into land. When the tide approaches land, the amount of water rushing forwards can be measured in terms of millions of gallons. This is an immense force of moving water. The largest tidal energy is the Bay of Fundy.

Part 2 Types of Tidal Power

Tidal power can be classified into three main types:

(1) **Tidal stream systems** make use of the kinetic energy of moving water to power turbines, in a similar way to windmills that use moving air. This method is gaining in popularity because of the lower cost and lower ecological impact compared to other methods.

Tidal stream generators draw energy from currents in much the same way as wind turbines. The higher density of water, 800 times the density of air, means that a single generator can provide significant power at low tidal flow velocities (compared with wind speed). Given that power varies with the density of medium and the cube of velocity, it is simple to see that water speeds of nearly one-tenth of the speed of wind provide the same power for the same size of turbine system.[5] However this limits the application in practice to places where the tide moves at speeds of at least 2 knots (1m/s) even close to neap tides.

Since tidal stream generators are an immature technology (no commercial scale production facilities are yet routinely supplying power), no standard technology has yet emerged as the clear winner, but a large variety of designs are being experimented with, some very close to large scale deployment. [6] Several prototypes have shown promise with many companies making bold claims, some of which are yet to be independently verified, but they have not operated commercially for extended periods to establish performances and rates of return on investments.[7]

(2) **Barrages** make use of the potential energy in the difference in height (or head) between high and low tides. Barrages are essentially dams across the full width of a tidal estuary, and suffer from very high civil infrastructure costs, a worldwide shortage of viable sites, and environmental issues.

Fig.24.2 Rance Tidal Power Plant

The barrage method of extracting tidal energy involves building a barrage across a bay or river as in the case of the Rance tidal power plant in France (as shown in Fig.24.2). Turbines installed in the barrage wall generate power as water flows in and out of the estuary basin, bay, or river. These systems are similar to a hydro dam that produces static head or pressure head (a height of water pressure). When the water level outside of the basin or lagoon changes relative to the water level inside, the turbines are able to produce power. The largest such installation has been working on the Rance River, France, since 1966 with an installed (peak) power of 240 MW, and an annual production of 600 GWh (about 68 MW average power).

The basic elements of a barrage are caissons, embankments, sluices, turbines, and ship locks. Sluices, turbines, and ship locks are housed in caissons (very large concrete blocks). Embankments seal a basin where it is not sealed by caissons.The sluice gates applicable to tidal power are the flap gate, vertical rising gate, radial gate, and rising sector.Barrage systems are affected by problems of high civil infrastructure costs associated with what is in effect a dam being placed across estuarine systems, and the environmental problems associated with changing a large ecosystem.

(3) **Tidal lagoons**, are similar to barrages, but can be constructed as self contained structures, not fully across an estuary, and are claimed to incur much lowercost and impact overall. Furthermore, they can be configured to generate continuously which is not the case with barrages.

Modern advances in turbine technology may eventually see large amounts of power generated from the ocean, especially tidal currents using the tidal stream designs. Tidal stream turbines may be arrayed in high-velocity areas where natural tidal current flows are concentrated such as the west and east coasts of Canada, the Strait of Gibraltar, the Bosporus, and numerous sites in Southeast Asia and Australia. Such flows occur almost anywhere where there are entrances to bays and rivers, or between land masses where water currents are concentrated.

Part 3 Off-Shore Turbines

Research is being carried out on the possibility of using off-shore turbines, driven by the power of the tides. The powerful tides could be used to drive underwater turbines. The turbines could rotate in either a clockwise or anticlockwise direction and as the tide is always moving, electricity production would be continuous. Off-shore turbines work together, rather like an underwater wind farm, as shown in Fig.24.3.

Fig.24.3 Offshore Turbines (http://www.marineturbines.com/)

This has the advantage of being much cheaper to build than a tidal barrage, and does not have the environmental problems that a tidal barrage would bring. As they are based well away from land they are out of sight and, unlike wind powered turbines, they do not create noise. [8]

The major difficulties with this type of system is that the off-shore turbines cost more money than land/wind-based turbines and they are also more expensive to maintain as they function under water. Furthermore, sea water is corrosive to steel and other metals because of the salt content.

The propellers would be maintained using large maintenance ships. When maintenance is required, the propellers would be extended above the water. The plan is for the turbines to be sited together similar to wind farms on land and the generated electricity to be cabled to the land.

* * * * * * * * * * * * * * * * * Explanations * * * * * * * * * * * * * * * *

* * * * * * * * * * * * * * * * * New Words and Expressions * * * * * * * * * * * * * * *

1. dissipation [ˌdisiˈpeiʃn] n. 消散，分散，挥霍，浪费
2. immense [iˈmens] adj. 极广大的，无边的
3. inexhaustible [ˌinigˈzɔ:stəbl] adj. 无穷无尽的
4. lesser [ˈlesə] adj. 较小的，更少的，次要的
5. periodic [ˌpiəriˈɔdik] adj. 周期的，定期的
6. relative motion 相对运动
7. turbulence [ˈtə: bjuləns] n. 湍流，紊流，（液体或气体的）紊乱
8. via [ˈvaiə] prep. 经，通过，经由
9. viscous [ˈviskəs] adj. 黏性的，黏滞的
10. annual production 年发电量
11. barrage [ˈbærɑ:ʒ] n. 拦河坝，堰
12. caisson [ˈkeisən] n. 沉箱（桥梁工程）
13. ecological [ˌekəˈlɔdʒikəl] adj. 生态学的，社会生态学的
14. flap [flæp] n. 折叠板，活板
15. immature [ˌiməˈtjuə] adj. 不成熟的，未完全发展的
16. incur [inˈkə:] v. 招致，带来
17. infrastructure [ˈinfrəˌstrʌktʃə] n. 下部构造，基础建设
18. knot [nɔt] n. 速度单位（每小时 1 海里，大约合每小时 1.85km）
19. lagoon [ləˈgu:n] n. 泻湖
20. land mass 大陆块，大陆板块
21. neap tide 小潮
22. popularity [ˌpɔpjuˈlæriti] n. 普及，流行
23. prototype [ˈprəutətaip] n. 原型，样机
24. seal [si: l] vt. 封，密封
25. viable [ˈvaiəbl] adj. 可行的
26. windmill [ˈwindmil] n. 风车
27. anticlockwise [ˌæntiˈlɔkwaiz] adj./adv 逆时针的（地）

28. clockwise [ˈklɔkwaiz] adj./adv 顺时针的

29. corrosive [kəˈrəusiv] adj. 腐蚀的，腐蚀性的

30. off-shore 离岸的，在近海处的

* * * * * * * * * * * * * * * * * * * Complicated Sentences * * * * * * * * * * * * * * * * * * *

1. Tidal power, sometimes called tidal energy, is the only form of energy which derives directly from the relative motions of the earth–moon system, and to a lesser extent from the earth–sun system.

【译文】潮汐能是直接从地球－月球系统相对运动（也有少量来自地球－太阳系统）中提取能量的唯一形式。

【说明】句中 to a lesser exten 表示“程度较小”，relative motion 意思是“相对运动”。由 which 引导的定语从句直到句子结束。sometimes called tidal energy 是对 tidal power 的补充说明，在英文中有 tidal energy 和 tidal power 的两种说法，在汉语中统称潮汐能，因而该部分可以不翻译。

2. The stronger the tide, either in water level height or tidal current velocities, the greater the potential for tidal energy generation.

【译文】潮汐越强（不论表现在水位高度还是潮流速度），潮汐能发电的潜力越大。

【说明】the stronger … the greater 的结构，相当于汉语中的“越强……越大”。

3. Tidal movement causes a continual loss of mechanical energy in the earth–moon system due to pumping of water through the natural restrictions around coastlines, and due to viscous dissipation at the seabed and in turbulence.

【译文】由于要泵水跨越海岸线周围的自然约束，以及海床上和湍流中的黏滞损耗，潮汐运动造成地球－月球系统中机械能的持续损耗。

【说明】句中两个 due to 结构并列，共同解释原因。

4. During the last 620 million years the period of rotation has increased from 21.9 hours to the 24 hours we see now; in this period the Earth has lost 17% of its rotational energy.

【译文】在过去的 6.2 亿年中，（地球的自转）周期已经从 21.9h 增加到我们目前所知的 24h；在这段时期中，地球已经损失了 17%的旋转能量。

【说明】句中第一个 period 表示“周期，循环变化的时间长度”，第二个 period 表示“时期，较长的时间段”。

5. Given that power varies with the density of medium and the cube of velocity, it is simple to see that water speeds of nearly one-tenth of the speed of wind provide the same power for the same size of turbine system.

【译文】假设功率随着介质密度和速度的立方变化，那么很容易知道，几乎相当于风速十分之一的水流速度可以为同等尺寸的轮机提供相同的动力。

【说明】given 表示“假设”。one-tenth 是一种分数表示形式，分子用基数词，分母用序数词。

6. Since tidal stream generators are an immature technology (no commercial scale production facilities are yet routinely supplying power), no standard technology has yet

emerged as the clear winner, but a large variety of designs are being experimented with, some very close to large scale deployment.

【译文】因为潮汐发电机是尚未成熟的技术（还没有商业化规模的生产设备日常供电），还没有明显的优胜者作为标准技术出现，但是大量的各种设计方案正在试验中，某些也已很接近大规模开发。

【说明】close to 表示“接近”。句子前面两个分句之间是因果关系，since 表示“因为，由于”。最后一个分句与前面是转折关系，由 but 来实现转折的连接。

7. Several prototypes have shown promise with many companies making bold claims, some of which are yet to be independently verified, but they have not operated commercially for extended periods to establish performances and rates of return on investments.

【译文】随着很多公司的大胆尝试，已生产出的若干样机也显示出了希望，当然，其中一些技术还有待独立验证，在较长的时间内，它们也都还没有商业化运行，也没有确定样机的性能和投资回收率。

【说明】with many companies making bold claims 表示“带着（承载着）很多公司的大胆尝试”。

8. As they are based well away from land they are out of sight and, unlike wind powered turbines, they do not create noise.

【译文】由于它们建立在远离大陆的位置，因而一般不会出现在人们的视野中，也不会像风力机那样产生噪声。

【说明】base 表示“以……为基础，建立基础”，away from 表示“离开，远离”，well 起强调作用。

*** Summary of Glossary ***

1. tidal power 潮汐能
2. tidal energy 潮汐能
3. tide 潮，潮汐
4. tital current 潮流
5. tidal stream 潮流
6. water level 水位
7. neap tide 小潮
8. civil infrastructure 土建基础工程
9. static head 静止水头
10. pressure head 压力水头
11. annual production 年发电量
12. average power 平均功率
13. ship lock 船闸
14. Strait of Gibraltar 直布罗陀海峡
15. Bosporus 博斯普鲁斯海峡

*** * * * * * * * * * * * * * * * * * Exercises * * * * * * * * * * * * * * * * * ***

(1) ________ that power varies with the density of medium and the cube of velocity, it is simple to see that water speeds of nearly one-tenth of the speed of wind provide the same power for the same size of turbine system.

(2) ________ tidal stream generators are an immature technology (no commercial scale production facilities are yet routinely supplying power), no standard technology has yet emerged as the clear winner, but a large variety of designs are being experimented with, some very close to large scale deployment.

(3) Periodic changes of water levels, and associated tidal currents, are ______ the gravitational attraction by the sun and moon.

*** * * * * * * * * * * * Word Building (24) anti-; cycl- * * * * * * * * * * * ***

1. anti- [前缀]，表示：反对，抵抗

clockwise	adj./adv.	anticlockwise	adj./adv.	逆时针的
dote	vi.	antidote	n.	解毒剂，校正方法
septic	n./adj.	antiseptic	n.	防腐剂，杀毒剂
biotics	adj.	antibiotics	n.	抗生素，抗生学
terrorism	n.	antiterrorism	n.	反恐怖主义

2. cycl- [词根]，表示：圆，环，轮

-ic（…的）	adj	cyclic	adj	循环的
mega	n.	megacycle	n.	兆周
		cyclone	n.	旋风

*** * * * * * * * * * * * * * * * Text Translation * * * * * * * * * * * * * * * ***

第二十四章　潮　汐　电　站

第一节　潮汐能概述

潮汐能是直接从地球—月球系统相对运动（也有少量来自地球—太阳系统）中提取能量的唯一形式。月球和太阳产生的潮汐力与地球的转动相结合，是潮汐形成的原因。（单词个数：56）

换言之，潮汐能的产生是通过相互作用的地球、太阳和月亮之间的引力相对运动的结果。水位的周期性变化，还有相应的潮汐，都是太阳和月亮引力作用的结果。某一处潮汐的高度

取决于太阳和月亮与地球相对位置的变化、地球自转影响，以及当地海底和海岸线的形状（见图 24.1）。（单词个数：82）

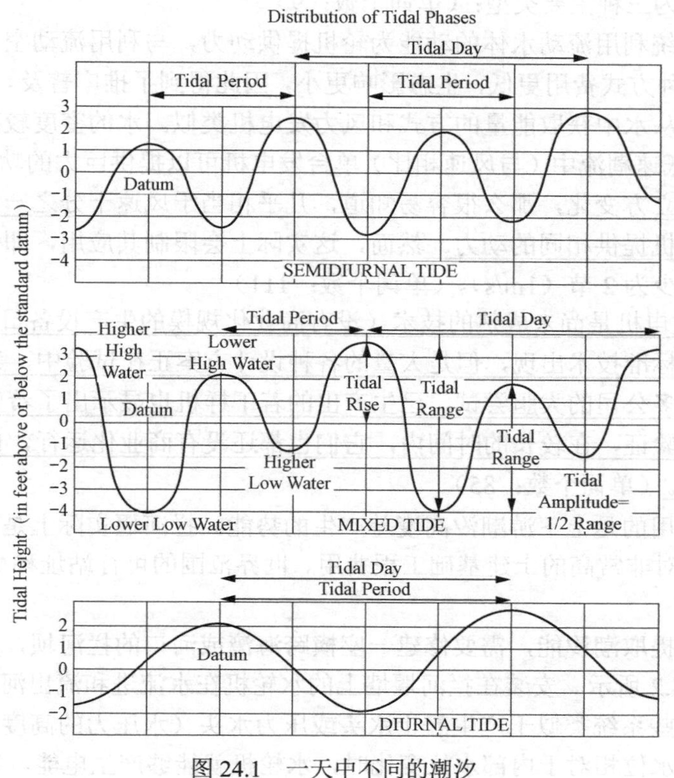

图 24.1　一天中不同的潮汐

因为地球上的潮汐是由月亮、太阳的引力作用以及地球自转引起的潮汐力产生的，所以潮汐能几乎是无尽的，并被列为可再生能源。（单词个数：35）

潮汐越强（不论表现在水位高度，还是潮流速度），潮汐能发电的潜力越大。（单词个数：31）

由于要泵水跨越海岸线周围的自然约束，以及海床上和湍流中的黏滞损耗，潮汐运动造成地球—月球系统中机械能的持续损耗。这一能量损耗导致了地球自形成以来 45 亿年间自转速度的减慢。在过去的 6.2 亿年中，（地球的自转），周期已经从 21.9h 增加到我们目前所知的 24h；在这段时期中，地球已经损失了 17% 的旋转能量。虽然潮汐能可能从地球系统中消耗了额外的能量，加快了地球自转速率的降低，但是这一影响在数百万年后才会表现出来，因此是可以忽略不计的。（单词个数：118）

潮汐能实质上是一种水力发电形式，它把潮汐能量转换成电能或其他有用能源。虽然没有得到广泛应用，但潮汐能在未来的电力生产中很有潜力。潮汐能比风能和太阳能更易于预测。历史上，在欧洲和大西洋沿岸北美洲都有潮汐磨坊的使用。最早开始利用潮汐能的时间，可能是在中世纪，甚至是罗马时期。（单词个数：53）

海洋往返的潮汐可以被用来生产电能。潮汐通常蕴有丰富能量，在它到达陆地时，海水流动速度非常快。当潮水接近陆地时，涌起水的数量能以百万加仑为单位来计量。这些流动的水流具有巨大的能量。潮汐能最大的地方在芬地湾。（单词个数：71）

第二节 潮汐能的类型

潮汐能可以分为三种主要类型：（单词个数：9）

（1）**潮汐流系统**利用流动水体的动能为轮机提供动力，与利用流动空气的风车类似。和其他方式相比，这种方式费用更低，生态影响更小，因此得到了推广普及。（单词个数：44）

潮汐流发电机从水中获取能量的方式和风力发电机类似。水的密度较高（是空气密度的800倍）意味着在低速潮流中（与风速相比）单台发电机可以提供巨大的功率。假设功率随着介质密度和速度的立方变化，那么很容易知道，几乎相当于风速十分之一的水流速度，可以为同等尺寸的水轮机提供相同的动力。然而，这实际上会限制其应用，即使在接近小潮时，潮水移动速度也至少为2节（1m/s）。（单词个数：111）

因为潮汐流发电机是尚未成熟的技术（没有商业化规模的生产设备日常供电），还没有明显的优胜者作为标准技术出现，但是大量的各种设计方案正在试验中，某些也已很接近大规模开发。随着很多公司的大胆尝试，已生产出的若干样机也显示出了希望，当然，其中一些技术还有待独立验证，在较长的时间内，它们也都还没有商业化运行，也没有确定样机的性能和投资回收率。（单词个数：85）

（2）**拦河堰**利用的是上下游潮汐高度差产生的势能。拦河堰实际上是横跨整个潮汐河口宽度的大坝，须面对非常高的土建基础工程费用、世界范围的可行站址稀少及环境问题。（单词个数：48）

用拦河堰方式提取潮汐能，需要修建一座横跨海湾或河口的拦河坝，类似于法国朗斯潮汐发电站，如图24.2所示。安装在拦河堰墙上的水轮机在水流进和流出河口盆地、海湾或河流时产生电能。这些系统类似于产生静止水头或压力水头（水压力的高度）的水电大坝。当盆地或泄湖外面的水位相对于内部水位变化时，水轮机就能够产生电能。1966年安装在法国朗斯河上的潮汐水轮机是最大的，它的安装（峰值）总功率是240MW，年发电量600GW·h（平均功率大约68MW）。（单词个数：131）

拦河堰的基本组成有沉箱、筑堤、水闸、水轮机以及船闸。水闸、水轮机和船闸装在沉箱（非常大的水泥箱）中。在没有沉箱密封的盆地处，由筑堤进行密封。用于潮汐利用的水闸是折叠活板门、垂直升降门、弧形闸门、升降扇形门。拦河堰系统受到土建基础设施成本高和环境问题的困扰，前者是因为横跨河口修建水坝，后者是因为改变了大范围的生态系统。（单词个数：91）

图24.2 法国朗斯潮汐发电站

（3）**潮汐泄湖**，类似于拦河堰，区别是可以建成自包含系统，而非完全地横跨河口，据称它可以降低成本和对环境的整体影响。而且，它们能够进行配置实现连续发电，这在拦河堰方式中是不可能的。（单词个数：45）

水轮机技术的现代研究进展，可能最终会实现从海洋中产生巨额能量，特别是当潮水发电应用了潮汐流设计时，潮汐流水轮机可以排列在自然潮流流向集中的高流速区，如加拿大的西部和东部海岸、直布罗陀海峡、博斯普鲁斯海峡，以及东南亚和澳洲的众多地区。这样

的潮流遍及海湾和河流的入口，以及水流集中的大陆板块之间。(单词个数：102)

第三节 离岸水轮机

关于使用由潮汐能驱动的离岸水轮机的可行性研究正在开展。蕴含能量的潮汐可以用来驱动水下轮机。这种轮机可以按顺时针或逆时针方向旋转，而且由于潮流的连续运动，它发出的电能也是连续的。离岸水轮机就像是水中的风电场（见图24.3)。(单词个数：60)

利用离岸水轮机发电的优势在于建筑成本比潮汐拦河堰低得多，也没有后者带来的环境问题。由于它们建立在远离大陆的位置，因而一般不会出现在人们的视野中，也不会像风力机那样产生噪声。(单词个数：51)

这一系统的主要困难是离岸型轮机与陆地或风力上的轮机相比更贵，而且由于它们在水下工作，维护费用也更多。另外，由于含有盐分，海水对钢铁和其他金属具有腐蚀作用。(单词个数：50)

螺旋桨将靠大型维修船来维修，需要维修的时候螺旋桨会被提至水面以上，这类轮机的布置方案类似于陆地上的风电场，发出的电能通过电缆输向陆地。(单词个数：47)

图 24.3 离岸水轮机

Chapter 25 Small Hydropower

Part 1 General Introduction of Small Hydropower

Small hydro is the development of hydroelectric power on a scale suitable for local community and industry, or to contribute to distributed generation in a regional electricity grid.[1] The definition of a small hydro project varies, but a generating capacity of 1 to 20 megawatts (MW) is common. In contrast, many hydroelectric projects are of enormous size, such as the generating plant at the Three Gorges Dam at 22 500 megawatts or the vast multiple projects of the Tennessee Valley Authority. Small hydro projects may be built in isolated areas that would be uneconomic to serve from a national electricity grid, or in areas where a national grid does not exist.

The use of the term "small hydro" varies considerably around the world, the maximum limit is usually somewhere between 10 and 30 MW. While a minimum limit is not usually set, the US National Hydropower Association specifies a minimum limit of 5 MW. The "small hydro" description may be stretched up to 50 MW in the United States, Canada and China. Small hydro can be further subdivided into mini hydro, usually defined as 100 to 1000 kilowatts (kW), and micro hydro which is 5 to 100 kW. Micro hydro is usually the application of hydroelectric power sized for smaller communities, single families or small enterprise. The smallest installations are pico hydro, below 5 kW.

Since small hydro projects usually have correspondingly small civil construction work and little or no reservoir, they are seen as having a relatively low environmental impact compared to large hydro.[2]

Part 2 Project Design of Small Hydropower

Many companies offer standardized turbine generator packages in the approximate size range of 200 kW to 10 MW. These "water to wire" packages simplify the planning and development of the site since one vendor looks after most of the equipment supply. Because non-recurring engineering costs are minimized and development cost is spread over multiple units, the cost of such package systems is reduced. While synchronous generators capable of isolated plant operation are often used, small hydro plants connected to an electrical grid system can use economical induction generators to further reduce installation cost and simplify control and operation.[3]

Small "run of the river" projects do not have a conventional dam with a reservoir, only a weir to form a headpond for diversion of inlet water to the turbine. Unused water simply flows over the weir and the headpond may only be capable of a single day's storage, not enough for dry summers or frozen winters when generation may come to a halt. A preferred scenario is to have the inlet in an existing lake.

Countries like India and China have policies in favor of small hydro, and the regulatory process allows for building dams and reservoirs. In North America and Europe the regulatory process is too long and expensive to consider having a dam and a reservoir for a small project.

Small hydro projects usually have faster environmental and licensing procedures, and since the equipment is usually in serial production, standardized and simplified, and the civil works construction is also reduced, the projects may be developed very rapidly.[4] The physically smaller size of equipment makes it easier to transport to remote areas without good road or rail access.

One measure of decreased environmental impact with lakes and reservoirs depends on the balance between stream flow and power production. Reducing water diversions helps the river's ecosystem, but reduces the hydro system's return on investment. The hydro system design must strike a balance to maintain both the health of the stream and the economics.

Part 3 Advantages of Small Hydropower

As a green and renewable energy resource, small hydropower has a lot of advantages, and some of them are unique in China. The main are as follows:

(1) It is clean and green. The small hydropower station is generally built on medium or small rivers with a small basin area and without water concentration. It doesn't change water quality and volume, and basically has no effect on the survival and reproduction of biological species of the rivers. Exploitation of small hydropower doesn't produce greenhouse gas and other noxious gases, and thus won't pollute the environment.

(2) The small hydropower technology is very mature. The power station can be designed and completed within one or two years, and has a long lifetime. The dam and related control engineering can work over 100 years with simple maintenance. Moreover, with technology development the turbine efficiency is getting higher and higher, and some can even reach 90%.

(3) Small hydropower plays a very important role in resisting natural disasters, emergency service and disaster relief because it supplies power distributedly. So that it can provide power rapidly in isolated regions when the main power network splits and cannot recover in time.

In January, 2008, the disaster of snow rainfall and freezing happened in South China, which was rare in history. It led to large area blackout in the main power network, severely affecting local people's production and life, and even the economic development. At this time, small hydropower quickly recovered to supply power, effectively reducing disaster loss. In May, 2008, the large earthquake occurred in Wenchuan, Sichuan province. The small hydropower fully took the advantage of isolated operation, which was vital to the disaster rescue.

(4) The electricity price of small hydropower for fuel is relatively low, which is very beneficial to reduce the peasants' burden in China. This advantage is very obvious especially in China.

According to the planning data and typical investigation, the average investment of small hydropower is about ¥6000 per kW • h, and the comprehensive network power price is about ¥0.25 per kW • h. With national support, the average consumer price can decrease to ¥0.23 per kW • h, which is much lower than the mean consumer price of ¥0.56 per kW • h in the rural areas around the country at the present time.[5]

* * * * * * * * * * * * * * * * * Explanations * * * * * * * * * * * * * * * * *

* * * * * * * * * * * * * * * * New Words and Expressions * * * * * * * * * * * * * * * *

1. small hydro 小水电
2. megawatt [ˈmegə.wɑt] n. 百万瓦特，兆瓦
3. considerably [kənˈsɪdərəblɪ] adj. 大幅度的，很大的
4. specify [ˈspɛsəˌfaɪ] v. 具体说明，详述，详列
5. subdivide [ˌsʌbdɪˈvaɪd, ˈsʌbdɪˌvaɪd] n. 细分，把……再分
6. kilowatt [ˈkɪlə.wɑt] n. 千瓦
7. weir [wɪr] n. 堰，拦河坝，导流坝
8. diversion [daɪˈvɜrʒn] n. 转移，转向，消遣，偏移
9. halt [hɔlt] n.&v. 停止，暂停，中断
10. scenario [səˈnɛrioʊ] n. 方案，前景
11. medium [ˈmidiəm] adj.&n. 中等的，中号的，媒介，手段
12. basin [ˈbesɪn] n. 盆地，流域，海盆
13. noxious [ˈnɑkʃəs] adj. 有毒的，有害的
14. split [splɪt] n.&v. 分歧，分裂，使分裂，分开

* * * * * * * * * * * * * * * * * * Complicated Sentences * * * * * * * * * * * * * * * * * *

1. Small hydro is the development of hydroelectric power on a scale suitable for local community and industry, or to contribute to distributed generation in a regional electricity grid.

【译文】小水电是水力发电在适当规模下的发展应用，它可以给社区和工业供电，或者把电能提供给分布式的区域电网。

【说明】此句虽是简单句，但信息量很多，可以自主断句，按汉语习惯翻译。

2. Since small hydro projects usually have correspondingly small civil construction work and little or no reservoir, they are seen as having a relatively low environmental impact compared to large hydro.

【译文】因为小水电项目通常只有相对较小的土建工程和极少数水库，所以相比于大的水电来说，它对环境的影响比较低。

【说明】此句为含有原因状语从句的复合句，句中 compared to large hydro. 为过去分词短语做状语。

3. While synchronous generators capable of isolated plant operation are often used, small hydro plants connected to an electrical grid system can use economical induction generators to further reduce installation cost and simplify control and operation.

【译文】当经常使用能够隔离电站操作的同步电机时，连接在大的电力网的小水电站可以使用比较经济的感应电机，从而更大限度地减少了安装费用，简化了控制与操作。

【说明】此句为含有时间状语从句的复合句，句中 conncted to…systme. 为过去分词短语做定语，修饰句子主语 small hydro plants。

4. Small hydro projects usually have faster environmental and licensing procedures, and since the equipment is usually in serial production, standardized and simplified, and the civil works construction is also reduced, the projects may be developed very rapidly.

【译文】小水电项目的环境许可过程一般很快，因为设备通常是连续、标准和简化生产的，并且土建工程也减少了，所以小水电项目建设得很快。

【说明】此句为含有 and（后面 2 个）连接的两个状语从句的复合句，前一个表示原因，后一个为结果，句中第一个 and 为连接词，连接 aster environmental 和 licensing procedures。

5. With national support, the average consumer price can decrease to ¥0.23 per kW・h, which is much lower than the mean consumer price of ¥0.56 per kW・h in the rural areas around the country at the present time.

【译文】因为有国家的支持，小水电的平均消费价格可以降低到 0.23 元每千瓦・时，这个价格远远低于目前国家偏远地区的 0.56 元每千瓦・时的电能消费价格。

【说明】此句为含有定语从句的复合句，句中 which 引导的定语从句进一步解释了价格低的程度。

****************** Summary of Glossary ******************

1. Small hydro	小水电
2. regional electricity grid	区域电网，地方电网
3. non-recurring engineering costs	一次性工程费用，临时工程费用
4. development cost	开发成本
5. remote areas	边远地区
6. power production	发电量
7. greenhouse gas	温室气体
8. snow rainfall	暴雪
9. disaster rescue	灾害救援
10. electricity price	电价

******************** Exercises ********************

(1) The use of the term "________" varies considerably around the world, the maximum limit is usually somewhere between 10 and 30 MW.

(2) Countries like India and China have policies in favor of small hydro, and the regulatory process allows for building dams and ________.

(3) Small hydropower plays a very important role in ________natural disasters, emergency service and disaster relief because it supplies power distributedly.

*************** Word Building (25)mal-;sub- **************

1. mal- [前缀]，表示：坏，错误

conduct	n/v	malconduct	n.	错误行为，恶行
development	n.	maldevelopment	n.	畸形，变形
distribution	n.	maldistribution	n.	分配不均，分配不公
observation	n.	malobservation	n.	观测误差
position	n/v	malposition	n.	位置不正

2. sub- [前缀] 表示：在……下；低于

marine	n.	submarine	adj.	水下的，海底的
merge	v.	submerge	v.	浸没，淹没；潜水

sub- [前缀]，表示：下级，分支，子，副，亚，次

subsystem	n.	子系统
subclasse	n.	子集
subdivide	v.	再分，细分

****************** Text Translation ******************

第二十五章 小 水 电

第一节 小水电概况

小水电是水力发电在适当规模下的发展形式，可以给社区和工业供电，或者把电能提供给区域分布式电网。小水电项目的定义多种多样，但是普遍功率在 1 MW 到 20MW。对比许多大型的水电项目，比如三峡电站的 22 500MW 的水电厂或者田纳西州流域管理局的多机组项目。小水电项目可以建设在国家电网无法供电的经济不发达的孤立地区，或者国家电网还没有覆盖到的地区。（单词个数：110）

“小水电”这个术语的定义在世界范围内多种多样，最大功率限制在 10MW 到 30MW 不等，最小值的限定没有一个普遍的规定，美国国家水电协会指定了一个最小限定值为 5MW。在美国、加拿大和中国，形容小水电的范围可以达到 50MW。小水电可以再被细分为微水电和小小水电，微水电一般容量在 100～1000kW，小小水电容量一般在 5～100kW。小小水电一般用来给较小的社区，单个家庭或者较小企业来供电。最小的装置是微微水电，容量低于 5kW。（单词个数：114）

因为小水电项目通常只需一个相对小的土建工程和极少数水库，所以相比于大的水电来说，它对环境的影响比较低。（单词个数：30）

第二节 小水电的项目设计

许多公司提供标准值大约在 200kW 至 10MW 范围内的涡轮发电机组。这些水电转换机组简化了原水电厂的规划和发展，因为一个供应商可以保证大多数设备供应。由于临时工程费用成本减至最低，并且开发成本分散到各个机组上，所以减少了这些机组系统的花费。当经常使用能够隔离电站操作的同步电机时，连接在大的电力网的小水电站可以使用比较经济的感应电机，从而更大限度地减少了安装费用，简化了控制与操作。（单词个数：98）

小的河流水电工程没有传统的大坝与水库，只有一个拦河堰来升高水位，从而使水流进入机组。没有使用的水，则直接流过拦河堰，高水位时，可能仅能蓄存一天的水量，这对干燥的夏季和冰冷的冬季来说并不够，这时候发电可能会暂停。更好的方案是让水流蓄存到已经有的湖里。（单词个数：76）

类似于中国和印度这些国家，有政策支持小水电的发展，并且管理过程中允许建造大坝和水库。在北美和欧洲，由于管控过程太长、花费太大，而不能考虑在小的项目上建造大坝和水库。（单词个数：47）

小水电项目的环境许可过程一般很快，因为设备通常连续、标准和简化生产，并且土建工程也减少了，所以小水电项目建设得很快。物理尺寸较小的设备更容易运送到没有公路和铁路的偏远地区。（单词个数：57）

在有湖和水库的情况下减少环境影响的有效措施是如何平衡河流流量和发电量。减少河流分流可以保持河流附近的生态系统，但是会减少小水电的投资回报。小水电系统设计必须做到二者平衡，以保持河流和经济的健康。（单词个数：54）

第三节 小水电的优点

作为一个绿色的可再生能源，小水电有很多优势，而且其中的一些优势是在中国独有的，主要的优点如下：（单词个数：27）

（1）清洁绿色。小水电厂通常建在一个中等的或小的盆地里，并且水能不富集的河流上。它不会改变水的质量和数量，并且基本上对河流附近的生物多样性没有影响。小水电的开发不会产生温室气体和有毒气体，所以不会污染环境。（单词个数：68）

（2）小水电技术非常成熟。小水电站可以在 1~2 年内建设完成，它同样可以使用很长的生命周期。在简单维护的条件下，大坝及其水库工程可以工作超过 100 年。另外，随着技术的发展，涡轮机效率越来越高，一些甚至可以达到 90%。（单词个数：60）

（3）小水电在抵抗自然灾害，紧急服务和减灾等方面扮演一个重要角色。由于它可以提供分布式能源，在一些边远地区，当主电网分离，且不能及时恢复的情况下，小水电可以快速提供电能。（单词个数：43）

2008 年 1 月，中国南方发生历史上罕见的雪灾和冰冻灾害，它导致了主电网的大面积区域停电，严重影响了当地人民的生产生活，甚至是经济发展。但与此同时，当地的小水电快速恢复供电，有效减少了灾害损失。2008 年 5 月，四川省汶川县发生大地震，小水电同样发挥了可以独立运行的优势，对灾害救援至关重要。（单词个数：84）

（4）小水电的电价相对较低，有益于减小中国农民的负担。这在中国的优势尤其明显。计划数据和典型调查显示，小水电的平均投资大约是每千瓦·时 6000 元，且综合电网的电价是每千瓦·时 0.25 元。因为有国家的支持，小水电的平均消费价格可以降低到 0.23 元每千瓦·时，这个价格远远低于目前国家偏远地区的 0.56 元每千瓦·时的电能消费价格。（单词个数：71）

Chapter 26 Navigation and Recreation

Part 1 Navigation and Service for Ships

Proper design of hydropower installations may provide other benefits besides hydroelectric production. Many of them have multipurpose dam. Now let's take the Three Gouges Dam as an example for describing the use of dam in navigation and service for ships.

1. Navigation

Backwater of the Three Gouges reservoir goes as far as to the southwest metropolis Chongqing, therefore it improves 660 km waterway, and largely enables10000-tonnage fleets to navigate between Shanghai and Chongqing. The annual one-way navigation capacity of the Yangtze at the dam will be upgraded from ten million tons to fifty million tons.

In the Three Gorge area, the mountains are high and footways are dangerous, and most of the local residents build their homes in the mountains and beside the river. Because of the sterile land and inconvenient transportation, economic development in this area is very slow. The construction of the Three Gorges Dam will bring innumerable chances to over a million relocatees in the reservoir region.

The Three Gorges featured numerous shoals, rapids and narrow river-course, in which navigation at night was at stake since ancient times. After the founding of the People's Republic of China, navigation was improved by restoring the river course and constructing the Gezhouba project. After the reservoir impoundment, the 660-km-long river course will be further improved and become golden waterway in real sense.

2. Ship Locks

As shown in Fig.26.1, the installation of ship locks is intended to increase river shipping from ten million to 100 million tonnes annually, with transportation costs cut by 30% to 37%.[1] Shipping will become safer, since the gorges are notoriously dangerous to navigate. Each of the two ship locks is made up of five stages, taking around four hours in total to transit, and has a vessel capacity of 10 000 tons.[2] Critics argue, however, that heavy siltation will clog ports such as Chongqing within a few years based on the evidence from other dam projects.[3]

Fig.26.1 Ship Locks for River Traffic to Bypass the Three Gorges Dam (http://en.wikepedia.org)

The locks are designed to be 280 m long, 35 m wide, and 5 m deep (918×114×16.4 ft). That is 30 m longer than those on the St Lawrence Seaway, but half as deep. Before the dam was constructed, the maximum freight capacity of the river at the Three Gorges site was 18.0 million tonnes per year. From year 2004 to 2007, a total of 198 million tonnes of freight passed through the Three Gorges Dam ship locks. The freight capacity of the river increased six times and the cost of shipping was reduced by 25%, compared to the previous years. The total capacity of the ship locks is expected to reach 100 million tonnes.

3. Ship Lifts

Fig.26.2 Ship Lifts (http://flickr.com)

In addition to the canal locks, the Three Gorges Dam will be equipped with a ship lift, a kind of elevator for vessels, as shown in Fig.26.2. The ship lift is designed to be capable of lifting ships of up to 3000 tons, having been reduced from the original plans where the ship lift was going to have the capacity to lift vessels of up to 11500 tons displacement. The vertical distance travelled will be 113 metres, and the size of the ship lift's basin will be 120×18×3.5 metres. The ship lift, when completed, will take 30 to 40 minutes to ascend or descend, as opposed to the three to four hours for stepping through the main locks.[4] One of the factors complicating the design is that the water level can vary dramatically. The ship lift had to be designed to work properly even if the water levels varied by 12 metres on the lower side, and 30 metres on the upper side.

The ship lift was not yet complete when the rest of the project was officially opened on May 20, 2006. Construction of the ship lift started in October 2007 and is anticipated to be completed in 2014.

Part 2 Recreation and Scenery

Reservoirs created by hydropower dams may also act as recreational areas. The large reservoirs created by dams create open bodies of water that are much more suitable for boating, water skying, fishing, and other activities. These recreational activities are good themselves, but can also act as an important economic stimulus for communities.

Water bodies account for 30%-40% relative to other recreational enterprises in satisfying human recreational needs. Intense recreational use of reservoirs is presently being observed. For instance, in a certain country, of the 200 large hydroelectric station reservoirs, more than 60 are used for recreation.[5] In contrast to seashores usedmainly for long recreation, reservoirs play an important role in creating conditions for brief recreation of the population. In cities located directly on the shores of reservoirs there lives 27 million persons, and within a 2-h drive an additional 50 million persons having the oppurtunity to recreate at reservoirs.[6] These data do not included those arriving from other places in the indicated regions for a comparatively long time. With consideration of the broad program of hydropower and water-management constraction in that country, the planned measures on the multipurpose development of small rivers, the recreation use of reservoirs has great prospects. At the same time, when developing water-management and hydropower constraction projects and schemes of the multipurpose use of individual streams and river basins, the section related to the recreational use of waters is often compiled formally, in connection with which the possibilities of organizing recreation on them are not fully realized. The creation of recreational systems on reservoirs is related to a number of charateristics.

We must consider both the engineering-technical problems and an economic substantiation of the proposed recreational measures with an evaluation of their positive and possible negative consequences with estimation not only of costs but also an estimation of the payback period and effectiveness from the recreational development of the territory.

However, there are still some critics argue that the construction of dams destroys previous recreational opportunities. In order to create a large vertical drop for power generation, dams are often built on rivers that have a significant gradient and whitewater rapids. Because of this, many reservoirs flood rivers that previously provided whitewater rafting and kayaking opportunities. Furthermore, fishing is often better in free-flowing rivers and reservoirs.

Dams are often viewed as magnificent and beautiful. There is a reason why dams are major tourist attractions. They are a magnificent and awe-inspiring site to see. But, sometimes, hydropower and their reservoirs destroy natural landscapes and impose man-made creations upon them. These tarnishes the beauty of pristine river ecosystems.

* * * * * * * * * * * * * * * * * Explanations * * * * * * * * * * * * * * * * *

* * * * * * * * * * * * * * * * New Words and Expressions * * * * * * * * * * * * * * * *

1. ascend [ə'send] v. 攀登，上升
2. at stake 危如累卵，危险
3. backwater ['bækwɔ: tə] n. 逆流，回水（被水坝或急流止住或往回流的水）
4. descend [di'send] v. 下来，下降
5. displacement [dis'pleismənt] n. 移置，转移，排水量
6. fleet [fli:t] n. 舰队
7. Gezhouba 葛洲坝
8. innumerable [i'nju: mərəbl] adj. 无数的，数不清的
9. lift [lift] n. 无数的，数不清的
10. metropolis [mi'trɔpəlis] n. 主要都市，都会，大城市
11. multipurpose ['mʌlti'pə: pəs] adj. 综合利用，多种用途的，多目标的
12. notoriously [nəu'tɔ: riəsli] adv.声名狼藉地
13. one-way adj. 单行道的，单程的
14. rapid ['ræpid] n. 急流
15. relocatee ['ri: ləu'keiti:] n. 迁至新址者
16. shoal [ʃəul] n. 浅滩，沙洲
17. sterile ['sterail] adj. 贫瘠的，不育的，不结果的
18. Yangtze (=Yangtze River, Changjiang) 扬子江，长江
19. awe-inspiring ['ɔ:in, spaiəriŋ] n. 令人敬畏的
20. kayak ['kaiæk] n. 皮船
21. rafting ['rɑ:ftiŋ] n. 筏运
22. stimulus ['stimjuləs] n. 刺激物，刺

激，促进因素

23. substantiation [sʌbsˈtænʃieiʃən] n. 实体化，证明，证实

24. tarnish [ˈtɑːniʃ] vt. 玷污；（使）失去光泽，（使）变灰暗

* * * * * * * * * * * * * * * * * * **Complicated Sentences** * * * * * * * * * * * * * * * * * *

1. The installation of ship locks is intended to increase river shipping from ten million to 100 million tonnes annually, with transportation costs cut by 30% to 37%.

【译文】船闸的安装将每年江上的船舶吨位由 1 千万吨增加到 1 亿吨，而运输成本却下降 30%～37%。

【说明】句中 intend 表示"计划，打算；为特殊目的而设计"，shipping 意思是"海运，(总称)运输船只，船舶吨数"。

2. Each of the two ship locks is made up of five stages, taking around four hours in total to transit, and has a vessel capacity of 10 000 tons.

【译文】两个船闸的每一个都由 5 级构成，总共需要 4 小时左右的时间驶过船闸，闸室有万吨级容量。

【说明】in total 表示"总计，一共"。两个逗号后面的内容都是对每个船闸的说明。

3. Critics argue, however, that heavy siltation will clog ports such as Chongqing within a few years based on the evidence from other dam projects.

【译文】然而，一些评论员根据其他大坝项目的证据提出质疑，主要由于严重的泥沙淤积，会在若干年内阻塞重庆等港口。

【说明】however 往往出现在句子中间，并用前后两个逗号分隔，在翻译时应该提到句首。Base on 的意思是"基于，在……基础上"。

4. The ship lift, when completed, will take 30 to 40 minutes to ascend or descend, as opposed to the three to four hours for stepping through the main locks.

【译文】船舶升降系统完成后，需要 30～40 分钟的时间实现上升或下降，明显少于通过主船闸分段提升所需的 3～4 小时。

【说明】when completed 是个状语从句，插入到了句子的主语和谓语之间。oppose to 表示"相对的，相反的"。

5. For instance, in a certain country, of the 200 large hydroelectric station reservoirs, more than 60 are used for recreation.

【译文】例如，在某国的 200 座大型水电站的水库中，有 60 多座用于娱乐休闲。

【说明】for instance 意思是"例如"，certain 表示"某个"。

6. In cities located directly on the shores of reservoirs there lives 27 million persons, and within a 2-h drive an additional 50 million persons having the oppurtunity to recreate at reservoirs.

【译文】直接居住于水库岸边的城市人口有 2700 万人，在 2 小时车程范围内，还有另外的 5000 万人有机会到水库来娱乐休闲。

【说明】and 是连词，介绍了两种条件下的人口情况，located 过去分词短语做定语，修饰 cities; having…为现在分词短语修饰 persons。

* * * * * * * * * * * * * * * * * * Summary of Glossary * * * * * * * * * * * * * * * * * *

1. Yangtze (=Yangtze River, Changjiang）扬子江，长江
2. the Three Gouge Dam 三峡大坝
3. Gezhouba 葛洲坝
4. the St Lawrence Seaway 圣劳伦斯海道
5. shoal 浅滩，沙洲
6. rapid 急流
7. river-course 河道
8. navigation 航行
9. ship lock 船闸
10. ship lift 升船机，船舶升降系统
11. whitewater 浪花，浪端的白色泡沫

* Exercises *

(1) The Three Gorges ________ numerous shoals, rapids and narrow river-course, in which navigation at night was at stake since ancient times.

(2) Each of the two ship locks is made up of five stages, ________ around four hours in total to transit, and has a vessel capacity of 10 000 tons.

(3) Water bodies ________ 30%-40% relative to other recreational enterprises in satisfying human recreational needs.

* * * * * * * * * * * Word Building (26)micro-; tri- * * * * * * * * * * * *

1. micro-[前缀]，表示：“极微小，仪器或工具用以扩大者”之义

microelement	n.	微型元件，微量元素
microcosmic	adj.	微观世界的, 微观的
microeconomic	n.	微观经济
microanalysis	n.	微量分析

2. tri- [前缀]，表示：三，三重，三次，三倍

triangle	n.	三角形
triangular	n.	三角形的, 三人间的
trigonometry	n.	三角学
triplywood	n.	三夹板
triode	n.	三极管
trisect	v.	三等分

****************** Text Translation ******************

第二十六章 航运与娱乐旅游

第一节 航运和船运服务

对许多综合利用枢纽大坝来说，合理的水电设施设计，除了可以水力发电外，还可以提供其他用途。这里，我们以三峡水库为例，来说明大坝在航运和船运服务方面的作用。（单词个数：40）

1. 航运

三峡水库的回水远到西南的都市重庆，因此，改善了660km的水路，使1万吨级的舰队可以在上海和重庆之间航行。长江三峡大坝的年平均单程航行容量也由1000万吨增加到5000万吨。（单词个数：54）

三峡地区的山很高，大多数居民在山里或河边建立家园，所以人们出行非常艰难。由于土地贫瘠，交通不便，这个地区的经济发展十分缓慢。三峡大坝的建设将会给三峡水库附近的一百多万居民带来很多机会。（单词个数：65）

三峡地区以无数的浅滩、急流和狭窄河道而著称，那里自古以来夜间航行就很危险。中华人民共和国成立以后，河道的恢复和葛洲坝的建设使航运得到了改进。三峡水库蓄水后，660km长的河道将会进一步改善，并将会变成真正意义上的黄金水道。（单词个数：62）

2. 船闸

图26.1 三峡修建的船闸

如图26.1所示，三峡船闸的规划是每年江上的通航量由1000万吨增加到1亿t，而运输成本下降30%～37%。船运将变得更安全，因为峡谷是出了名的危险。两个船闸的每一个都由5级构成，总共需要4个小时左右的时间驶过船闸，闸室可容纳万吨级船。但是，一些评论员根据其他大坝项目的证据提出质疑，主要源于严重的泥沙淤积会在若干年内阻塞重庆等港口。（单词个数：91）

该船闸长280m、宽35m、深5m（918×114×16.4英尺），比圣劳伦斯河海道上的船闸要长30m，但只有其一半深。在大坝建成前，三峡每年的最大货运容量为1800万吨，2004～2007年，一共有19800万吨货物通过三峡船闸。相比前几年，货运容量增长了6倍，而航运成本降低了25%。船闸总容量预计达到1亿吨。（单词个数：119）

3. 升船机

除了船闸，三峡大坝还将配备升船机，一种船只的升降机，如图 26.2 所示。设计的升船机能够升起重达 3000 吨的船只，比原计划（升船机将升起重达 11 500 吨的船只）降低了规模。升起距离是 113m，升船机的尺寸为 120m×18m×3.5m。船舶升降系统完成后，需要 30～40 分钟的时间实现上升或下降，时间明显少于通过主船闸分段提升所需的 3～4 小时。使升船机设计复杂的一个因素是水位落差很大。升船机必须保证在水位落差达到 12m 时，也能在水位低的一侧正常运行，相应的，水位落差达到 30m 时，能在水位高的一侧正常工作。（单词个数：160）

图 26.2　升船机

2006 年 5 月 20 日，升船机还未完全竣工，配套部分已经开工。升船机从 2007 年 10 月开始施工，预计于 2014 年完成。（单词个数：37）

第二节　娱乐和旅游

建造水坝所形成的水库，还可作为休闲娱乐的地方。由水坝形成的大型水库提供了一个更适合划船、滑水、钓鱼或其他活动的开放空间。这些活动本身很好，但也可以作为一项发展地区经济的重要措施。（单词个数：52）

相对于其他娱乐事业，水体可以提供人类娱乐需求的 30%～40%。近来，利用水库进行一些刺激的娱乐活动引起关注。例如，在某国，200 个大型水电站的水库中，有 60 多个用于娱乐休闲。相比海岸用来长期休憩，水库在为人们提供简单娱乐条件方面发挥着重要作用。直接在居住水库岸边的城市人口有 2700 万人，在 2 个小时车程范围内，还有另外的 5000 万人，也可以到水库来休憩。这些数据不包括从其他地方来到这里游览名胜的人们。经过对该国水电和水资源管理建设的整体调查，发现对中小河流进行有计划综合开发，水库休憩利用有着很大的前景。同时，在利用个别小溪和河流流域发展水资源管理和水电建设工程时，与水的娱乐应用有关的地区，连同在那里组织娱乐不能完全实现的可能性，也往往被正式编列汇总。水库上娱乐设施的建立和水库的一些特性相关。（单词个数：214）

我们必须通过估计提出的娱乐项目的利益和可能带来的负面影响，来考虑工程技术问题和经济评估，不仅做成本估计，而且要预计投资回收期和地区发展娱乐的效益。（单词个数：49）

然而，有一些评论家对于大坝建设对目前的旅游事业造成的影响还有争议。为发电要创造很大的落差，大坝通常建造在有显著梯度或激流险滩的河流上。因为这样，许多水库就淹没了原来提供泛舟和划艇机会的河流，而且渔业往往是在自由流动的河流和水库中发展更好。（单词个数：66）

水坝往往被视为宏伟，美观的。水坝成为主要的旅游景点还有一个原因，它们是一个宏伟和令人惊叹的景点。但是，有时候水电和水库破坏自然景点，在自然景观上施加人为创作。这损害了河流生态系统的原始之美。（单词个数：50）

Chapter 27 Bids

Part 1 Introduction

Once the bidder's list is fixed, there is a formal procedure to follow. Variation will depend on the preferences of the project manager, but generally most of these steps will be followed to achieve a satisfactory bidding process. In many instances the bidding process is regulated to insure fair and equal treatment of all bidders. Few deviations or exceptions to the rules are permitted or encouraged.

In project management, flexibility in bidding is a major factor in obtaining effective results. This freedom allows the project manager to select the best system for each circumstance. In certain cases, the project manager may recognize and recommend that the owner should not bid at all, but would be better served through negotiation. There is almost no end to the possibilities for variation from the basic bidding procedure, as long as the objective of owner's benefit and fairness to the contractors are observed.

Part 2 Bid Procedures

The first step in the bidding process is to inform all selected or available bidders that bids will be sought. A notice to bidders is published, or otherwise disseminated, which identifies the work, the design team, and other pertinent facts. The purpose of the notice is to alert potential bidders in order to allow them to make the preliminary decision regarding their interest and to reserve the time and resources needed to bid. Generally, the notice does not require or encourage a response. It is usually a legal requirement in public work following the method of giving notice that is strictly enforced by the governing authority; such notices are often useful in private work.

The next step after the notice is the actual invitation to bidders, which solicits bids. The invitation may be mailed to an individual bidder or may be made available for all bidders to read and respond. The invitation is more specific regarding the work and the mechanics of bidding and established the time and place for the receipt of bids. The availability of documents is explained and the prescribed bid forms are disseminated.

1. Bid Forms

The formulation of the bid forms is a critical step in the bidding. To obtain fair and meaningful bid, which can be compared and analyzed without confusion, it is necessary to design the bid form with clarity and specificity. [1]The work that is subject to bid should be careful defined with reference to contract documents, location, timing, and any other major limitations or criteria. Alternates or options should be separately designated so that their relation to the basic work is clear, thus allowing the bidder to evaluate his costs without ambiguity.

Bid forms that are carefully prepared in this manner contribute to the simplicity of bid analysis and fairness in contract award. It is an obligation to the bidders, who invest too much time and money in responding, to provide forms that are equitable. The project manager should review all proposed bid forms with the Architect-Engineer, the client, and with the contractors during prebid meeting to eliminate ambiguity and to establish the clear meaning and use of the forms.

2. Bid Bonds

As a part of their bid submission, all bidders are required to submit a bond, or the acceptable substitute for the same, to ensure that the bidder will enter into the formal bidding process. This is necessary for two reasons:

(1) Unforeseen circumstances. Prior to the signing of the contract, the successful bidder may encounter circumstances, which will preclude his fulfilling the contract.

(2) Bid spread. When the successful bid is much lower than the competitive bids, the difference, referred to as "money left on the table", often has a disquieting effect on the low bidder. His first reaction is to carefully review his bid for errors. If an honest substantial error is found, he may ask to be excused from signing a contract, if no error is found, the low bidder also still feels uncertain of a profitable outcome, he may again petition for release of the bid. He may choose to surrender the bonded amount rather than undertake the work.

The bid bond is generally issued by the contractor's insurer in the amount of 5% of the bid. An acceptable substitute is a certified check to the owner in the same amount.

3. Performances bonds

A performance bond guarantees that the contractor will perform the obligation according to the contract. This bond is effective within the construction period and all guarantee-warranty period. Should any portion of the work prove unacceptable to the owner and should the contractor refuse to provide a remedy, the client has recourse to the insurer to correct the deficiency. In case of contractor default, the insurer will undertake to complete the work or retain a suitable contractor to fulfill the contract, at no expense to the owner. Default usually results in delay of progress. Extensive delays in honoring the performance bond may render the insurer liable to additional claims by the owner and other parties to the work.

The performance bond is written in the full amount of the contract. In case of default, retention money and the unpaid balance of the contract are available for the completion of the work. This fact places a heavy responsibility on the project manager to closely monitor payouts to the contractor to assure that payments do not exceed the value of acceptable work in place, materials, and equipment delivered to the site.[2]

4. Pre-bid Conference

At least one week before bids are due a pre-bid conference is assembled in order to respond to all questions that the bidders wish to ask regarding the project. The timing of the conference should allow for the issuance of any agenda, which might develop, applicable to the conference. This should be a comprehensive and wide-ranging question-and-answer session with no holds barred. However, this does not rule out an agenda for the conference. Bidders, including subcontractors, should be invited to the following groups be present to respond to all questions:

(1) Representatives of each discipline in the design group.

(2) The owner's representative and staff members involve in financing, insurance, operation, and maintenance.

(3) Representatives of each discipline in the project management group.

(4) Local inspection, code, and zoning authorities.

In addition, and particularly where outside agencies are involved, it may be judicious to invite:

(1) Local authorities.

(2) Major suppliers.

(3) Police, fire, and street department representatives.

(4) Utility representatives.

Part 3 Bidding, Bid Opening and Award of Contract

The bidding documents should state clearly whether contracts will be awarded on the basis of unit process (for work performed or goods supplied) or of a lump sum of the contract, according to the nature of goods or works to be provided.

The size and scope of individual contracts will depend on the magnitude, nature, and location of the project. For projects requiring a variety of works and equipment such as power, water supply, or industrial projects, separate contracts are normally awarded for the civil works, and for the supply and erection of different major items of plant and equipment.[3]

On the other hand, for a project requiring similar but separate civil works or items of equipment, bids should be invited under alternative contract options that would attract the interest of both smaller and larger firms. Contractors or manufacturers, small and large, should be allowed to bid for individual contracts or for a group of similar contracts at their option, and all bids and combinations of bids should be opened band evaluated simultaneously so as to determine the bid or combination of bids offering the most advantageous solution for the borrower.[4]

Detailed engineering of the works or goods to be provided, including the preparation of technical specification and other bidding documents, should precede the invitation to bid for the contract. However, in the case of turnkey contracts or contracts for large complex industrial projects, it may be undesirable to prepare technical specifications in advance. In such a case, it will be necessary to use a two-step procedure inviting unpriced technical bids subject to technical clarification and adjustment, followed by the submission of priced proposals.

The time allowed for preparation of bids should depend on the magnitude and complexity of the contract. Generally, not less than 45days from the date of invitation to bid should be allowed for international bidding. Where large civil works are involved, generally, not less than 90 days from the date of invitation should be allowed to enable prospective bidders to conduct investigations at the site before submitting their bids. The time allowed, however, should be governed by the particular circumstances of the project.

The date, hour, and place for latest delivery of bids by the bidder, and of the bid opening, should be announced in the invitation to bid, and all bids should be opened at the stipulated time. Bids delivered after the time stipulated should be returned unopened unless the delay was not due to any fault of the bidder and its late acceptance would not give him any advantage over other bids. Bids should normally be opened in public. The name of the bidder and total amount of each bid, and, of any alternative bids if they have been requested or permitted, should, when opened, be read aloud and recorded.

Extension of validity of bids should normally not be requested; if, in exceptional circumstances, and extension is required, it should be requested of all bidders before the expiration date. Bidders should have the right to refuse to grant such an extension without forfeiting their bid bond, but those who are willing to extend the validity of their bid should be neither required nor permitted to modify their bids.

It is undesirable that information related to the examination, clarification, and evaluation of bids and recommendations concerning awards be communicated after the public opening of bids to bidders or to persons not officially concerned with these procedures until the award of a contract to the successful bidder is announced.

No bidder should be permitted to alter his bid after bid has been opened. Only clarification not changing the bid may be accepted. The borrower may ask any bidder for a clarification of his but should not ask any bidder to change the substance of price of his bid.

Following the opening, it should be ascertained whether material errors in computation have been in the bids, whether the bids are substantially responsive to the bidding documents, or contains inadmissible reservation, it should be rejected, unless it is an alternative bid permitted, or requested, or requested, under the bidding documents. A technical analysis should then be evaluated each responsive bid and to enable bids to be compared.

A detailed report on the evaluation and comparison of bids setting forth the specific reasons on which the decision for the award of the contract, or rejection of all bids is based, should be prepared by the borrower or by its consultants.

The award of a contract should be made, within the period specified for the validity of bids, to the bidder whose responsive bid has been determined to be the lowest evaluated bid, and who meets the appropriate standards of capability and financial resources.[5] Generally, such bidder should not be required, as a condition of award, to undertake responsibilities or work not stipulated in the specification or to modify his bid.

Part 4 Bidding Documents

The bidding documents shall furnish all information necessary for a prospective bidder to prepare a bid for the goods and works to be provided. While the detail and complexity of these documents may vary with the size and nature of the proposed bid package and contract, they generally include: invitation to bid; instructions to bidders; form of bid; form of contract; conditions of contract, both general and special; specifications and drawings; relevant technical data (including of geological and environmental nature); list of goods or bill of quantities; delivery time or schedule of completion; and necessary appendices, such as formats for various securities.[6] The basis for bid evaluation and selection of the lowest evaluated bid shall be clearly outlined in the instructions to bidders and/or the specifications. If a fee is charged for the bidding documents, it shall be reasonable and reflect only the cost of their printing and delivery to prospective bidders, and shall not be so high as to discourage qualified bidders. The Borrower may use an electronic system shall be secure to distribute bidding documents, provided that the Bank is satisfied with the adequacy of such system. If bidding documents are distributed electronically, the electronic system shall be secure to avoid modifications to the bidding documents and shall not restrict the access of Bidders to the bidding documents. Guidance on critical components of the bidding documents is given in the following paragraphs.

Borrowers shall use the appropriate Standard Bidding Documents (SBDs) issued by the Bank with minimum changes, acceptable to the Bank, as necessary to address project-specific conditions. Any such changes shall be introduced only through bid or contract data sheets, or through special conditions of contract, and not by introducing changes in the standard wording of the Bank's SBDs. Where no relevant standard bidding documents have been issued, the Borrower shall use other internationally recognized standard conditions of contract and contract forms acceptable to the Bank.

Part 5 Validity of Bids and Bid Security

Bidders shall be required to submit bids valid for a period specified in the bidding documents which shall be sufficient to enable the Borrower to complete the comparison and evaluation of bids, review the recommendation of award with the Bank (if required in the Procurement Plan), and obtain all the necessary approvals so that the contract can be awarded within that period.[7]

Borrowers have the option of requiring a bid security. When used, the bid security shall be in the amount and form specified in the bidding documents and shall remain valid for a period of four weeks beyond the validity period for the bids, in order to provide reasonable time for the Borrower to act if the security is to be called. Bid security shall be released to unsuccessful bidders once the contract has been signed with the winning bidder. In place of bid security, the Borrower may require bidders to sign a declaration accepting that if they withdraw or modify their bids during the period of validity or they are awarded the contract and they fail to sign the contract or to submit a performance security before the deadline defined in the bidding documents, the bidder will be suspended for a period of time from being eligible for bidding in any contract with the Borrower.[8]

Bidding documents shall be worded as to permit and encourage international competition and shall set forth clearly and precisely the work to be carried out, the location of the work, the goods to be supplied, the place of delivery or installation, the schedule for delivery or completion minimum performance requirements, and the warranty and maintenance requirements as well as any other pertinent terms and conditions.[9] In addition, the bidding documents, where appropriate, shall define the tests, standards, and methods that will be employed to judge the conformity of equipment as delivered, or works as performed, with the specifications. Drawings shall be consist with the text of the specifications, and an order of precedence between the two shall be specified.

The bidding documents shall specify any factors, in addition to price, which will be taken into account in evaluating bids, and how such factors will be quantified or otherwise evaluated. If bids based on alternative designs, materials, completion schedules, payment terms, etc., are permitted, conditions for their acceptability and the method of their evaluation shall be expressly stated.

Part 6 Competitive Bids

Competitive bids offers extended by businesses in which they detail proposed compensation that they will receive in exchange for executing a specific task or tasks. These tasks can range from providing a service for a set period of time to manufacturing and transporting a certain quantity of goods or materials. Competitive bidding differs from other pricing strategies in that with bid pricing, a specific price is put forth for each possible job rather than a generic price that applies to all customers. "The big problem in bid pricing," wrote E. Jerome McCarthy and William D. Perrault Jr. in Basic Marketing, is estimating all the costs that will apply to each job. This may sound easy, but a complicated bid may involve thousands of cost components. Further, management must include an overhead charge and a charge for profit. Competition must be considered when adding in overhead and profit. Usually, the customer will get several bids and accept the lowest one. So unthinking addition of 'typical' overhead and profit rates should be avoided. Some bidders use the same overhead and profit rate on all job—regardless of competition—and then are surprised when they don't get some jobs.

Competitive bidding is an especially common practice with government buyers, many of whom have instituted mandatory procedures. Government buyers are typically required to accept the lowest bid that receive, but it is important to note that low bids can often be disregarded if they are judged to be lacking in meeting minimum job specifications.

1. Small Business and Private-Sector Procurement

"Most small-company owners chalk up the time and energy spent bidding on customer contracts to the cost of getting new business," wrote Stephanie Gruner in Inc. However, some customers may simply be using the small business to push down the price demands of already established vendors. Given this reality, there are several steps that small business owners can take to 1) improve their changes of securing a contract, and 2) minimize loss of time and energy on the competitive bidding process.

"Ask for information," wrote Gruner. "Most companies won't share confidential data, but it doesn't hurt to ask for specifics, such as current costs. The way the customer handles your requests will also tell you whether the company seriously wants to work with you." Gruner also counseled small business operators to try and establish contract with a variety of people within the targeted company. Making contracts with several people within an organization can be valuable not for information-gathering purposes, but also because, as Gruner put it, "one might become your advocate."

Another key to success in competitive bidding is simply recognizing long-term trends, both in your company's industry and in the larger would of commerce. For example, writer Matthew S. Scott pointed out in Black Enterprise in 1995 that many entrepreneurial ventures have the potential to reap huge benefits from the ever-expanding telecommunications industry. "The key to linking up with the billions of dollars in procurement contracts that will potentially become available over the next decade is to remember that the information superhighway is still under constructions ... Over the next decade, these high-tech telecommunications firms will spend billions of dollars to develop the new technologies and information transmission systems that will make up the infrastructure of the information superhighway.[10] So, considering the type of small business you own, if you can help the telecommunications giants build these new systems quickly, and in a cost-efficient manner, you can benefit from the coming golden age of telecommunications."

Other business experts indicate that a spectrum of business trends are also increasing the importance of competitive bidding in ensuring small business success. One of the most dramatic of these trends is the one toward increased outsourcing by companies of all sizes. As responsibilities that were previously attended to in-house continue to be shipped outside, opportunities for establishing your company as a valued producer of goods and/or services will also increase.

One way to minimize the time and cost associated with preparing a competitive bid is to use a computerized bidding program. "If you're putting in a number and a factor changes, everything changes. You used to have to make all the changes by hand," consultant Elizabeth Jeppesen told Susan Craig of the New Mexico Business Journal. "On bid day you need not only the base bid but any alternates plus all the paperwork to go with the bid, such as lists of subcontractors, qualifications, insurance carriers, and more. You must cross your T's and dot your I's. Bid day is organized chaos."

2. Negotiated Bids

Not all competitive bidding situations end with the customer's acceptance of one of the bids offered. In some instances, a subsequent negotiation step may take place. "Some buying situations (including much government buying) require the use of bids-and the purchasing agent must take the lowest bid," said Mc-Carthy and Perrault. "In other cases, however, the customer asks for bids and then singles out the company that submits the most attractive bid-not necessarily the lowest-for further bargaining. The list price or bidding price the seller would like to charge is sometimes only the starting point for discussion with individual customers. What a customer will buy-if the customer buys at all-depends on the negotiated price, a price set based on bargaining between the buyer and seller." McCarthy and Perrault go on to note that negotiated pricing, like simple bid pricing, "is most common in situations where the marketing mix is adjusted for each customer-so bargaining may involve the whole marketing mix, not just the price level."

* * * * * * * * * * * * * * * * * * Explanations * * * * * * * * * * * * * * * * * *

* * * * * * * * * * * * * * * * * New Words and Expressions * * * * * * * * * * * * * * * *

1. disseminate [dɪˈsɛmɪnet] v. 播撒，宣传，散播

2. pertinent [ˈpɝtnənt] adj.相关的，相干的；中肯的；切题的

3. prescribed [prɪˈskraɪbd] adj. 规定的，指定的

4. formulation [ˌfɔrmjəˈleʃən] n. 构想，规划；公式化；简洁陈述

5. criteria [kraɪˈtɪrɪə] n. 标准，条件（criterion 的复数）

6. ambiguity [ˌæmbɪˈɡjuəti] n. 含糊；不明确；暧昧；模棱两可的话

7. obligation [ɔbliga(ɑ)sjɔ̃] n. 责任，义务；债务，债券；必须，需要

8. subcontractors [ˌsʌbkənˈtræktə] n. 转包商，次承包商；分承包方

9. magnitude [ˈmæɡnɪtud] n. 大小；量级；[地震] 震级；重要；光度

10. precede [prɪˈsid] vi. 领先，在前面
    vt. 领先，在……之前；优于，高于

11. forfeit [ˈfɔrfət] vt.（因犯罪、失职、违约等）丧失（权利、名誉、生命等）
    adj. 因受罚而丧失的；被没收的
    n. 罚金；没收物；丧失的东西

12. clarification [ˌklærəfəˈkeʃən] n. 澄清，说明；净化

13. ascertain [ˈæsɚˈten] vt. 确定；查明；探知

14. stipulate [ˈstɪpjulet] vi. 规定；保证
    vt. 规定；保证
    adj. 有托叶的

15. compensation[ˌkɑmpɛnˈseʃən] n. 补偿，赔偿，抵消；清算，划账

16. overhead [ˌovɚˈhɛd] adv. 在头顶上；在空中；在高处
    n. 天花板；[会计] 经常费用；间接费用；吊脚架空层

17. procurement[prəˈkjʊrmənt] n. 采购；获得，取得

18. vendor [ˈvɛndɚ] n. 卖主；小贩；供应商；[贸易] 自动售货机

19. confidential[ˌkɑnfɪˈdɛnʃl] adj. 机密的；表示信任的；获信任的

20. counsel [ˈkaʊnsl] n. 法律顾问；忠告；商议；讨论；决策
    vt. 建议；劝告
    vi. 商讨；提出忠告

21. reap [rip] vt. 收获，获得；收割
    vi. 收割，收获
    n. (Reap)人名；(英)里普

22. outsourcing [ˈaʊtˌsɔrsɪŋ] n. 外包；外购；外部采办

23. specification [ˈspɛsəfəˈkeʃən] n. 规格；说明书；详述

24. generic [dʒəˈnɛrɪk] adj. 类的；一般的；属的；非商标的

25. entrepreneurial[ˌɑntrəprəˈnjʊrɪəl] adj. 企业家的，创业者的；中间商的

26. telecommunications [ˈtɛləkəˈmjʊnəˈkeʃənz] n. 通信行业；服务类型变更，缴纳话费

27. subsequent [ˈsʌbsɪkwənt] adj. 后来的，随后的

28. mandatory [ˈmændətori] adj. 强制的；托管的；命令的
    n. 受托者（等于 mandatary）

* * * * * * * * * * * * * * * * * * * **Complicated Sentences** * * * * * * * * * * * * * * * * * * *

1. The formulation of the bid forms is a critical step in the bidding. To obtain fair and meaningful bid, which can be compared and analyzed without confusion, it is necessary to design the bid form with clarity and specificity.

【译文】对招标程序的设计是招投标过程中的一个重要步骤。为了实现公平有效的招标，经过冷静的比较和分析，设计出简洁、具体的招标程序十分有必要。

【说明】To obtain fair and meaningful bid 作目的状语，which 引导的从句起补充说明作用。

2. This fact places a heavy responsibility on the project manager to closely monitor payouts to the contractor to assure that payments do not exceed the value of acceptable work in place, materials, and equipment delivered to the site

【译文】这说明一个事实，项目经理负有很大的责任，来密切监督承包商的费用支出，以确保承包商在工程场地、材料、设备、运输费上不超过可接受范围。

【说明】assure 后跟从句表目的，用以修饰"监督"这个行为。

3. For projects requiring a variety of works and equipment such as power, water supply, or industrial projects, separate contracts are normally awarded for the civil works, and for the supply and erection of different major items of plant and equipment.

【译文】对于包含种类繁多的工程建筑和设备的项目，如电力、供水或者工业项目，通常是就土木工程和不同的工厂厂房、设备等主要项目的供应商和安装方分别签订合同。

【说明】主句是"separate contracts are normally awarded for…"即"……合同需要单独签订"，"For projects requiring…"是句中的状语成分。

4. Contractors or manufacturers, small and large, should be allowed to bid for individual contracts or for a group of similar contracts at their option, and all bids and combinations of bids should be opened band evaluated simultaneously so as to determine the bid or combination of bids offering the most advantageous solution for the borrower.

【译文】应该允许各大小承包商或制造商，对单个合同或一组类似的合同随意进行投标。所有的单标或组标应该同时开标和进行估价，以决定哪个单标或组标对借款人最合算。

【说明】"small and large,"是插入语部分，修饰说明"Contractors or manufacturers"。"and all bids and combinations of…"该从句与主句是并列关系。"simultaneously"表示"同时地"。

5. The award of a contract should be made, within the period specified for the validity of bids, to the bidder whose responsive bid has been determined to be the lowest evaluated bid, and who meets the appropriate standards of capability and financial resources.

【译文】要在规定的投标有效期内，把合同授予以最低价投标的，而且在能力和财力方面符合相应标准的投标人。

【说明】"within the period specified for the validity of bids"作插入成分，起补充说明作用。"whose responsive bid…"和"and who meets…"这是两个并列的定语从句修饰"bidder"。

6. While the detail and complexity of these documents may vary with the size and nature of the proposed bid package and contract, they generally include: invitation to bid; instructions to bidders; form of bid; form of contract; conditions of contract, both general and special;

specifications and drawings; relevant technical data (including of geological and environmental nature); list of goods or bill of quantities; delivery time or schedule of completion; and necessary appendices, such as formats for various securities.

【译文】尽管这些文件的细节和复杂性可能因提议的投标包和合同的规模和性质不同而有所不同，但它们通常包括:招标、投标人说明、投标形式、合同形式、合同的一般和特殊条款、规范和图纸；有关技术资料(包括地质和环境性质)、货物清单或数量清单、交付时间或完成计划、以及必要的附加条件，例如保证金的格式。

【说明】“complexity”是“复杂性”的意思，“vary with the size and nature”表示“因规模和性质不同而产生差异”，“appendices”是“附件”的意思。

7. Bidders shall be required to submit bids valid for a period specified in the bidding documents which shall be sufficient to enable the Borrower to complete the comparison and evaluation of bids, review the recommendation of award with the Bank (if required in the Procurement Plan), and obtain all the necessary approvals so that the contract can be awarded within that period.

【译文】投标人应按要求在招标文件规定的期限内提交有效的投标书，以使借款人有充足的时间完成对投标的比较和评价、同银行一起审查授标建议(如果采购计划要求的话)，并获得一切必要的批准，以便能在该期限内授予合同。

【说明】valid 作后置定语修饰“bids”,which 后引导一个较长的目的状语从句，so that 后引导一个内嵌从句，也是表目的。

8. In place of bid security, the Borrower may require bidders to sign a declaration accepting that if they withdraw or modify their bids during the period of validity or they are awarded the contract and they fail to sign the contract or to submit a performance security before the deadline defined in the bidding documents, the bidder will be suspended for a period of time from being eligible for bidding in any contract with the Borrower.

【译文】如果不要求提交投标保证金，借款人可以要求投标人签署一项声明，声明如果在投标有效期内撤回或修改其投标书，或者在授予合同之后不能签订合同，或不能在招标文件规定的截止日期前提交履约保证金，该投标人的投标资格将被暂停一段时间，从而不得参与借款人有关的任何合同的投标。

【说明】that 后引导定语从句，修饰“declaration”，对“声明”的内容起补充说明作用。If 后引导一个内嵌的条件状语从句，“eligible for...”表示“……的资格”。

9. Bidding documents shall be worded as to permit and encourage international competition and shall set forth clearly and precisely the work to be carried out, the location of the work, the goods to be supplied, the place of delivery or installation, the schedule for delivery or completion minimum performance requirements, and the warranty and maintenance requirements as well as any other pertinent terms and conditions.

【译文】招标文件的措辞应允许和鼓励国际竞争，并且应清楚准确地阐述需要开展的工作、工作的地点、需要提供的货物、交货或安装的地点、交货或完工的时间表、最低性能要求、质量保证和维修要求以及其他任何有关的条款和条件。

【说明】“word”在这用作动词，意思为“措词”，“set forth”是“陈述”的意思。“and shall

set forth…”与主句并列，该句内大量宾语并列，详细说明需“陈述”的内容。

10. “The key to linking up with the billions of dollars in procurement contracts that will potentially become available over the next decade is to remember that the information superhighway is still under constructions…” Over the next decade, these high-tech telecommunications firms will spend billions of dollars to develop the new technologies and information transmission systems that will make up the infrastructure of the information superhighway.

【译文】“在接下来十年可能会出现的数十亿美元的采购合同中，关键是要记住，信息高速公路仍在建设中……”在接下来的十年中，这些高科技电信公司将花费数十亿美元开发构成信息高速公路基础设施的新技术和信息传输系统。

【说明】“The key to…is to remember…”是主句的主要结构，remember 后引导一个内嵌宾语从句，“high-tech telecommunications firms”意思是“高科技电信公司”，“infrastructure”意为“基础设施”。

********************* Summary of Glossary *********************

1. no end to	没有终止
2. prebid meeting	招标会
3. bid spread	买入差价
4. certified check	保付支票
5. recourse to	求助于；依赖
6. no holds barred	无拘无束
7. civil works	土木工程
8. stipulated time	约定期限
9. bidding documents	招标文件
10. rule out	排除；取消；划去；反对；阻止
11. set forth	阐明，陈述
12. private-Sector	私营企业
13. chalk up	归咎于
14. cross your T's and dot your I's	一丝不苟
15. negotiated bid	投标谈判
16. procurement contract	采购合同
17. bid security	投标保证金
18. procurement plan	采购计划
19. expressly stated	明文规定的
20. eligible for	合格；够资格

********************* Abbreviations (Abbr.) *********************

SBD Standard Bidding Document 标准招标文件

******************** Exercises ********************

(1) In project management, _________ is a major factor in obtaining effective results, because there is almost no end to the possibilities for _________ from the basic bidding procedure, as long as the objective of ______ and ________ to the contractors are observed.

(2) Generally, the notice does not ________ or _______ a response, however, the invitation may be ___________ to an individual bidder or may be made available for all bidders to _______.and _______.

(3) Bidders are required to submit a _______ , or the acceptable _________ for the same for two reasons: _________ and _________.

(4) Contractors or manufacturers can bid for ________ contracts or for ________ similar contracts at their option, and all bids and combinations of bids should be opened band evaluated ________.so as to determine the bid or combination of bids offering the most _________ for the borrower.

(5) Electronic system is valid to distribute bidding documents if the ________ is satisfied with the adequacy of such system. If bidding documents are distributed electronically, the electronic system shall be secure to avoid ________ to the bidding documents and shall not ________ the access of Bidders to the bidding documents.

(6) Bidders shall be required to submit bids valid for a period which shall be sufficient to enable the Borrower to complete the___________and ___________of bids, the recommendation of award with the Bank, and obtain all the necessary so that __________ within that period.

(7) Competitive bidding differs from other _________ in that with bid pricing, a specific price is___________for each possible job rather than a generic price that applies to _________ . So the big problem in bid pricing is estimating all the costs that will apply to _________.

(8) Not all competitive bidding situations end with the _________of one of the bids offered. In some instances, a subsequent__________step may take place. In some cases, the customer will singles out the company that submits the most bid but not necessarily____________.

************* Word Building (27) -tion;-ity *************

1. -tion [词尾]，表示：形容词或动词转名词

formation	n.	形成，构成
deposition	n.	沉淀物，沉积作用
gravitation	n.	地心引力，引力作用
portion	n.	一部分

2. -ity [名词词尾]，由形容词演变而来，构成抽象名词，表示“……性，……度”

feasibility	n.	可行性
flexibility	n.	灵活性
reliability	n.	可靠性

security	n.	安全性
stability	n.	稳定性

* * * * * * * * * * * * * * * * * * **Text Translation** * * * * * * * * * * * * * * * * * *

第二十七章 招 投 标

第一节 概 述

一旦投标人的名单被确定下来，随后将有一个正式的程序。根据项目经理的意愿，招标会发生一定的变化。但一般情况下多数步骤都是为了完成一个令人满意的招标过程。通常情况下，招标过程受招投标法的约束，以确保所有参与竞标的投标人得到公平公正的对待。通常对规则少许的偏离或例外是被允许和鼓励的。（单词个数：66）

在项目管理中，为了取得有效的结果，招标的灵活性是一个主要的因素。这种特权允许项目经理为每一环节选择最佳方案，并且能够缩短那些烦冗、耗时的程序。在特定情况下，项目经理可以认可或建议业主不进行投标，而是通过协商以得到更好的服务。在基本的招标程序中，既然要保证业主的利益和承包商公平，改变就不可避免。（单词个数：83）

第二节 投 标 程 序

在招标的过程中第一步是通知入选的投标人竞标，发布或散布招标通知，确认招标事项、设计小组和其他相关的事情。通知的目的是提醒那些潜在的投标人，以便使他们就投资收益以及参加所必需的时间和财力做初步的决定。一般来说，通知不要求也不鼓励回应，在公共工程中这通常是一个法令性条款，即政府机构强制要求的发布通知的方法。这样的条款在非公共工程中也十分有用。（单词个数：114）

通知的下一步就是对投标人发出邀请，即请求其投标。这种邀请可以通过发邮件的方式通知投标人或者是通过发布公告，让所有有意竞标的投标人知晓并可以做出回应。这种邀请更加具体，主要是涉及竞标工作事项，标书的构成、时间的确定、地点的确认等。同时，要说明相关文件的获取途径以及发布规定的投标程序。（单词个数：73）

1. 招标程序

对招标程序的设计是招投标过程中的一个重要步骤。为了实现公平有效的招标，经过冷静的比较和分析，设计出简洁、具体的招标程序十分有必要。与招标相关的工作需详细说明，包括合同文件、招标地点、时间以及其他重要的限制或标准。变更或选择部分应该单独标明，以便明确它们与基础工作之间的关系，进而便于投标人精确估计投标成本。（单词个数：91）

这样精心设计的招标程序有助于简化投标分析和授予合同的公平性。对于投入了大量时间和金钱并对招标做出回应的投标人来说，这一举措是有益的，它为投标人提供了公平的竞标方式。项目经理应该与建筑工程师、投标人的委托人以及承包人在招标会议上就所有建议的招标程序进行协商，以消除歧义并明确招投标程序的用意。（单词个数：78）

2. 投标保函

作为投标的一部分，所有的投标人都被要求交纳保证金或是可接受的等值替代物，以确保投标人能够参与竞标。这是必需的，其理由如下：（单词个数：40）

（1）不可预见的情况。在签订合同之前，中标人可能会遇到阻碍其完成合同的情况。（单词个数：24）

（2）买入差价。当中标人的标价远远低于有竞争力的标价时，这一差价，即“差额”，通常让出价低的竞标者感到不踏实。对于标价低的投标者来说，他的第一反应就是认真检查投标中出现的失误。如果发现低价中标是由于自己的重大失误造成的，他就会在签订合同时请求原谅。如果发现失误，中标人也会因为对最终的收益结果没有把握，而要求撤销中标。他们宁愿放弃保证金也不愿意承揽工程。（单词个数：100）

保证金通常由承包商投保的保险公司按照标价报价的5%预付，也可以递交同等数额的保付支票。（单词个数：32）

3. 履约保函

履约保函用来保证承包商按照合同规定正常履行其义务。履约保函在工期和保修期内都是有效的。工程的任何部分如果达不到业主的要求，并且承包商拒绝采取补救措施，委托人将要求担保人纠正其过失行为。一旦承包商毁约，则保险公司将承担完成工程的任务或指定合适的承包商来履行合同，且不对业主提出任何收费要求。承包商毁约通常会导致工期延误。长时间的工期延误不仅使担保人要兑现履约保证金，而且可能会遭到业主或第三方就工程提出额外索赔要求。（单词个数：117）

履约保证金要在合同中写明。一旦承包商毁约，留存款项及未结算价款可用于工程的完工。事实上，项目经理负有很大的责任，来密切监督承包商的费用支出，以确保承包商在工程场地、材料、设备、运输费上不超支可接受范围。（单词个数：71）

4. 标前会议

至少在招标前一个星期要召开一次会议来解答所有投标者关于工程的问题。会议时间的确定要考虑会议议程的发布，这是一个广泛、全面、无拘无束的咨询会，然而，不排除会议议程。投标者包括分包商，应被邀请参加并且弄清楚合同文件的内容。为使会议更有成效，以下各方应出席会议并对所有由投标人以及分包商提出的问题做出解答：（单词个数：90）

（1）业主代表。

（2）全体职员。（单词个数：10）

（3）项目管理班子代表，包括融资、保险、施工、维修代表。（单词个数：11）

（4）地方监督、法规、职能部门。（单词个数：8）

除此之外，要特别邀请相关的外部机构，以下是明智的做法：（单词个数：15）

（1）地方政府代表。（单词个数：4）

（2）主要供应商。（单词个数：4）

（3）警察、消防、街道管理代表。（单词个数：8）

（4）公共事业代表。（单词个数：4）

第三节　招标、开标和授予合同

招标文件应该清楚地说明，根据所提供的货物或者工程建筑的性质，合同是以分项价格（完成的建筑或者提供的货物），还是以总价格为基础授予的。（单词个数：42）

单个合同的规模和范围取决于项目的大小、性质和特点。对于包含种类繁多的工程建筑和设备的项目，如电力、供水或者工业项目，通常是就土木工程和不同的工厂厂房、设备等主要项目的供应和安装分别签订合同。（单词个数：58）

另一方面，如果一个项目需要有一些相似的，独立的土木工程或设备，合同的方式应该多样化，供投标人选择，这样能吸引规模大小不同的公司。应该允许各大小承包商或制造商，对单个合同或一组类似的合同随意进行投标。所有的单标或组标应该同时开标和进行估价，以决定哪个单标或组标对借款人最合算。（单词个数：91）

在招标之前，要把所需工程建筑或货物的详细设计包括技术要求和其他招标文件准备就绪。然而，对于包括规划、设计和管理的整套施工合同或者大的综合性工业项目的合同，事先准备好的技术规范，也许是不合乎需要的。这时，就需要采用一种分为两步的程序，先是进行不确定费用技术招标，以便于进行技术性说明及安排，随后再提供带价格的标书。（单词个数：83）

投标所需要的准备时间应该视合同的规模和复杂程度而定。一般来说，国际投标从招标之日起，应有不少于45天的时间。如果涉及大型土木工程，一般来说，从招标之日起，应该给予不少于90天的时间，以使有可能中标的投标人在投标前能到施工现场进行调查。而到底需要给多少时间，应该视工程项目的具体情况而定。（单词个数：83）

投标人投标截止的日期、时间和地点，以及开标的日期、时间和地点都要在招标时公布，所有的投标都要在规定的时间打开。在截止时间以后收到的投标，都要原封不动地退回。除非迟到的原因不在于投标人，而且在截止期后接受其标书并不使他比其他投标占便宜。投标应该在公开场合打开，投标人的名字，每个投标的总金额，包括要求他们或者允许他们提出的可供选择的投标，都要在开标时诵读并记录在案。（单词个数：109）

一般来说，不应该延长投标的有效期限。如果在特殊情况下，需要延长投标期限，就要在终止日期之前，向所有投标人提出要求。投标人有权拒绝延长期限，并不会因此而损失其投标保证金。对那些同意延长投标期限的投标人，既不要求也不允许修改他们的投标。（单词个数：68）

在公开开标以后，宣布合同中标者之前，不允许把有关对投标进行审查、说明和评价以及关于中标的推荐意见等情况，告诉投标人或者与这些程序没有正式关系的人。（单词个数：49）

开标之后，任何投标人都不允许修改投标，只接受不改动投标实际内容的说明。借款人可以要求投标人对其投标加以说明，但不能要求投标人改动实际内容或报价。（单词个数：49）

开标之后，就应该查明投标中有无计算错误，投标是不是基本上回答了招标文件提出的要求，有没有提供所需要的保证，文件是否都签了字，投标文件在其他方面是否完整。如果投标基本上没有回答招标文件所提出的要求，或者包含不能允许的保留，就要加以拒绝，除非这是根据招标文件的要求，或者文件所允许的供选择的参考投标。随后，就要进行技术分析，对每个投标进行评价，并进行比较。（单词个数：67）

借款人或者其顾问要准备一个对投标的评价和比较的详细报告，在里面要写明其所依据的具体理由。并应写明将合同授予谁，或者所有投标都不能接受。（单词个数：42）

要在规定的投标有效期内，把合同授予以最低价投标的，而且在能力和财力方面符合相应标准的投标人。通常，不应当要求投标人承担技术要求中没有规定的责任或者工作，或者要求其修改投标，并以此作为授予合同的条件。（单词个数：70）

第四节 投 标 文 件

招标文件应提供潜在投标人的一切必要信息，以便投标人准备投标货物和提供的工作。尽管这些文件的细节和复杂性可能因提议的投标包和合同的规模和性质不同而有所不同，但它们通常包括：招标、投标人说明、形式合同、合同形式、合同的一般和特殊条款、规范和图纸、有关技术资料(包括地质和环境性质)、货物清单或数量清单、交付时间或完成计划以及必要的附加条件，例如各种证券的格式。投标评估的基础和最低报价的选择应当清楚地列在投标人和规格的说明中。对招标文件收取的费用，应当合理，只反映其印刷和交付给潜在投标人的成本，不应过高，使合格的投标人受到阻碍。如果银行对该系统的妥善性满意，借款人可以使用电子系统来分发投标文件。如果招标文件通过电子系统发布，那么，该电子系统应有安全保障措施，以避免对招标文件进行修改，并且不应限制投标人进入该系统获得招标文件。招标文件的关键组成部分将在下面几段中给予说明。（单词个数：234）

借款人应使用银行发布的适当的标准招标文件(简称 SBD)。银行可接受为适应项目具体情况而做的必要的最小改动。任何此类改动只能放在投标资料表或合同资料表中，或放在合同专用条款中，而不得对银行标准招标文件(SBD)中的标准文字进行修改。如果银行还没有发布相关的标准招标文件(SBD)，那么，借款人应使用银行可接受的其他在国际上公认的标准合同条件和合同格式。（单词个数：85）

第五节 投标及投标保证金有效期

投标人应按要求在招标文件规定的期限内提交有效的投标书，以使借款人有充足的时间完成对投标的比较和评价、同银行一起审查授标建议(如果采购计划要求的话)，并获得一切必要的批准，以便能在该期限内授予合同。（单词个数：62）

借款人可以选择是否要求投标人提交投标保证金。如果要求提交投标保证金，投标保证金应该按照招标文件中规定的金额和格式提交。投标保证金应在投标有效期期满后的 4 周内保持有效，以使借款人在需要索取保证金时，有合理的时间采取行动。在与中标的投标人签订合同之后，应及时退还未中标的投标人的投标保证金。如果不要求提交投标保证金，借款人可以要求投标人签署一项声明，声明如果在投标有效期内撤回或修改其投标书，或者在授予合同之后不能签订合同，或不能在招标文件规定的截止日期前提交履约保证金，该投标人的投标资格将被暂停一段时间，从而不得参与借款人有关的任何合同的投标。（单词个数：158）

招标文件的措辞应允许和鼓励国际竞争，并且应清楚准确地阐述需要开展的工作、工作的地点、需要提供的货物、交货或安装的地点、交货或完工的时间表、最低性能要求、质量保证和维修要求以及其他任何有关的条款和条件。另外，在适当的情况下，招标文件还应规定用于判定所提交的货物或所完成的工程是否和技术规格相一致所需的检验、标准和方法。图纸应与技术规格的文字内容相一致，并应规定两者之间的优先顺序。（单词个数：119）

除价格因素外，招标文件应规定评标时将考虑的全部评标因素以及如何将这些因素量化或进行评价。如果允许采用替代的设计方案、材料、完成时间和付款条件等进行投标，那么，应对接受它们的条件和对它们进行评价的方法应做出明确的说明。（单词个数：58）

第六节 竞争性投标

竞争性投标提供商业扩展，他们详细说明他们将收到的补偿，以换取执行特定任务或多个任务。这些任务可以从提供一段时间的服务到制造和运输一定数量的货物或材料。竞争性投标与其他定价策略不同之处在于，使用出价定价，针对每个可能的工作提供特定价格，而不是适用于所有客户的通用价格。（单词个数：82）

基准营销中 E. Jerome McCarthy 和 William D. Perrault Jr.写道："投标价格中的大问题是估计将适用于每项工作的所有成本"。这可能听起来很简单，但复杂的投标可能涉及数千个成本组件。此外，管理层必须包括间接费用和利润费用。增加开销和利润时，必须考虑竞争。通常情况下，客户将收到几个投标并接受最低价。因此，应避免增加"典型"间接费用和利润率。一些投标人在所有工作上使用相同的开销和利润率，无论竞争如何，然后当他们没有得到一些工作时，他们会感到惊讶。（单词个数：82）

竞争性投标是政府买家特别常见的做法，其中许多人已经制定强制性程序。政府采购员通常需要接受最低的出价，但重要的且要注意是，如果被判定缺乏满足最低工作规格的要求时，低标价往往会被忽视。（单词个数：114）

1. 小企业和私营部门的采购

Stephanie Gruner 在公司中写道："大多数小公司的老板都把时间和精力都花在了客户合同上，以获得新业务的成本。然而，一些客户可能只是在利用小企业的价格来压低已经建立的供应商的价格。"考虑到这一现实，小企业所有人可以采取以下几个步骤：① 改善他们的合同的变更；② 在竞争投标过程中减少时间和精力的损失。（单词个数：54）

"询求信息，" Gruner 写道，"大多数公司不会分享机密数据，但要求细节具体，比如当前的成本，也不会有什么坏处。"客户处理你的请求的方式也会告诉你该公司是否真的想要和你一起合作。Gruner 还建议小型企业运营商尝试与目标公司内的多个人建立合同。与组织内的几个人签订合同不是为了获取信息，而是因为，正如格伦所说，"任何人都可能会成为你的拥护者"。（单词个数：81）

在竞争性招标中，成功的另一个关键是要认识到长期趋势，无论是在公司的行业，还是在更大的商业领域。例如，作家马修斯科特 1995 年在黑人企业中指出，许多创业企业有潜力从不断扩张的电信行业中获得巨大利益。"在接下来十年可能会出现的数十亿美元的采购合同中，关键是要记住，信息高速公路仍在建设中……在接下来的十年中，这些高科技电信公司将花费数十亿美元开发构成信息高速公路基础设施的新技术和信息传输系统。因此，考虑到你所拥有的小企业类型，如果你能帮助电信巨头迅速建立这些新系统，并且是以一种合算的方式，你就可以从即将到来的电信黄金时代中获益。”（单词个数：155）

其他业务专家表示，一系列业务趋势也增加了竞争性投标在确保小企业成功方面的重要性。其中最引人注目的趋势之一就是各种规模的公司外包的增加。由于以前在内部履行的职责继续在外部输送，因此建立贵公司作为有价值的货物或服务生产商的机会也将增加。（单词个数：71）

一种减少与准备竞争投标相关的时间和成本的方法是使用一个计算机化的投标程序。“如果你输入一个数字，一个因素改变，一切都会变。你过去必须手工做所有的改变，”Elizabeth Jeppesen 对新墨西哥商业杂志的苏珊克雷格说，“在竞标日，你不仅需要基本的报价，还需要任何附加条件，加上所有的投标文件，比如分包商、资格证书、保险公司等。你必须一丝不苟，竞标日是有组织的混乱”。（单词个数：102）

2. 谈判出价

并非所有的竞争性招标都是在客户接受其中一个报价的情况下结束的。在某些情况下，可能会发生后续的协商步骤。Mmccarthy 和 Perrault 说：“一些购买情况(包括大量的政府购买)需要使用投标书，而采购代理必须采取最低的报价。”不过，在其他情况下，客户会要求报价，然后选出最具吸引力的公司——不一定是最低的价格。卖方想要收取的标价或出价，有时只是与个别客户讨论的起点。顾客购买的东西完全取决于协商的价格，一种基于买卖双方讨价还价的价格。麦卡锡和 Perrault 继续指出：谈判的定价，就像简单的投标定价一样，最常见的情况是，在市场营销组合为每个客户所调整时，讨价还价可能涉及整个营销组合，而不仅仅是价格水平。（单词个数：164）

Chapter 28 How to Write Your First Research Paper

Part 1 Schedule Your Writing Time in Outlook

Whether you have written 100 papers or you are struggling with your first, starting the process is the most difficult part unless you have a rigid writing schedule. [1]Writing is hard. It is a very difficult process of intense concentration and brain work. As stated in Hayes' framework for the study of writing: "It is a generative activity requiring motivation, and it is an intellectual activity requiring cognitive processes and memory". In his book *How to Write a Lot: A Practical Guide to Productive Academic Writing*, Paul Silvia says that for some, "it's easier to embalm the dead than to write an article about it". Just as with any type of hard work, you will not succeed unless you practice regularly. If you have not done physical exercises for a year, only regular workouts can get you into good shape again. The same kind of regular exercises, or I call them "writing sessions," are required to be a productive author. Choose from 1- to 2-hour blocks in your daily work schedule and consider them as non-cancellable appointments. When figuring out which blocks of time will be set for writing, you should select the time that works best for this type of work. For many people, mornings are more productive. One Yale University graduate student spent a semester writing from 8 a.m. to 9 a.m. when her lab was empty. At the end of the semester, she was amazed at how much she accomplished without even interrupting her regular lab hours. In addition, doing the hardest task first thing in the morning contributes to the sense of accomplishment during the rest of the day. This positive feeling spills over into our work and life and has a very positive effect on our overall attitude.

Part 2 Start With an Outline

Now that you have scheduled time, you need to decide how to start writing. The best strategy is to start with an outline. This will not be an outline that you are used to, with Roman numerals for each section and neat parallel listing of topic sentences and supporting points. This outline will be similar to a template for your paper. Initially, the outline will form a structure for your paper; it will help generate ideas and formulate hypotheses. Following the advice of George M. Whiteside, "…start with a blank piece of paper, and write down, in any order, all important ideas that occur to you concerning the paper". Use Table 1 as a starting point for your outline. Include your visuals (figures, tables, formulas, equations, and algorithms), and list your findings. These will constitute the first level of your outline, which will eventually expand as you elaborate.

The next stage is to add context and structure. Here you will group all your ideas into sections: Introduction, Methods, Results, and Discussion/Conclusion (Table 2). This step will help add

coherence to your work and sift your ideas.

Table 1	**Outline-level 1**
1. What is the topic of my paper?	
2. Why is this topic important?	
3. How could I formulate my hypotheses?	
4. What are my results?	
5. What is my major finding?	

Table 2	**Outline-level 2**
Introduction	
1. Why is your research important?	
2. What is known about the topic?	
3. What are your hypotheses?	
4. What are your objectives?	
Materials and Methods	
1. What materials did you use?	
2. Who were the subjects of your study?	
3. What was the design of your research?	
4. What procedure did you follow?	
Results	
1. What are your most significant results?	
2. What are your supporting results?	
Discussion and Conclusions	
1. What are the studies major findings?	
2. What is the significance/implication of the results?	

Part 3 Continue With Drafts

After you get enough feedback and decide on the journal you will submit to, the process of real writing begins. Copy your outline into a separate file and expand on each of the points, adding data and elaborating on the details. When you create the first draft, do not succumb to the temptation of editing. Do not slow down to choose a better word or better phrase; do not halt to improve your sentence structure. Pour your ideas into the paper and leave revision and editing for later. As Paul Silvia explains, "Revising while you generate text is like drinking decaffeinated coffee in the early morning: noble idea, wrong time".[2]

Many students complain that they are not productive writers because they experience writer's block. Staring at an empty screen is frustrating, but your screen is not really empty: you have a template of your article, and all you need to do is fill in the blanks. Indeed, writer's block is a logical fallacy for a scientist — it is just an excuse to procrastinate. When scientists start writing a research paper, they already have their files with data, lab notes with materials and experimental designs, some visuals, and tables with results. All they need to do is scrutinize these pieces and put them together into a comprehensive paper.

If you still struggle with starting a paper, then write the Materials and Methods section first. Since you have all your notes, it should not be problematic for you to describe the experimental design

and procedures. Your most important goal in this section is to be as explicit as possible by providing enough detail and references. In the end, the purpose of this section is to allow other researchers to evaluate and repeat your work.

Interestingly, recent studies have reported that the Materials and Methods section is the only section in research papers in which passive voice predominantly overrides the use of the active voice.[3] For example, Martínez shows a significant drop in active voice use in the Methods sections based on the corpus of 1 million words of experimental full text research articles in the biological sciences. According to the author, the active voice patterned with "we" is used only as a tool to reveal personal responsibility for the procedural decisions in designing and performing experimental work. This means that while all other sections of the research paper use active voice, passive voice is still the most predominant in Materials and Methods sections.

Writing Materials and Methods sections is a meticulous and time consuming task requiring extreme accuracy and clarity. This is why when you complete your draft, you should ask for as much feedback from your colleagues as possible. Numerous readers of this section will help you identify the missing links and improve the technical style of this section.

For many authors, writing the Results section is more intimidating than writing the Materials and Methods section. If people are interested in your paper, they are interested in your results. That is why it is vital to use all your writing skills to objectively present your key findings in an orderly and logical sequence using illustrative materials and text.[4]

Your Results should be organized into different segments or subsections where each one presents the purpose of the experiment, your experimental approach, data including text and visuals (tables, figures, schematics, algorithms, and formulas), and data commentary. For most journals, your data commentary will include a meaningful summary of the data presented in the visuals and an explanation of the most significant findings. This data presentation should not repeat the data in the visuals, but rather highlight the most important points. In the "standard" research paper approach, your Results section should exclude data interpretation, leaving it for the Discussion section. However, interpretations gradually and secretly creep into research papers: "Reducing the data, generalizing from the data, and highlighting scientific cases are all highly interpretive processes. It should be clear by now that we do not let the data speak for themselves in research reports; in summarizing our results, we interpret them for the reader". As a result, many journals including the *Journal of Experimental Medicine* and the *Journal of Clinical Investigation* use joint Results/Discussion sections, where results are immediately followed by interpretations.

Now that you are almost half through drafting your research paper, it is time to update your outline. While describing your Methods and Results, many of you diverged from the original outline and re-focused your ideas. So before you move on to create your Introduction, re-read your Methods and Results sections and change your outline to match your research focus. The updated outline will help you review the general picture of your paper, the topic, the main idea, and the purpose, which are all important for writing your introduction.

The best way to structure your introduction is to follow the three-move approach shown in Table 3.

Table 3 **Moves in research paper introductions**

Move 1. Establish a research territory
show that the general research area is important, central, interesting, and problematic in some way.
Move 2. Find a niche
indicate a gap in the previous research, or extend previous knowledge in some way.
Move 3. Occupy the niche
a. outline purposes or state a nature of the present research;
b. list research questions or hypotheses;
c. announce principle findings;
d. state the value of present research;
e. indicate the structure of research paper

The moves and information from your outline can help to create your Introduction efficiently and without missing steps. Some academic writers assume that the reader "should follow the paper" to find the answers about your methodology and your findings. As a result, many novice writers do not present their experimental approach and the major findings, wrongly believing that the reader will locate the necessary information later while reading the subsequent sections.[5] However, this "suspense" approach is not appropriate for scientific writing. To interest the reader, scientific authors should be direct and straightforward and present informative one-sentence summaries of the results and the approach.

Part 4 Discussion of the Results

For many scientists, writing a Discussion section is as scary as starting a paper. Most of the fear comes from the variation in the section. Since every paper has its unique results and findings, the Discussion section differs in its length, shape, and structure. However, some general principles of writing this section still exist. Knowing these rules, or "moves," can change your attitude about this section and help you create a comprehensive interpretation of your results.

The purpose of the Discussion section is to place your findings in the research context and "to explain the meaning of the findings and why they are important, without appearing arrogant, condescending, or patronizing". The structure of the first two moves is almost a mirror reflection of the one in the Introduction. In the Introduction, you zoom in from general to specific and from the background to your research question; in the Discussion section, you zoom out from the summary of your findings to the research context, as shown in Table 4.

Table 4 **Moves in research paper discussions**

Move 1 The study's major findings
a. state the study's major findings;
b. explain the meaning and importance of your finding;
c. consider alternative explanations of the findings.
Move 2 Research context
a. compare and contrast your findings with those of other published results;
b. explain any discrepancies and unexpected findings;
c. state the limitations, weaknesses, and assumptions of your study.

Table 4 (continued)

Move 3 Closing the paper
a. summarize the answers to research questions;
b. indicate the important of work by stating applications, recommendations, and implications

The biggest challenge for many writers is the opening paragraph of the Discussion section. Following the moves in Table 4, the best choice is to start with the study's major findings that provide the answer to the research question in your Introduction. The most common starting phrases are "Our findings demonstrate…," or "In this study, we have shown that …," or "Our results suggest …" In some cases, however, reminding the reader about the research question or even providing a brief context and then stating the answer would make more sense. This is important in those cases where the researcher presents a number of findings or where more than one research question was presented. Your summary of the study's major findings should be followed by your presentation of the importance of these findings. One of the most frequent mistakes of the novice writer is to assume the importance of his findings. Even if the importance is clear to you, it may not be obvious to your reader. Digesting the findings and their importance to your reader is as crucial as stating your research question.

* * * * * * * * * * * * * * * * * Explanations * * * * * * * * * * * * * * * * *

* * * * * * * * * * * * * * * * * New Words and Expressions * * * * * * * * * * * * * * *

1. motivation [ˌmotəˈveʃən] n. 动机；积极性；推动
2. cognitive [ˈkɑɡnətɪv] adj. 认知的，认识的
3. interrupt [ˈɪntəˈrʌpt] vi. 打断；打扰
   vt. 中断；打断；插嘴；妨碍
   n. 中断
4. hypotheses [haɪˈpɑθəsiz] n. 假定；臆测
5. elaborate [ɪˈlæbəret] adj. 精心制作的；详尽的；煞费苦心的
   vt. 精心制作；详细阐述；从简单成分合成
   vi. 详细描述；变复杂
6. sift [sɪft] vt. 筛选；撒；过滤；详查
   vi. 筛；详查；撒下；细究
7. succumb [səˈkʌm] vi. 屈服；死；被压垮
8. frustrating [ˈfrʌstretɪŋ] adj. 令人沮丧的
9. procrastinate [proˈkræstɪnet] vt. 耽搁，延迟
   vi. 耽搁，延迟
10. scrutinize [ˈskrutənaɪz] vi. 详细检查；细看
    n. 仔细或彻底检查
    vt. 细阅；作详细检查
11. meticulous [məˈtɪkjələs] adj. 一丝不苟的；小心翼翼的；拘泥小节的
12. intimidating [ɪnˈtɪmɪdetɪŋ] adj. 吓人的

*** * * * * * * * * * * * * * * * * * Complicated Sentences * * * * * * * * * * * * * * * * * ***

1. Whether you have written 100 papers or you are struggling with your first, starting the process is the most difficult part unless you have a rigid writing schedule.

【译文】无论你是已经撰写了 100 篇论文，还是你正在第一次写论文上挣扎，除非你有一个严格的写作计划，开始这个过程是最难的部分。

【说明】whether...or...表示“无论……还是……”，first 后面省略了“paper”一词。“starting the process”是动名词结构，作为该句主语。

2. As Paul Silvia explains, “Revising while you generate text is like drinking decaffeinated coffee in the early morning: noble idea, wrong time”

【译文】正如保罗・西尔维亚（Paul Silvia）解释的那样，“撰写初稿时的修改就像在清晨喝去掉咖啡因的咖啡一样：高贵的想法，错误的时间”。

【说明】decaffeinated coffee 是指去咖啡因的咖啡，这句话的含义是，修改文字固然很好，但在写初稿的时候修改文字，容易陷入繁杂的细节，反而影响论文的表达。就像喝去掉咖啡因的咖啡虽然很好，但在清晨喝不能起到使人清醒的作用，因此是“高贵的想法，错误的时间”。

3. Interestingly, recent studies have reported that the Materials and Methods section is the only section in research papers in which passive voice predominantly overrides the use of the active voice.

【译文】有趣的是，最近的研究报道说，材料与方法部分是论文中唯一的被动语态使用显著超过主动语态的部分。

【说明】interestingly 为 interesting 的副词形式；in which 引导的从句是用来修饰 section 的，passive voice 和 active voice 分别表示被动语态和主动语态；predominantly override 的意思是显著超过。

4. That is why it is vital to use all your writing skills to objectively present your key findings in an orderly and logical sequence using illustrative materials and text.

【译文】这就是为什么使用所有的写作技巧，以一个有序的、条理分明的顺序并用说明性的材料和文本来客观地呈现你的关键发现至关重要。

【说明】why it is vital to...表示为什么……是至关重要的。objectively present 表示客观的呈现。in an orderly and logical sequence 表示有序并富有逻辑的。

5. As a result, many novice writers do not present their experimental approach and the major findings, wrongly believing that the reader will locate the necessary information later while reading the subsequent sections.

【译文】因此，许多新手作家并没有提出他们的实验方法和主要发现，错误地相信之后读者会在阅读随后的章节时会找到必要的信息。

【说明】novice writers 指初学者；wrongly believing that 是主句的补充分句，以动名词结构引导，意思是“错误的认为”；locate 此处是定位或找到的意思；subsequent section 意指后面的章节

****************** **Summary of Glossary** ******************

1. struggling with 与……挣扎
2. physical exercises 体育锻炼
3. spills over 溢出
4. logical fallacy 逻辑谬论
5. novice writer 写作新手
6. supporting points 支撑观点
7. passive voice 被动语态
8. active voice 主动语态

****************** **Exercises** ******************

(1) Doing the hardest task first thing in the morning contributes to the sense of accomplishment during the rest of the day, because the ________ feeling spills over into our work and life and has a very positive effect on our overall __________.

(2) The outline of an academic paper is not an outline with Roman numerals for each section. It is more similar to a ____________ for your paper which forms a ________ for your paper and helps generate ________ and formulate __________.

(3) When you create the first draft, do not __________ to the temptation of editing, such as choose a better ________ or better ________ or halt to improve your ____________. Pour your ideas into the paper and leave ________ and __________ for later.

(4) In the Materials and Methods section, the most important goal is to be as ________ as possible by providing enough __________ and ________. In the end,the purpose of this section is to allow other researchers to ________ and ________ your work.

************* **Word Building (28) tele-; terr-** *************

1. tele- [前缀]，表示：远的

television	n.	电视
telescope	n.	望远镜
telegraph	n.	电报
telecommunication	n.	电讯

2. terr- [词根]，表示："土地"

terra	n.	土地
terrace	n.	梯田
territory	n.	领土
mediterranean	n.	地中海

******************** **Text Translation** ********************

第二十八章 怎样写出你的第一篇学术论文

第一节 预先安排好写作时间

无论你是已经撰写了 100 篇论文，还是你在第一次写论文上挣扎，除非你有一个严格的写作计划，开始这个过程是最难的部分。写作是很困难的。这是一个十分困难的集中注意力和脑力劳动的过程。正如在海斯的写作研究框架中所述：“这是一种需要动机的生成活动，它是一种需要认知过程和记忆的智力活动。”在他的书《如何写作：一本实用的学术写作指南》中，Paul Silvia 说，对于一些人来说，“铭记死者比写一篇关于他的文章更容易。”就像任何类型的辛勤工作一样，除非你经常练习，否则你不会成功。如果你一年没有进行体育锻炼，只有经常的锻炼，才能使你恢复良好的体型。同样的常规练习，或者称之为“写作会议”，是成为一个多产作者所需要的。在你的日常工作日程中选择 1～2 小时的时间，并把它们视作不可取消的预定时间。当哪块时间确定为写作时间时，你应该选择最适合这类工作的时间。对很多人来说，早晨的效率更高。一位耶鲁大学的研究生在早上 8 点到 9 点写作了一个学期，那时她的实验室是没有人的。在学期结束时，她惊讶地发现自己竟然完成了如此多的工作，甚至都没妨碍到她正常的实验时间。此外，在早上做第一件事的时候，做一件最困难的事情，会在接下来的一天中给自己带来成就感。这种积极的感觉会渗透到我们的工作和生活中，对我们的整体态度有非常积极的影响。（单词个数：293）

第二节 从 大 纲 开 始

既然你已经预定了时间，你需要决定如何开始写作。最好的策略是从大纲开始。这将不是你习惯的一个大纲，每个部分都有罗马数字以及主题句和支持点的整齐并列。这个大纲将类似于你的论文模板。首先，大纲将为你的论文形成一个结构;它将有助于产生想法和制定假设。遵循乔治・怀特塞德斯的建议“……从一张空白的纸张开始，以任何顺序写下所有你想到的关于文章的重要想法”。使用表 1 作为大纲的起点。 包括你的图形部分（图、表、公式、方程式和算法），并列出你的发现。这些将构成你大纲的第一级，最终会随着你的详细阐述而展开。（单词个数：151）

下一个阶段是添加上下文和结构。 在这里，你将所有的想法分为几个部分：介绍、方法、结果和讨论/结论（表 2)。这一步将有助于增加您的工作的一致性并筛选你的想法。（单词个数：38）

表 1	大纲——第 1 级（单词个数：36）
1. 我论文的主题是什么？	
2. 这个主题为什么重要？	
3. 我可以怎样制定我的假设？	
4. 我得到的结果是什么？	
5. 我主要的发现是什么？	

表 2	大纲——第 2 级（单词个数：91）
介绍	
1. 你的研究为什么重要？	
2. 对于这个主题你知道些什么？	
3. 你的假设是什么？	
4. 你的目标是什么？	
材料与方法	
1. 你使用了什么材料？	
2 你研究的科目是什么？	
3. 你的研究设计是什么？	
4. 你遵循什么程序？	
结果	
1. 你最重要的结果是什么？	
2. 你所支持的结果是什么？	
讨论与结论	
1. 研究的主要发现是什么？	
2. 结论的意义是什么	

第三节 继续写草稿

当你得到足够的反馈，并决定你将要提交的期刊后，真正的写作过程就开始了。把你的大纲复制到一个单独的文件中，并在每个点上展开延伸，添加数据并详细说明细节。当你创作你的初稿时，不要屈从于编辑的诱惑。不要放慢速度去选择一个更好的词或短语，不要停下来去改进你的句子结构。将你的想法倾注在纸上，之后再进行修改和编辑。正如保罗·西尔维亚（Paul Silvia）解释的那样，“撰写初稿时的修改，就像在清晨喝去掉咖啡因的咖啡一样：高贵的想法，错误的时间”。（单词个数：110）

许多学生抱怨说，他们不是富有成效的作者，因为他们经历了写作的瓶颈。盯着一个空白的屏幕是令人沮丧的，但是你的屏幕并不是空的：你有一篇文章的模板，所有你需要做的就是填补空白。事实上，写作的瓶颈是科学家的逻辑谬误 ——这只是拖延的借口。当科学家开始撰写论文时，他们已经有文件、有数据、有材料和实验笔记，有一些图形效果和具有结果的表格。他们所需要做的就是仔细检查这几块资料，然后把它们放到一篇综合的论文里。（单词个数：108）

如果你还在为如何开始写论文而纠结，那就先写材料和方法部分。既然你已经有了所有的笔记，那么描述实验设计和程序，就应该不成问题了。你在这部分中，最重要的是通过提供足够的细节和参考文献来尽可能地明确目标。最后，这个部分的用途是允许其他研究人员评估和重复你的工作。（单词个数：75）

有趣的是，最近的研究报道说，材料与方法部分是论文中唯一被动语态使用显著超过主动语态的部分。例如，在基于 100 万字的生物科学实验全文论文语料库中，Martínez 在“方法”部分中就表现出主动语态使用的显著下降。根据作者的说法，“我们”的主动语态仅被用作一种工具，用来揭示在设计和执行实验工作过程中程序决策的个人责任。这意味着，虽然论文的所有其他部分都使用主动语态，但被动语态在材料和方法部分中仍然占主导地位。（单词个数：119）

写作材料和方法部分是一个细致而耗时的任务，需要极高的准确性和清晰性。 这就是为什么当你完成草稿时，你应该尽可能多地向同事询求反馈意见。本部分的许多读者将帮助你识别缺少的链接并改进这部分的技术风格。（单词个数：57）

对于许多作者来说，编写“结果”部分比编写“材料和方法”部分更令人生畏。如果人们对你的论文感兴趣，他们会对你的结果感兴趣。这就是为什么使用所有的写作技巧，以一个有序的、条理分明的顺序并用说明性的材料和文本来客观地呈现你的关键发现至关重要。（单词个数：59）

你的“结果”部分应该分为不同的段落或子部分，各个部分介绍实验的目的、实验方法，以及包括文本和图表内容（表格、图形、原理图、算法和公式）的数据以及数据说明。对于大多数期刊，你的数据说明将包括图表形式呈现的数据摘要和对最重要发现的解释说明。这个数据呈现不应该重复图形部分中的数据，而是强调最重要的一点。在“标准”论文写作方法中，你的结果部分应排除数据解释，将其留作讨论部分。然而，解释说明是渐渐地悄然地渗透到论文的：“减少数据，从数据中归纳，突出科学案例都是高度解释性的过程。现在应该清楚的是，我们不可以让研究报告中的数据自言自语；在总结我们的结果时，我们要向读者说明它们”。因此，包括《实验医学杂志》和《临床调查杂志》在内的许多期刊都使用联合“结果/讨论”部分，文中解释紧随结果其后。（单词个数：181）

现在你几乎是起草了论文的一半，现在是更新你的大纲的时候了。在描述你的方法和结果部分的同时，你们中的许多人偏离了原本的大纲，并重新组织了你的想法。因此，在继续创作介绍部分之前，请重新阅读“方法和结果”部分，并根据研究重点改变大纲。 更新后的大纲将帮助你审查你的论文的总体情况、主题、主要思想和目的，这些对于撰写你的介绍都很重要。（单词个数：89）

撰写介绍部分的最佳方法是遵循表 3 所示的三步法。（单词个数：17）

表 3　　进行论文的简介部分（单词个数：87）

步骤 1　建立一个研究领域
表明一般研究领域是重要的、核心的，某种程度上也是存在问题的。
步骤 2　找到一个定位
表明与之前研究存在的差异，或以某种方式拓展之前的知识。
步骤 3　充实这个定位
a. 概述目的或阐述现今研究的本质；
b. 列出研究问题或假设；
c. 宣布主要发现；
d. 声明该项研究的价值；
e. 指出论文的结构

你的大纲步骤和信息可以有效地帮助你创作简介，而且不会丢失任何步骤。一些学术作家认为读者“应该遵循论文”来找出你个人的方法论和你发现的答案。因此，许多新手作家并没有提出他们的实验方法和主要发现，错误地相信之后读者会在阅读随后的章节时会找到必要的信息。然而，这种“悬念”的做法不适合科学写作。为了让读者感兴趣，科学作者应该是直接而坦率的，并提供有关结果和方法的一句话信息性总结。（单词个数：103）

第四节 结 果 讨 论

对于许多科学家来说，撰写论文的讨论部分与开始部分一样可怕。大部分恐惧来自该部分的变化。由于每篇论文都有其独特的结果和发现，所以讨论部分的篇幅，框架和结构不同。但是，编写本节的一般原则仍然存在。了解这些规则或“步骤”可以改变你对这部分的态度，并帮助你创建对结果的全面解释。（单词个数：76）

讨论部分的目的是将你的发现放在研究环境中去，并“解释这些发现的意义以及它们为什么重要，而不是显得傲慢、居高临下或屈尊俯就”。前两个步骤的结构几乎是引言中的一个真实写照。在介绍部分，你可以从一般放大到特殊，从背景深入到你研究的问题；在“讨论”部分，你将从研究结果总结中缩小到研究背景，如表 4 所示。（单词个数：93）

表 4　　进行论文的讨论部分（单词个数：93）

步骤 1　研究的主要发现
a. 说明研究的主要发现；
b. 交代你的发现的意义和重要性；
c. 考虑对这些发现的其他解释。
步骤 2　研究背景
a. 把你的发现与已发表的研究结果对比比较；
b. 解释任何不一致和出乎意料的发现；
c. 陈述你的研究的局限性、劣势和假设。
步骤 3　论文收尾
a. 总结一下研究问题的答案；
b. 通过说明应用、建议和启示来阐释工作的重要性

许多作者面临的最大挑战是“讨论”部分的开篇。按照表 4 中的步骤，最好的选择是从研究的主要发现开始，为你的简介中的研究问题提供答案。最常见的起始词是“我们的发现表明……”或“在这项研究中，我们已经表明……”或“我们的结果表明……”然而，在某些情况下，把有关研究问题对读者进行提示，甚至提供一个简短的背景，然后再说明答案会更有意义。这在研究人员提出了一些研究结果，或提出了多个研究问题的情况下很重要。你对研究结果的总结应该放在对这些发现重要性的介绍之后。新手最常见的错误之一就是假定他的发现的重要性。即使你自己对你的发现的重要性是清楚的，但你的读者可能不是很清楚。帮助读者领会结果及其重要性，与说明研究问题一样至关重要。（单词个数：184）

Chapter 29 EST Translation

EST stands for English of Science and Technology. The translation of scientific and technical texts is considerably different from that of literature works and religious books. In the translation process, different criterion and rules are applied to different types of texts. As a result, translators must understand the characteristics of the source text before he sets his pen to paper, and adopt proper translation techniques in translation.

Part 1 Lexical Features of EST—Wide Use of Technical Terms

In the *A Textbook of Translation,* Newmark writes, "Technical translation is primarily distinguished from other forms of translation by terminology, although primarily distinguished from other forms of translation by terminology, although terminology usually only makes up about 5-10% of the text." (Newmark, A Textbook of Translation, 151). The more professional the source text is, the more technical terms it has. Therefore, it is necessary to analyze the characteristics of EST in terms of vocabulary. It is known that a large number of technical terms have emerged and been introduced into English vocabulary in today's information age.

Generally speaking, most of the technical terms are used in a certain scientific field with a unique meaning. For example:

Polypropylene is the second resin in the family of polyolefin composed of long chain saturated hydrocarbons.

译：聚丙烯是含有长链饱和烃的聚烯烃系列的第二个树脂。

The underlined terms above are only used in a certain scientific field—Chemistry. They seldom appear in other fields and daily language. At the same time, there are many words which are used both in daily life and certain scientific fields. These words are called semi-technical words. For example:

1) We missed the morning bus.

译：我们没赶上早上的公共汽车。

2) The bus runs to the connecting lug over there.

译：母线连接到那边的线鼻子上。

In sentence 1), the word "bus" refers to a transportation tool in daily English. However, in the field of Electronics, it means a circuit that connects the main parts of a computer so that signals can be sent from one part of the computer to another.

Part 2 Syntactical Characteristics of EST

1. Wide Use of Passive Voice

EST differs from other types of English in terms of voice, tense, and sentence structure. The predominance of the passive voice is one of the greatest features of EST. It is estimated that one third of the verbs in EST are in passive voice. This is because, in English language, the subject carries the most important information and catches the reader's attention first. In EST, which emphasizes the phenomenon itself, "what is done" is more important than "who does it". Sentences in passive voice highlight the phenomenon to be described and help to avoid subjectivity. For example:

1) If an oscillatory motion was superimposed on steady shearing, the maximum torque on the top plate during the combined motion was scarcely more than the torque in steady shearing alone; the minimum was considerably less.[1]

译：如果往稳恒剪切上叠加振荡运动，则在联合运动过程中作用于顶板上的最大转矩很少大于仅只稳恒剪切时的转矩，而最小转矩则比后者小得多。

2) The hydrographs from each *X* increment of rain are determined from the *X*-hour unit hydrograph. The ordinates are then added at corresponding times to determine the total hydrograph.

译：每*X*历时降雨增量的单位过程线将由*X*小时单位过程线确定，然后在相应时间叠加到纵坐标，以确定总的过程线。

2. Wide Use of Nominalization Structure

Nominalization structure is an important stylistic feature of EST. Nominalization, just as its name suggests, changes the verbs or adjectives into nouns or noun phrases to achieve grammatical function of nouns, while keeping the meaning of verbs and adjectives.[2] The nominalization part can be used as subject, object, prepositional object, predicative, etc. With nominalization structure, sentences can be more concise and objective, which is needed in scientific and technological texts.

3. Wide Use of Long Sentences

In EST, long sentences are often used to explain terms, describe an object, or illustrate technical process. These long sentences often contain more than two coordinate clauses or subordinate clauses, which make the translation harder. Therefore, translators need to analyze the sentence structure before translating, find out the main sentence, participle phrases, clauses etc., and understand the logical relationship between each part. Only by doing this could a translator do the translation job effectively and efficiently.

Part 3 EST Translation Techniques

To master certain translation techniques and skills is a must for translators. Consciously or not, we use translation skills to achieve a better translation. Actually, some of the mistranslation is caused by the inappropriate words or structures used, not by incorrect understanding of source text.

1. Conversion of Part of Speech

a. Nouns into verbs

Nominalization of verbs is common in EST. while translating nominalization structures, we can choose to convert some nouns or noun phrases into verbs to have a more coherent Chinese expression.

The application of these theories to the study of water resources development has yielded good results.

1st version：这些理论的运用研究水资源规划，已经收到了良好的效果。

2nd version：运用这些理论研究水资源规划，已经收到了良好的效果。

The use of the noun—“application “is appropriate in the original text. However, when translated into Chinese，it should be converted into a verb. Otherwise, as the example shows, the translation could be confusing for the target readers.

b. Adjectives into verbs

With this technology, it is possible to develop large scale power generation from water in this country, especially in wet mountainous regions.

译：有了这种新技术，就可以在这个国家进行大规模水力发电，尤其是山区水能资源丰富的地区。

c. Nouns into adjective

At present, increasing water use efficiency is an absolute necessity.

译：目前提高水资源的使用效率是绝对必要的。

d. Prepositions into verbs

In general, positive or negative rake tools can be used on stainless steel.

译：通常，正前角和负前角的刀具都可以用来加工不锈钢。

2. Translation of the Passive sentences

As discussed in the previous part, about one third of the EST sentences are in passive voice. However, Chinese language does not take such a number. While translating the passive sentences, we need take the habit of Chinese expression into consideration.[3] There are three ways to translate passive voiced sentences.

a. Translating into the active voice

Today different measures are taken to prevent corrosion.

译：今天，为预防腐蚀人们采取了各种措施。

b. Omitting the subject

Gully erosion should be paid attention to.

译：应该注意沟道侵蚀。

c. Remaining the passive voice

The circuit is broken by the in sulating material.

译：电路被绝缘材料隔断了。

3. Amplification, Omission and Repetition

Since Chinese differs greatly from English, translation skills like amplification，omission and repetition of words are often used in the English-Chinese translation. Translators add, reduce or repeat words with the purpose to express the meaning more accurately and completely. When you use these skills in translating EST, remember that what is added or reduced is not the meaning but words only.

a. Amplification

Amplification helps to clarify the meaning and yield better translation result.

1) It proves that world fresh water is becoming substantially stressed nowadays.

译：事实证明地球淡水正在变得越来越紧张。

2) It turns out that many of China's river reaches are so polluted that they are unsuitable for direct human contact.

译：人们终于弄明白许多河段污染严重，不适合人类直接接触。

b. Omission

In translating EST, word by word translation is not encouraged and omission is one of the most frequently-used translation methods. In the translation process, the pronoun and conjunction words are often omitted. If a word's meaning is already expressed in another word, we can reduce it in translation to achieve conciseness in language. For example:

1) Up and down motion can be changed to circular motion.

译：上下运动可以改变为圆周运动。

2) Automatic lathes perform basically similar functions but appear in a variety of forms.

译：各种自动机床的作用基本相同，但是形式不同。

c. Repetition

We want such material as can bear high temperature and pressure.

译：我们需要能耐高温高压的材料。

4. Translation of long sentences

As for the long and complex sentences, it is suggested to separate some words, phrases from the whole sentence and translate them into two or more sentences. For example:

1) Lower temperature is associated with lower growth rates.

译：温度一低，生长速度就慢下来。

2) The atom consists of a nucleus in the center，formed of protons and neutrons，and electrons moving round the nucleus in circular or elliptical orbits.[4]

译：原子由一个位于中心的原子核和一些电子组成。原子核又由质子和中子组成，电子则以圆形或椭圆形轨道绕原子核运行。

3) Application an *X*-hour unit graph to design rainfall excess amounts other than 10 mm is accomplished simply by multiplying the rainfall excess amount by the unit graph ordinates, since the runoff ordinates for a given duration are assumed to be directly proportional to rainfall excess.

译：采用 *X*-h 的单位线计算并非 10mm 的雨量过程时，可简单采用净雨深乘以单位线的纵坐标。因为对一个既定的时段来说，单位线假定径流与净雨直接成比例。

Part 4 Conclusion

With the increasing development of science and technology, EST translation has become increasingly important in the exchange of scientific and technological information. As a practical variety of English, EST has its own characteristics in vocabulary, grammar and style, which make it harder to translate. In order to translate EST in a satisfying way, translators should broaden their knowledge in the field of science and technology, understand the features of EST, improve their bilingual capacity and properly apply some translation techniques and skills.

* * * * * * * * * * * * * * * * * Explanations * * * * * * * * * * * * * * * * *

* * * * * * * * * * * * * * * * New Words and Expressions * * * * * * * * * * * * * * *

1. terminology [ˌtɜːmɪˈnɒlədʒɪ] n.术语，术语学
2. professional [prəˈfɛʃənl] adj.专业的；职业的；职业性的
   n. 专业人员；职业运动员
3. analyze [ˈænəˌlaɪz] vt. 对…进行分析，分解（等于 analyse）
4. vocabulary [vəˈkæbjəlɛri] n. 词汇；词表；词汇量
5. tense [tens] adj. 紧张的；拉紧的
   v. 变得紧张；使拉紧
   n. 时态
6. stainless [ˈsteɪnlɪs] adj. 不锈的；纯洁的，未被玷污的；无瑕疵的
7. highlight [ˈhaɪlaɪt] vt. 突出；强调；使显著
   n. 最精彩的部分；最重要的事情；加亮区
8. steady [ˈstedɪ] adj.稳定的；不变的；沉着的
   vt. 使稳定；稳固；使坚定
9. nominalization [ˌnɔminəlaiˈzeiʃən] n. 名词化；名物化
10. feature [ˈfiːtʃə] n.特色，特征；容貌；特写或专题节目
    vi. 起重要作用
11. subjective [səbˈdʒektɪv] adj. 主观的；个人的；自觉的
12. concise [kənˈsaɪs] adj. 简明的，简洁的
13. objective [əbˈdʒektɪv] adj. 客观的；目标的；宾格的
    n. 目的；目标；
14. illustrate [ˈɪləstreɪt] vt. 阐明，举例说明；图解
15. conscious [ˈkɒnʃəs] adj.意识到的；故意的；神志清醒的
16. convert [kənˈvɜːt] vt. 使转变；转换……
17. broaden [ˈbrɔːd(ə)n] vt. 使扩大；使变宽
18. lexical [ˈleksɪk(ə)l] adj. 词汇的
19. syntactical [sɪnˈtæktɪkəl] adj. 句法的

* * * * * * * * * * * * * * * * * Complicated Sentences * * * * * * * * * * * * * * * * *

1. If an oscillatory motion was superimposed on steady shearing, the maximum torque on the top plate during the combined motion was scarcely more than the torque in steady shearing

alone.

【译文】如果往稳恒剪切上叠加振荡运动，则在联合运动过程中作用于顶板上的最大转矩很少大于仅只稳恒剪切时的转矩。

【说明】句子前部分是由 if 引导的条件状语从句。主句是主系表结构（the maximum torque was more than…），on the top plate 和 during the combined motion 分别做地点状语和时间状语修饰主语 the maximum torque。

2. Nominalization, just as its name suggests, changes the verbs or adjectives into nouns or noun phrases to achieve grammatical function of nouns, while keeping the meaning of verbs and adjectives.

【译文】名词化，顾名思义，是将动词或者形容词转变为名词或者名词词组，从而使名词或者名词词组具有动词或形容词的含义，同时又保留了名词的语法功能。

【说明】just as its name suggests 做句子的插入语成分。To achieve grammatical…为不定式做目的状语。

3. While translating the passive sentences, we need take the habit of Chinese expression into consideration.

【译文】在翻译被动句时，译者需要考虑中文的表达习惯。

【说明】Take…into consideration 意为“将……纳入考虑”。

4. The atom consists of a nucleus in the center, formed of protons and neutrons, and electrons moving round the nucleus in circular or elliptical orbits.

【译文】原子由一个位于中心的原子核和一些电子组成。原子核又由质子和中子组成，电子则以圆形或椭圆形轨道绕原子核运行。

【说明】此句的核心结构为 The atom consists of…。 Formed of…是过去分词做定语修饰 nucleus。and 连接 a nucleus 和 electrons，并列结构做 consist of 的宾语。Moving round…是现在分词结构做定语修饰 electrons。

*********************** Summary of Glossary ***********************

1. atom 原子
2. proportional 成比例的
3. insulating 绝缘的；隔热的
4. erosion 侵蚀
5. positive 正的
6. negative 负的
7. rainfall 降雨
8. sentence structure 句子结构
9. participle phrases 分词短语
10. saturated 饱和的；渗透的
11. maximum 最大限度
12. minimum 最小限度

13. superimpose 添加，附加
14. lug 接线片
15. circuit 电路

********************* Abbreviations (Abbr.) *********************

EST	English for Science and Technology	科技英语

*********************** Exercises ***********************

Find answers to the following questions from the text:

1. What is the lexical feature of EST? Please give some examples.
2. What are the syntactical characteristics of EST? Please give some examples.
3. What are the translation techniques we often use in EST translation?

************* Word Building (29) poly-; sur- ***************

1. poly- [前缀]，表示：多的，各种的

technic	adj.	polytechnic	adj.	各种工艺的，综合技术的
morphism	n.	polymorphism	n.	多态性，多形性
polyester	n.	聚酯		

2. sur- [前缀]，表示：超，外加

surface	n.	外加
surtax	n.	附加税
surplus	n.	过剩，剩余

******************** Text Translation *********************

第二十九章 科技英语翻译

EST 是 English for Science and Technology 的简称，意为科技英语。翻译科技英语和翻译文学作品、宗教著作有着较大的差异。因为不同体裁的文本有不同的语言特点，翻译实践中采用翻译方法和标准也不尽相同。因此，译者需要在提笔翻译之前理解源语的特点和规律，并且在翻译实践中使用一定的翻译技巧和方法。（单词个数：87）

第一节 科技英语的词汇特点——大量使用专业词汇

Newmark 在他的著作 *A Textbook of Translation* 一书中指出“科技类翻译和其他类别翻译的最显著差异就是专业词汇翻译，尽管这些专业词汇只占整个文本的 5%~10%”(Newmark, *A Textbook of Translation*, 151)。而且，原文越专业，科技术语就越多。因此，在学习科技英语的过程中，译者有必要了解科技英语的词汇特点。我们都知道，当今信息爆炸的时代涌现了不计其数的科技词汇。总的来说，大多数科技词汇仅用于某一专业领域，这类词汇被称作专业特有词汇。例如：（单词个数：63）

Polypropylene is the second resin in the family of polyolefin composed of long chain saturated hydrocarbons. 聚丙烯是含有长链饱和烃得聚烯烃系列的第二个树脂。（单词个数：16）

该例中划线部分的词汇仅用在化学领域，很少出现在其他专业领域或者日常生活中。同时，有一部分词汇既是科技专业词汇，又是日常生活中的普通词汇。例如：（单词个数：49）

1) We missed the morning bus. 我们没赶上早上的公共汽车。（单词个数：6）

2) The bus runs to the connecting lug over there. 母线连接到那边的线鼻子上。（单词个数：11）

在这两个句子中，bus 有不同的含义：① 句中意为“公共汽车”；② 中意为母线。（单词个数：47）

第二节 科技英语的句法特点

1. 大量使用被动语态

科技英语在语态、时态、和句子结构上有其自身的特点。大量被动语态的运用是科技英语最显著的特征之一。据估计，科技英语文本中 1/3 的动词都是用的被动语态。在英语语言中，主语是很重要的成分，传递着句子重要的信息，而且读者一眼就能看到。而科技英语主要是讨论事物的发展过程或者阐述科学规律，关注的是论证的结果，而动作的执行者显得不太重要。在这种情况下，句子的核心应该是陈述事物或者是结果。为满足这些要求，科技英语往往采用被动语态。例如：（单词个数：198）

1) If an oscillatory motion was superimposed on steady shearing, the maximum torque on the top plate during the combined motion was scarcely more than the torque in steady shearing alone; the minimum was considerably less.

译：如果往稳恒剪切上叠加振荡运动，则在联合运动过程中作用于顶板上的最大转矩很少大于仅只稳恒剪切时的转矩，而最小转矩则比后者小得多。（单词个数：35）

2) The hydrographs from each *X* increment of rain are determined from the *X*-hour unit hydrograph. The ordinates are then added at corresponding times to determine the total hydrograph.

译：每 *X* 历时降雨增量的单位过程线将由 *X* 小时单位过程线确定。然后在相应时间叠加到纵坐标，以确定总的过程线。（单词个数：28）

2. 大量使用名词性结构

名词性结构是科技文章中的一个重要文体特征。名词化，顾名思义，是将动词或者形容词转变为名词或者名词词组，从而使名词或者名词词组具有动词或形容词的含义，同时又保留了名词的语法功能。名词化结构可做主语、宾语、介词宾语等。科技类文章要求行文简洁、表达客观，而名词化结构的使用可以更好地实现这一目标。（单词个数：70）

3. 大量使用长句

在科技类文本中，大量长句用于解释术语，描述事物及技术流程。一般而言，这些长句是由多个并列句或者从句构成，这给长句的翻译造成了不少的困难。因此，译者在翻译的过程中，首先要分析句子的结构。通过分析句子结构，确定主句、从句分词短语等句子构成部分，从而理解每个部分之间的逻辑关系。只有这样，才能对长句实现更有效的翻译。（单词个数：76）

第三节 科技英语翻译技巧

掌握一定的翻译技巧对任何译者来说是必须的。在翻译过程中，我们有意无意地使用一些翻译技巧，以达到更好的翻译效果。实际上，一些错译并不是因为对原文的错误理解而造成的，而是翻译过程中使用了不当的词汇或是语言结构。（单词个数：46）

1. 根据语境选择恰当的词性（单词个数：31）

a. 名词转换成动词

前面讲过，名词化结构被广泛用于科技英语中。但是，有时候动词更加符合中文的逻辑和表达习惯。例如：（单词个数：39）

The application of these theories to the study of water resources development has yield good results.

1st version：这些理论的运用研究水资源规划，已经收到了良好的效果。

2nd version: 运用这些理论研究水资源规划，已经收到了良好的效果。

名词“application”在源语中是十分贴切的。但是在翻译时，应该将其转换成动词，否则正如例子（译 1）所示，读者将难以理解译文。（单词个数：39）

b. 形容词转化成动词

例：With this technology, it is possible to develop large scale power generation from water in this country, especially in wet mountainous regions.

译：有了这种新技术，就可以在这个国家进行大规模水力发电，尤其是山区水能资源丰富的地区。（单词个数：22）

c. 名词转化成形容词

例：At present, increasing water use efficiency is an absolute necessity.（单词个数：10）

译：目前提高水资源的使用效率是绝对必要的。

d. 介词转化成动词

例：In general，positive or negative rake tools can be used on stainless steel. （单词个数：14）

译：通常，正前角和负前角的刀具都可以用来加工不锈钢。

2. 被动句的翻译方法

正如前面一部分所讲，科技英语中大约有三分之一的被动句型。然而，中文的被动句远少于三分之一。因此，在翻译被动句时，译者需要考虑中文的表达习惯。通常，我们有三种方式来处理被动句的翻译。（单词个数：50）

a. 将被动语态翻译成主动语态

例：Today different measures are taken to prevent corrosion.（单词个数：9）

译：今天，为预防腐蚀人们采取了各种措施。

b. 省略主语

例：Gully erosion should be paid attention to.（单词个数：7）

译：应该注意沟道侵蚀。

c. 保持被动语态

例：The circuit is broken by the insulating material.（单词个数：8）

译：电路被绝缘材料隔断了。

3. 增译、省译和重译法

由于中文和英文有很大的不同，在英译中过程中常常会用到增译法、省译法和重译法。译者为了更加准确完整地表达原文意义，会添加、省略或是重复词语。一般来说，增译、省译和重译需要遵从一条原则，即增加或省略的不是意义而是词语。（单词个数：62）

a. 增译

例 1：It proves that world fresh water is becoming substantially stressed nowadays.

译：事实证明地球淡水正在变得越来越紧张。（单词个数：11）

例 2：It turns out that many of China’s river reaches are so polluted that they are unsuitable for direct human contact.（单词个数：20）

译：人们终于弄明白许多河段污染严重，不适合人类直接接触。

b. 省译法

省译法是科技英语翻译中使用最多的方法之一。科技英语翻译是为了翻译原文意义而非逐字翻译原文单词。因此，翻译过程中我们常常省略原句中的代词和连词。另外，如果一个单词的意义已经在另一个单词中体现了出来，我们也可以在翻译时省略该词。这样，译文将更为简洁。例如：（单词个数：55）

例 1：Up and down motion can be changed to circular motion（单词个数：12）

译：上下运动可以改变为圆周运动。

例 2：Automatic lathes perform basically similar functions but appear in a variety of forms.（单词个数：16）

译：各种自动机床的作用基本相同，但是形式不同。

c. 重译

例：We want such material as can bear high temperature and pressure.（单词个数：14）

译：我们需要能耐高温高压的材料。

4. 长句拆译法

对于长句或者是复杂的句子结构，建议将单词或者短语从整个句子中分离出来单独翻译。例如：（单词个数：30）

1) Lower temperature is associated with lower growth rates. （单词个数：10）

译：温度一低，生长速度就慢下来。

2) The atom consists of a nucleus in the center，formed of protons and neutrons，and electrons moving round the nucleus in circular or elliptical orbits.（单词个数：28）

译：原子由一个位于中心的原子核和一些电子组成。原子核又由质子和中子组成，电子则以圆形或椭圆形轨道绕原子核运行。

3) Application an *X*-hour unit graph to design rainfall excess amounts other than 10 mm is accomplished simply by multiplying the rainfall excess amount by the unit graph ordinates, since the runoff ordinates for a given duration are assumed to be directly proportional to rainfall excess.（单词个数 45）

译：采用 *X* 小时的单位线计算并非 10mm 的雨量过程时，可简单采用净雨深乘以单位线的纵坐标。因为对一个既定的时段来说，单位线假定径流与净雨直接成比例。

第四节　结　　论

随着科技的迅速发展，科技英语翻译正吸引着越来越多人的关注。作为应用类语言的一支，科技英语在词汇、句法和文体上有着自身的特点。这些特点加大了科技英语的翻译难度。为了获得满意的翻译作品，译者应加强自己在科技领域的知识，了解科技英语的特点，提高自身的双语能力并且有意识地运用一些翻译技巧。（单词个数：82）

Appendix 1 Test Example

____学年 第____学期考试试卷（A）

课程名称	专业英语阅读（水利）	课程编号		考核日期时间	
专业班级		需要份数		送交日期	
考试方式	开卷	试卷页数		A B 卷齐全	是✓ 否
命题教师		主任签字		备　注	

班级：____________　姓名：____________　学号：____________

一、名词解释题（共 30 分，每题 5 分，＞15words）

1. Water resources
2. Process of water cycle
3. Type of river closure
4. Dams
5. Hydroelectricity
6. Climate change

二、词组及短语汉译英（共 30 分，每题 2 分）

1. 水源　　2. 河流径流量　　3. 可行性研究
4. 蒸散发　　5. 坝型　　6. 地表水
7. 污染指标　　8. 大坝截流　　9. 连拱坝
10. 地下水水位　　11. 潮汐能　　12. 分期施工
13. 水污染指标　　14. 冲击式水轮机　　15. 抽水蓄能电站

三、英文摘要或总结（共 10 分，15-30 words）

Water has been getting more attention in China as it becomes scarcer due to the growing demand and the recent droughts that have affected the southwestern provinces and the downstream Yangtze River, "wet areas" in common perception. Indeed, such attention is well deserved. In Yunnan province alone, the direct economic loss due to drought accounted for more than 9 percent of the province's GDP in 2010. In the same year, more than 13 million hectares of crops were affected by drought in China.

四、综合英译汉（共 30 分）

Two changes are essential for ensuring a sustainable water future for China. First, most of the previous water resources development efforts were focused on hardware such as construction of hydraulic structures. But in the future, more efforts should be directed toward the improvement of water management institutions. Sound water policy and regulations can create incentives for users to save water and protect the environment.

Second, most of the previous water management methods largely depended on supply augmentation to meet increasing demand. Future water management should rely more on managing water demand to avoid uncontrolled growth and promote water saving through properly designed policies.

____学年 第____学期考试试卷（B）

课程名称	专业英语阅读（水利）	课程编号		考核日期时间	
专业班级		需要份数		送交日期	
考试方式	开卷	试卷页数		A B 卷齐全	是✓ 否
命题教师		主任签字		备　注	

班级：____________ 姓名：____________ 学号：____________

一、英语解释题（共 30 分，每题 5 分，＞15words）

1. Water cycle
2. Planning
3. Hydroelectricity
4. Cradge
5. Climate changes
6. Tidal power

二、词组及短语汉译英（共 30 分，每题 2 分）

1. 地表径流　2. 土壤侵蚀　3. 漫堤
4. 蒸散发　5. 雨强　6. 洪水与干旱
7. 上游防渗铺盖　8. 水质　9. 抽水蓄能电站
10. 大坝导流　11. 分期施工　12. 水轮机
13. 年发电量　14. 土石坝　15. 反击式水轮机

三、英文摘要或总结（共 10 分，15-30 words）

The drought damages are just part of the water problems that challenge a sustainable water future for China. Despite huge achievements in water development, China faces a number of water problems that may unfavorably affect its socioeconomic development, if not addressed properly and timely. For instance, in the North China Plain, the breadbasket of China, the overdraft of groundwater keeps lowering the water table over large continuous areas, by as much as 1 meter a year in certain areas, with deep cones of depression forming under several cities.

四、综合英译汉（共 30 分）

In recent decades, the combined average economic growth rate of this region exceeds 6 per cent per year. This needs a large quantity of energy, making the region energy-starved. The Himalayan range of the region has the potential of more than 85 000 MW of hydro-power (combined potentiality of Bhutan and Nepal). This is the opportunity to generate electricity in the region.

This potential of hydro-power still remains untapped. The development of this potentiality hangs over the mistrust, suspicion and doubt throughout the region. For instance, it is deeper in Nepal over the utilization of water for mutual benefit. Rivers either originating in Tibet or in Nepal have a large potential for hydro-power, which could be a boon if harnessed to meet the growing demand of electricity. To some extent, Bhutan has been successful in utilizing it as it sales energy to India. But, the hydro-power potential of Nepal remains to be tapped, because of the lack of vision of the Nepalese rulers.

Appendix 2 Answers to Exercises

Chapter 1 Water Resources

(1) Converting (saltwater) to freshwater is generally too (expensive) to be used for (industrial), agricultural or (household) purpose.

(2) Only 2.5% of the world's water (supply) is fresh water and 68.7% of that is (frozen), forming the (polar) icecaps, (glaciers), and icebergs.

(3) Surface water is (visible) above the ground (surface), such as (creeks), river, (ponds) and lakes.

(4) Ground water is water that either (fills) the spaces between soil (particles) or (penetrates) the (cracks) and spaces within rocks.

Chapter 2 Planning for Water Resources Development

(1) Planning can be defined as the orderly consideration of a project from the original (statement) of purpose through the evaluation of (alternatives) to the final (decision) on a course of action.

(2) An overall regional water-management plan, (developed) with care and closely coordinated (with) other regional plans, may be a useful tool in determining which of many possible actions should be (taken)。

(3) There is no substitute for "(engineering) judgment" in the selection of the method of approach (to) project planning.

(4) It is the basis for the decision to proceed (with) (or to abandon) a (proposed) project and is the most important aspect of the engineering for the project.

Chapter 3 Water Resources for Sustainable Development

(1) Many of these meanings are (encapsulated) in term "(sustainable) development" which is being (broadly) used nowadays.

(2) In fact, "sustainable development" is an old (concept) that has been used in the management of (renewable) natural resources to ensure that the rate of (harvesting) a resource is smaller than the rate of its (renewal).

(3) The (availability) of water in adequate quantity and quality is a necessary condition for sustainable development. Water, the basic (element) of the life support system of the planet, is (indispensable) to sustain any form of life and (virtually) every human activity.

(4) At first sight this may look like a relative (abundance) of water. However, these apparently

comforting global figures are largely (misleading) as water (availability) at smaller scales is concerned.

Chapter 4 Hydrology and Water Cycle

(1) The cycle of movement of water between (atmosphere), (hydrosphere), (lithosphere) and (biosphere) is termed the hydrologic cycle.

(2) This information is essential to (design) and (evaluation) of natural and man-made channels, bridge openings and dams.

(3) Hydrology is the science that encompasses the (occurrence), distribution, movement and (properties) of the waters of the earth and their relationship with the environment within each phase of the hydrology cycle.

(4) As water goes through its cycle, it can be a solid (ice), a liquid (water), or a gas (water vapor).

(5) Water that (runs off) into rivers flows into ponds, lakes, or oceans where it (evaporates) back into the atmosphere.

(6) Water vapor (condenses) into millions of tiny (droplets) that form clouds.

(7) As a result of (evaporation), (condensation) and (precipitation), water travels from the surface of the Earth goes into the atmosphere, and returns to Earth again.

Chapter 5 Principle of Hydrology—Unit Hydrographs

(1) Ways to predict flood (peak discharges) and discharge hydrographs from (rainfall) events have been studied (intensively) since the early 1930s.

(2) The ordinates are then added at (corresponding) times to determine the (total) hydrograph.

(3) This is generally not true; (consequently), (variations) in ordinates for different storms of equal duration can be expected.

Chapter 6 Sediments

(1) Desert sand (dunes) and (loess) are examples of aeolian transport and deposition.

(2) (Arid and semi-arid) regions are characterized by relatively low and infrequent rainfall.

(3) The method of (measurement) of the size by its volume and fall velocity are based on the premise that measurements made can be expressed as the diameter of an (equivalent) sphere. The following definitions of sizes are used (in practice).

(4) Fall diameter of a particle is defined as the diameter of a sphere of relative (density) 2.65 and having the same standard fall velocity as that of the particle.

Chapter 7 River

(1) A river is a natural (watercourse), usually freshwater, flowing toward an ocean, a lake, a sea or another river.

(2) Small rivers may also be called by several other names, including (stream), creek, (brook), (rivulet), and rill; there is no general rule that defines what can be called a river.

(3) The water in a river is usually confined to a (channel), made up of a stream bed between banks.

Chapter 8 Flood

(1) When a (rainfall) does occur, it can sometimes result in a sudden (flood) of water filling dry streambeds known as a "(flash flood)".

(2) When heavy (rainfall) or (melting snow) causes the river's depth to increase and the river to (overflow) its banks, a vast (expanse) of shallow water can rapidly cover the adjacent flood plain.

(3) Floods may also cause millions of dollars worth of (damage) to a city, both evicting people from their (homes) and ruining (businesses).

Chapter 9 Water Pollution

(1) Converting (saltwater) to freshwater is generally too (impure) to be used for (industrial), agricultural or (household) purpose.

(2) Raw sewage, (garbage), and (oil spill) have begun to (overwhelm) the diluting capabilities of the oceans.

(3) Congress has passed laws to try to (combat) water pollution thus (acknowledging) the fact that water pollution is, indeed, a serious issue.

(4) We have to (preserve) existing trees and plant new trees and shrubs to help prevent soil erosion and promote (infiltration) of water into the soil.

(5) This, in turn, proves very (harmful) to aquatic organisms as it affects the (respiration) ability of fish and other (invertebrates) that reside in water.

(6) Under natural conditions, lakes, rivers, and other water bodies undergo (eutrophication), an aging (process) that slowly fills in the water body with (sediment) and organic matter.

Chapter 10 Dams

(1) A dam is a structure built across a (stream), river or (estuary) to retain water.

(2) The (failure) of a dam may cause serious loss of life and (property); consequently, the design and (maintenance) of dams are commonly under government (surveillance).

(3) The (failure) of the Teton Dam in Idaho in June 1976 added to the (concern) for dam (safety)

in the United States.

(4) Dams are (classified) on the basis of the type and materials of construction, as arch dams, (buttress) dams, (gravity) dams, and (embankment) dams.

Chapter 11 Levee

(1) A levee, levée, dike (or dyke), (embankment), floodbank or stopbank is a natural or artificial slope or wall to regulate water levels.

(2) Levees can be (permanent) earthworks or emergency constructions (often of sandbags) built (hastily) in a flood (emergency).

(3) The ability of a river to carry (sediments) varies very strongly with its speed. When a river floods over its banks, the water (spreads) out, slows down, and deposits its load of sediment.

Chapter 12 Hydropower Plants

(1) The (reservoir), basin formed by the construction of a (dam), holds back (a very large volume of water) so that the flow rate can be controlled. Top of dam, (upper part of the dam), rises above the water level of the reservoir by several yards.

(2) The water discharging is controlled by (spillway gate), a movable vertical panel; the reservoir's overflow is allowed to (pass through) when it is (opened). And the (cement crests) over which the reservoir's overflow discharge when the spillway gates are opened are called crests of spillway.

(3) Part of the reservoir immediately in front of the dam where the current originates is called (headbay). Channel that carries water under pressure to the power plant's turbines is called (penstock). And the area of the watercourse where water is discharged after passing through the turbines is called (afterbay).

(4) The water stored in such reservoirs may (have a residence time of several months) during which time (normal biological processes) are able to substantially reduce many contaminants and almost eliminate any turbidity.

Chapter 13 Water Environment

(1) The water quality in the (mainstreams) of the river systems is better than that in the (branched).

(2) Compared with the previous year, the order of the (pollution) level remained same and the pollution (extent) is quite close.

(3) The key (pollution) indicators are (ammonia nitrogen), (petroleum), potassium permanganate index and (BOD).

Chapter 14 Climate Change

(1) Predictions include higher incidences of severe weather events, a higher likelihood of flooding, and more (droughts).

(2) As a direct (consequence) of warmer temperatures, the hydrologic cycle will undergo significant (impact) with accompanying changes in the rates of (precipitation) and evaporation.

(3) There are a variety of climate change (feedbacks) that will either amplify or (diminish) the initial forcing.

(4) In their effect on climate, (orbital) variations are in some sense an (extension) of solar variability, because slight variations in the Earth's orbit lead to changes in the distribution and abundance of sunlight reaching the Earth's surface.

(5) Global sea level change for much of the last century has generally been estimated using (tide) gauge measurements (collated) over long periods of time to give a long-term average.

Chapter 15 Water Use Efficiency and Water Conservation: Definitions

(1) Benefit use has two calculation methods: 1. Single (ET) and 2. (ET) plus the amount of water required for (leaching salts) from the root zone.

(2) Usually, maximum average benefit value occurs at a lower infiltrated water value than the (economically optimal) quantity. Economically optimal input is at the point where the (marginal benefit) equals the (marginal costs).

(3) Irrigation scheduling refers to the (time), (duration), and (quantity) of an irrigation.

(4) (Neutron) probe can be used to measure the water content in the soil profile, but this method is (labor intensive).

Chapter 16 Irrigation culture

(1) Irrigation was regarded as culture in the (1930s) but now more and more regarded as a (business) that competes in a global economy.

(2) The priority rule for western America water use can be explained as (first in time), (first in right), which can sort out the (competing) claims to water.

(3) In the nineteenth century irrigation practices, (Chinese) immigrants played a central role in the reclamation of the (Sacramento) River floodplains and delta.

(4) In future efforts to reform irrigation institutions, irrigation is a combination of (culture), (economic) and (political) system.

Chapter 17 Irrigation Methods

(1) (Surface irrigation), which is also called (flood irrigation), has been the most common

method of irrigating agricultural land and still used in most parts of the world.

(2) In drip irrigation, water is delivered (drop by drop) to crop root, which can be the most water-efficient method of irrigation, since (evaporation) and runoff are minimized. The field water efficiency of drip irrigation is typically in the range of 80 to (90) percent when managed correctly.

(3) The spacing of emitters in the drip pipe of SSTI is not critical as the (geotextile) moves the water along the fabric up to 2 m from the dripper. The impermeable layer effectively creates an (artificial) water table.

(4) Irrigation by Lateral move system is less expensive to install than a center pivot, but much more (labor-intensive) to operate - it does not travel (automatically) across the field.

Chapter 18 Irrigation Performance Evaluation

(1) The area has about 7000 ha of irrigated lands distributed in 843 parcels and devoted to a (diverse) crop mix, with (cereals), sunflower, cotton, (garlic) and olive trees as principal crops.

(2) Irrigation is on demand from a (pressurized) system and hand-moved (sprinkler) irrigation is the most popular (application) method.

(3) The ancient irrigation (techniques) depended mostly on the (terrain), water supply, and the (engineering) skills of the civilization.

(4) Therefore, irrigation systems had to be simple in terms of (construction), and built high along the (riverbank) in order to deal with only the (peak) of the flood.

Chapter 19 River Closure and Diversion in Dam Construction

(1) It is consequently of interest to define safe and (realistic) conditions for closures and to estimate the (relevant) costs according to maximum water head in order to reach overall (optimization) of the whole scheme (closure, diversion, programme, etc...) or to give more flexibility for programming closure (beyond) the low flow season if suitable.

(2) The possibility of using diversion tunnels as a place for the bottom outlet is a clearly feasible solution in the case of (embankment) dams, but not so much as for concrete dams.

(3) In rockfill dams of medium or small height, the solution of providing an incorporated (bulk) of concrete housing spillways and (outlets), renders diversion operations much easier.

Chapter 20 Embankment and Fills

(1) In general, there are two types of embankment dams: earth (earthfill) dam and rock (rockfill) dam.

(2) The alignment of an earthfill dam should be such as to (minimize) construction costs but such alignment should not be such as to encourage (sliding) or (cracking) of the embankment.

(3) If necessary, test fills should be constructed with variations in placement water (content), lift

thickness, number of (roller) passes and (type) of rollers.

(4) Before fresh fill material is (placed), all such deteriorated fill or foreign material shall be removed to a (depth) at which material of an acceptable standard is (exposed).

Chapter 21 Main Equipment in Hydropower Plants

(1) A Pelton turbine has one or more (free jets) discharging water into an aerated space and impinging (on the buckets of a runner).

(2) The Francis turbine is also (similar to) a (waterwheel), as it looks like a spinning wheel with fixed blades (in between) two rims.

(3) Circuit breaker is a useful (mechanism) automatically cutting off the (power supply) in the event of overload (so that) the devices nearby can be (protected) from too much current.

Chapter 22 Hydroelectricity and Its Characteristics

(1) A hydroelectric power plant converts this energy into electricity by forcing water, often held at a (dam), through a (hydraulic turbine) that is connected to a (generator). The water exits the turbine and is returned to (a stream or riverbed) below the dam.

(2) The major (advantage) of hydroelectricity is elimination of the cost of fuel. The cost of operating a hydroelectric plant is nearly immune to (increases) in the cost of (fossil fuels) such as (oil, natural gas or coal), and no imports are needed.

(3) Dam failures have been some of the largest (man-made disasters) in history. Also, good design and construction are not an adequate (guarantee of safety). Dams are tempting industrial targets for (wartime attack), sabotage and terrorism.

Chapter 23 Pumped-Storage Plants

(1) Taking into (account) evaporation losses from the exposed water surface and conversion losses, approximately 70% to 85% of the electrical energy used to pump the water into the elevated reservoir can be (regained).

(2) In certain conditions, electricity prices may be close to zero or occasionally negative, indicating there is (more) generation than load available to absorb it; although at present this is rarely due to wind alone, (increased) wind generation may increase the likelihood of such occurrences.

(3) Plants that do not use pumped-storage are referred to as (conventional) hydroelectric plants; conventional hydroelectric plants that have significant storage capacity may be able to play a (similar) role in the electrical grid as pumped storage, by (deferring) output until needed.

Chapter 24 Tidal Power Station

(1) (Given) that power varies with the density of medium and the cube of velocity, it is simple to see that water speeds of nearly one-tenth of the speed of wind provide the same power for the same size of turbine system.

(2) (Since) tidal stream generators are an immature technology (no commercial scale production facilities are yet routinely supplying power), no standard technology has yet emerged as the clear winner, but a large variety of designs are being experimented with, some very close to large scale deployment.

(3) Periodic changes of water levels, and associated tidal currents, are (due to) the gravitational attraction by the Sun and Moon.

Chapter 25 Small Hydropower

(1) The use of the term "(small hydro)" varies considerably around the world, the maximum limit is usually somewhere between 10 and 30 MW.

(2) Countries like India and China have policies in favor of small hydro, and the regulatory process allows for building dams and (reservoirs).

(3) Small hydropower plays a very important role in (resisting) natural disasters, emergency service and disaster relief because it supplies power distributedly.

Chapter 26 Navigation and Recreation

(1) The Three Gorges (featured) numerous shoals, rapids and narrow river-course, in which navigation at night was at stake since ancient times.

(2) Each of the two ship locks is made up of five stages, (taking) around four hours in total to transit, and has a vessel capacity of 10 000 tons.

(3) Water bodies (account for) 30%-40% relative to other recreational enterprises in satisfying human recreational needs.

Chapter 27 Bids

(1) In project management, (flexibility) is a major factor in obtaining effective results, because there is almost no end to the possibilities for (variation) from the basic bidding procedure, as long as the objective of (owner's benefit) and (fairness) to the contractors are observed.

(2) Generally, the notice does not (require) or (encourage) a response, however, the invitation may be (mailed) to an individual bidder or may be made available for all bidders to (read) and (respond).

(3) Bidders are required to submit a (bond), or the acceptable (substitute) for the same for two

reasons: (unforeseen circumstances) and (bid spread).

(4) Contractors or manufacturers can bid for (individual) contracts or for (a group of) similar contracts at their option, and all bids and combinations of bids should be opened band evaluated (simultaneously) so as to determine the bid or combination of bids offering the most (advantageous) solution for the borrower.

(5) Electronic system is valid to distribute bidding documents if (the Bank) is satisfied with the adequacy of such system. If bidding documents are distributed electronically, the electronic system shall be secure to avoid (modifications) to the bidding documents and shall not (restrict) the access of Bidders to the bidding documents.

(6) Bidders shall be required to submit bids valid for a period which shall be sufficient to enable the (Borrower to complete the comparison) and (evaluation) of bids, (review) the recommendation of award with the Bank, and obtain all the necessary (approvals) so that (the contract can be awarded) within that period.

(7) Competitive bidding differs from other (pricing strategies) in that with bid pricing, a specific price is (put forth) for each possible job rather than a generic price that applies to (all customers). So, the big problem in bid pricing is estimating all the costs that will apply to (each job).

(8) Not all competitive bidding situations end with the (customer's acceptance) of one of the bids offered. In some instances, a subsequent (negotiation) step may take place. In some cases, the customer will single out the company that submits the most (attractive) bid but not necessarily (the lowest price).

Chapter 28 How to Write Your First Research Paper

(1) Doing the hardest task first thing in the morning contributes to the sense of accomplishment during the rest of the day, because the (positive) feeling spills over into our work and life and has a very positive effect on our overall (attitude).

(2) The outline of an academic paper is not an outline with Roman numerals for each section. It is more similar to a (template) for your paper which forms a structure for your paper and helps generate (ideas) and formulate (hypotheses).

(3) When you create the first draft, do not (succumb) to the temptation of editing, such as choose a better (word) or better (phrase) or halt to improve your (sentence structure). Pour your ideas into the paper and leave (revision) and editing for later.

(4) In the Materials and Methods section, the most important goal is to be as (explicit) as possible by providing enough (detail) and (references). In the end, the purpose of this section is to allow other researchers to (evaluate) and (repeat) your work.

Appendix 3 Vocabulary

abnormal [æb'nɔ: məl]	adj.	反常的，异常的	14.1
abrasion [ə'breiʒən]	n.	表面磨损	6.2
absolutely ['æbsəlu:tli]	adv.	绝对地，完全地；独立地；确实地	4.5
abutment [ə'bʌtmənt]	n.	拱座，桥台(支撑桥梁底部的结构)	10.2
accommodate [ə'kɔmədeit]	v.	适应	21.1
accompanied[ə'kʌmpənid]	adj.	伴随的	11.4
accomplish [ə'kʌmpliʃ]	vt.	达到(目的)，完成(任务)，实现(计划)	5.1
accrue [ə'kru]	vt.	产生；自然增长或利益增加	15.2
accumulate [ə'kju: mjuleit]	vi.	积聚，累积，堆积	6.1
activate ['æktiveit]	vt.	刺激，使活动	9.3
adjacent [ə'dʒeisnt]	adj.	邻近的，接近的	8.1
adjoin [ə'dʒɔin]	vt.	贴近，与……毗连	11.1
advent ['ædvənt]	n.	出现，到来	3.2
aerated ['eiəreitid]	adj.	充气（通气，通风， 鼓风）的	21.1
aerobic [ˌeiə'rəubik]	adj.	需氧的，有氧的，增氧健身法的	9.3
aesthetics [ɛs'θɛtɪks]	n.	美学	17.2
affordable [ə'fɔ: dəbl]	adj.	供应得起	23.3
Afghanistan		阿富汗	10.1
afterbay ['ɑ:ftəbei]	n.	下游池	12.1
aggravation [ˌægrə'veʃən]	n.	[口]激怒，惹恼；加重，加剧，恶化；	3.1
agrarian [ə'greəriən]	adj.	农业的	11.1
albeit [ɔ:l'bi:ɪt]	conj.	尽管，固然，即使	16.1
alfalfa [ælˌfælfə]	n.	[植] 紫花苜蓿	18.2
algae ['ældʒi:]	n.	藻类，海藻	9.2
alignment [ə'lainmənt]	n.	定线；（国家、团体间的）结盟；排成直线	20.1
alleviate [ə'li: vieit]	vt.	减轻，使（痛苦等）易于忍受	8.3
alluvial fans		冲积扇	6.1
alluvium [ə'lu: viəl]	n.	冲积层；冲积土，冲积物	6.1
Almanza Dam		阿尔曼扎坝	10.1
alter ['ɔ: ltə]	v.	改变，改动，变更	2.3
alternating current		交流（电）	12.4
alternative [ɔ: l'tə: nətiv]	n.	可供选择的办法，替代物	2.1
aluminum [ˌælju:'minjəm]	n.	铝	17.2

ambiguity [ˌæmbɪˈgjuəti]	n.	含糊；不明确；暧昧；模棱两可的话	27.2
ammonia [əˈməunjə]	n.	氨水，阿摩尼亚	13.1
amount [əˈmaunt]	n.	数（量），总额　vi. 合计；接近	1.1
anaerobic lagoon		厌氧生物礁湖	9.3
analyze [ˈænə,laɪz]	vt.	对……进行分析，分解（等于 analyse）	5.2
anastomosing [əˈnæstəməʊzing]	n.	网状河流	7.1
anchor [ˈæŋkə]	n.	铁锚	10.2
annual production		年发电量	24.2
anthropogenic [ˌænθrəpəuˈdʒenik]	n.	人为的	3.2
anticlockwise [ˌæntiˈklɔkwaiz]	ad.	逆时针的（地）	24.3
approximately [əˈprɔksimətli]	adv.	近似地	5.2
aquatic [əˈkwætik]	n.	水生动物，水草	7.3
aquifer [ˈækwəfə]	n.	含水土层，蓄水层；地下含水层	4.2
arch [ɑ:tʃ]	n.	拱门，弓形　v. 成拱形弯曲，成拱形	10.2
arch dam		拱坝	10.2
arid and semi-arid		干旱和半干旱	6.1
arid region		干旱地区	6.1
arrester [əˈrestə]	n.	避雷器	12.4
ascend [əˈsend]	v.	攀登，上升	26.1
ascertain [ˌæsəˈtein]	vt.	查明，弄清	5.2
assembly [əˈsembli]	n.	集合，装配	12.3
assumption [əˈsʌmpʃən]	n.	假定，臆断；担任，承担	2.1
asynchronous[eiˈsiskrənəs]	adj.	异步的	23.3
at stake		危如累卵，危险	26.1
atmosphere [ˈætməsfiə]	n.	大气圈	4.1
atmospheric [ˌætməsˈferik]	adj.	大气的	1.1
auxiliary [ɔ: gˈziljəri]	adj.	辅助的，补助的；备用的，后备的	11.1
awe-inspiring [ˈɔ:inˌspaiəriŋ]	n.	令人敬畏的	26.2
backwater [ˈbækwɔ: tə]	n.	逆流，回水（被水坝或急流止住或往回流的水）	26.1
bankfull [bæŋkfull]	n.	满水时期	8.3
	adj.	水位齐岸的	
banquette [ˈbæŋkwit]	n.	护坡道，凸部，人行道，弃土堆	11.1
barrage [ˈbærɑ:ʒ]	n.	拦河坝，堰	24.2
basin [ˈbeisn]	n.	（河川的）流域	2.1
bay [bei]	n.	海湾；狗吠声；绝路	8.1
bedrock [ˈbedˈrɔk]	n.	[矿]岩床，基础	4.1
beneath [biˈni:θ]	adv.	在……下面	4.1
	prep.	在……之下	

Word	POS	Meaning	Unit
beyond [biˈjɔnd]	prep.	远于；迟于；越出	3.2
	ad.	在更远处	
billabong [ˈbɪləˌbɔŋ]	n.	〈澳〉死河(指支流)，死水潭	7.1
biosphere [ˈbaiəsfiə]	n.	生物圈	4.1
blade [bleid]	n.	叶片，刀片	12.3
blower [ˈbləuə]	n.	送风机，吹风机	21.1
bluff [blʌf]	n.	断崖，绝壁，诈骗	11.1
	adj.	绝壁的，直率的	
	v.	诈骗	
border strip [ˈbɔrdə-strɪp]	n.	分区	17.1
braid[breid]	vt.	混合，交错，编织，编结	7.1
branch [bra: ntʃ]	n.	支流，分枝，树枝	2.2
	v.	分枝	
breach [bri: tʃ]	n.	缺口；破坏；不和	11.1
	vt.	攻破；破坏	
break through [ˈbreik,θru:]	n.	突破	22.2
brine [brain]	n.	盐水	23.3
brink [briŋk]	n.	(悬崖、河流等的) 边缘，边沿	11.3
broaden [ˈbrɔ:d(ə)n]	vt.	使扩大；使变宽	29.4
brook [bruk]	n.	小溪	7.1
	vt.	容忍[常用于否定句或疑问句]	
bucket [ˈbʌkit]	n.	桶状物，铲斗，叶片	21.1
bulk [bʌlk]	n.	体积，大部分，大多数，大块，大批，容积	9.3
	vi.	越来越大	
bund [bund]	n.	急坡堤岸	12.2
burrow [ˈbʌrəu]	n.	地洞	6.1
	vt.	挖掘(洞穴)；钻进	
	vi.	挖洞；翻寻	
bushing [ˈbuʃiŋ]	n.	[机]轴衬， [电工]套管	21.3
buttress [ˈbʌtrəs]	v.	支持	10.2
	n.	扶墙，拱壁，支墩	
buttress dam		支墩坝	10.2
caisson [ˈkeisən]	n.	沉箱(桥梁工程)	24.2
capacity [kəˈpæsiti]	n.	容量，生产量	10.2
carbon dioxide		二氧化碳（CO_2）	14.2
Carboniferous [ˌkɑ:bəˈnifərəs]	n.	石炭纪(时期)	14.2
cardiovascular [ˌkɑ:diəuˈvæskjulə]	adj.	心脏血管的	9.4
catalyst [ˈkætəlist]	n.	催化剂，促使事情发展的因素，刺激因素	11.1

Word			
cement [si'ment]	n.	水泥，结合剂	12.1
cemeteries ['semitəris]	n.	墓园	17.1
cereal ['siəriəl]	n.	谷类食物	7.3
chromosome ['krəuməsəum]	n.	[生物] 染色体	9.4
chute [ʃu:t]	n.	瀑布，斜道	12.1
circuit breaker		断路器	12.4
cistern ['sistən]	n.	蓄水池；储水器	12.2
civil work		土建工程	21.1
clarification [ˌklærəfə'keʃən]	n.	澄清，说明；净化	27.3
classify ['klæsifai]	vt.	把……分类，把……分级	6.2
clay [klei]	n.	黏土，泥土，肉体，人体，似黏土的东西，陶土制的烟斗	6.2
climate ['klaimit]	n.	气候，气候区；风气，气氛	3.2
climatological [ˌklaɪmətə'lɒdʒɪkəl]	adj.	与气候学有关的	14.2
clockwise ['klɔkwaiz]	adj./adv	顺时针的	24.3
clog [klɔg]	v.	障碍，阻塞	9.2
closure ['kləuʒə]	n.	截流，关闭，终止	19.1
coast [kəust]	n.	海岸，海滨	8.1
coastal ['kəustl]	adj.	海洋的，海岸的，沿海的，沿岸的	8.2
cofferdam ['kɔfədæm]	n.	围堰	19.2
cognitive ['kɑgnətɪv]	adj.	认知的，认识的	28.1
cohesion [kəu'hi: ʒən]	n.	内聚力	6.2
coil [kɔil]	n.	线圈	21.2
compensation[ˌkɑmpɛn'seʃən]	n.	补偿，赔偿，抵消；清算，划账	7.6
complex ['kɔmpleks]	n.	综合物；综合性建筑	2.1
compound ['kɔmpaund]	v.	混合，调和，妥协；	9.3
	n.	化合物，混合物，复合词	
concise [kən'saɪs]	adj.	简明的，简洁的	29.2
condensation [ˌkɔnden'seiʃən]	n.	凝结，凝聚；浓缩	4.3
conductor [kən'dʌktə]	n.	导体；导线	12.4
conduit ['kɔndit]	n.	管道，导管，沟渠	12.1
conduit joint		管道接头	20.1
confidential[ˌkɑnfɪ'dɛnʃl]	adj.	机密的；表示信任的；获信任的	27.6
connotation [ˌkɑnə'teɪʃ(ə)n]	n.	含义,含意	2.1
conscious ['kɒnʃəs]	adj.	意识到的；故意的；神志清醒的	29.3
consequence['kɑnsəkwɛns]	n.	结果；重要性；推论	2.2
considerably [kən'sɪdərəblɪ]	adj.	大幅度的，很大的	9.3
consistency [kən'sistənsi]	n.	连接，结合，坚固性，浓度，密度，一致性，连贯性	18.2

consolidation [kənˌsɔliˈdeiʃən]	n.	固结作用	20.1
construction [kənˈstrʌkʃən]	n.	建造，建设；建造物，建筑物；结构	2.3
contaminant [kənˈtæminənt]	n.	污染物	9.3
contamination [kənˌtæmiˈneiʃən]	n.	污染	4.1
contemporary [kənˈtɛmpərɛri]	n.	同时代的人，同时期的东西	14.2
	adj.	当代的，同时代的，属于同一时期的	
continuous [kənˈtinjuəs]	adj.	连续不断的，不断延伸的	4.2
contour [ˈkɔntuə]	n.	等高线，轮廓，周线，电路，概要；	4.5
	vt.	画轮廓(画等高线)	
controlling section		监测断面	13.1
controversial [ˌkɔntrəˈvəː ʃəl]	adj.	争议的，争论的	9.1
convert [kənˈvɜːt]	vt.	使转变；转换……	1.1
coordination [kəʊˌɔːdiˈneiʃn]	n.	协调，和谐	2.2
correspond [ˈkɔriˈspɔnd]	vi.	相符合；相类似；通信	3.1
corrosive [kəˈrəusiv]	adj.	腐蚀的，腐蚀性的	24.3
corrugate [ˈkɔrugeit]	v.	弄皱，起皱，（使）成波状	18.2
cost-effective [ˈkɔstiˈfektiv]	adj.	有成本效益的，划算的	23..2
counsel [ˈkaʊnsl]	n.	法律顾问；忠告；商议；讨论；决策	27.6
	vt.	建议；劝告	
	vi.	商讨；提出忠告	
couple [ˈkʌpl]	vt.	耦合，连接	8.3
course [kɔː s]	n.	路线，过程，课程，讲座，一道（菜）	2.1
crane [krein]	n.	起重机；鹤	12.3
	vt.	伸长（脖子等）	
crank up		加快，做好准备	23.1
creek [kriː k]	n.	小溪，小港，小湾	7.1
crest [krest]	n.	顶部，顶峰，浪头	10.2
	vi.	到达绝顶	
	vt.	加以顶饰	
criteria [kraɪˈtɪrɪə]	n.	标准，条件（criterion 的复数）	5.2
critical [ˈkritikəl]	adj.	决定性的，关键性的，危急的；批评（判）的	1.4
cross-flow		交叉流动，横向流动	21.1
cross-sectional	adj.	过流断面的	6.2
Cryptosporidium[ˌkriptəʊspɔːˈridiəm]	n.	隐孢子虫	9.4
cubic [ˈkjuː bik]	adj.	立方体的，立方的	23.2
cylinder [ˈsilində]	n.	圆柱体，圆筒；汽缸，泵(或筒)体	6.1
data acquisition		数据采集	18.1
decentralization [ˌdisɛntrələˈzeʃən]	n.	疏散化	3.2

decision factor		决定因素	23.2
decommission [ˌdi: kəˈmiʃən]	vt.	使退役	22.4
defence/defense [diˈfens]	n.	防御，保卫	
	[pl.]	防御工事；辩护	11.1
deflation [diˈfleiʃən]	n.	风力侵蚀	6.1
deforestation [ˌdifɔrəsˈteʃən]	n.	滥伐森林	3.2
delta [ˈdeltə]	n.	(河流的)三角洲	6.1
deluge [ˈdelju: dʒ]	n.	洪水，暴雨	8.1
	v.	使泛滥，淹，浸，压倒	
demographic [ˌdiməˈgræfik]	adj.	人口的，人口统计学的	3.2
densely [ˈdensli]	adv.	浓密地，浓厚地	1.3
density [ˈdensiti]	n.	密集，稠密；密度	6.2
deplete [diˈpli:t]	vt.	耗尽，使衰竭	1.4
derive [diˈraiv]	vt.	取得；追溯起源	5.1
	vi.	起源，衍生	
descend [diˈsend]	v.	下来，下降	13.3
destabilise [ˌdestiˈneiʃən]	vt.	使动摇	7.3
destructive [diˈstrʌktiv]	adj.	破坏（性）的	8.4
deviation [ˌdi: viˈeiʃən]	n.	背离，偏离；偏差，偏向；离题	14.2
device		起重设备	12.1
devolution [ˌdɛvəˈluʃən]	n.	移交，转让	3.2
diameter [daiˈæmitə]	n.	直径	6.2
diarrhea [ˌdaiəˈriə]	n.	痢疾，腹泻	3.1
dilemma [dɪˈlemə; daɪ-]	n.	困境，进退两难，两刀论法	16.1
dimension [diˈmenʃən]	n.	方面，特点；尺寸；	
	[pl.]	面积，规模，程度	3.2
diminish [diˈminiʃ]	vt.	减少，减小，降低	
	vi.	变少，变小，降低	9.4
diminished [diˈminiʃt]	adv.	减少了的	22.4
disinfect [ˌdisinˈfekt]	vt.	消毒	13.4
displacement [disˈpleismənt]	n.	移置，转移，排水量	26.1
dispose [disˈpəuz]	v.	处理，处置，销毁	9.3
dispute [ˈdɪsˈpjʊt]	n.	辩论，争吵	16.4
	vt.	辩论，怀疑，阻止，抗拒	
disrupt [disˈrʌpt]	v.	使中断，使分裂，使瓦解，使陷于混乱，破坏	9.1
disruptive [disˈrʌptiv]	adj.	分裂(性)的；破坏性的	22.4
disseminate [dɪˈsɛmɪnet]	v.	播撒，宣传，散播	27.2
dissipation [ˌdisiˈpeiʃn]	n.	消散，分散，挥霍，浪费	24.1

Word	POS	Meaning	Unit
dissolution [ˌdisəˈlu: ʃən]	n.	分解	23.3
dissolve [diˈzɔlv]	v.	溶解，解散，分解	4.1
distinctive [dɪˈstɪŋ(k)tɪv]	adj.	有特色的，与众不同的	16.2
disturbance [disˈtə: bəns]	n.	扰动，骚扰	20.2
ditch [ditʃ]	n.	沟，渠道	11.1
diversion [daiˈvə: ʃən]	n.	导流，分出，引出；转向，改道	12.1
doctrine [ˈdɔktrin]	n.	教义，教条，主义	19.2
downriver [daʊnˈrɪvər]	n.	向下游	7.1
downstream [ˈdaʊnˌstri:m]	n.	下游	2.2
	adv.	下游地	
draft tube	n.	尾水管	12.3
drain [drein]	vt.	排出沟外，喝干，耗尽	5.1
	vi.	排水，流干	
	n.	排水沟	
drainage [ˈdreinidʒ]	n.	排水，放水；排水系统，下水道；废水，污水，污物	5.1
drainage basin		排水区	6.1
dramatic [drəˈmætik]	adj.	激动人心的，引人注目的；戏剧的	3.1
drastically [ˈdræstikəli]	adv.	激烈地，彻底地	18.2
drift [drift]	n.	漂移，漂流物，观望，漂流，吹积物，趋势	4.4
	v.	漂移，漂流，吹积	
droplet [ˈdrɔplit]	n.	小滴	4.3
drought [draut]	n.	干旱，旱灾	4.1
dumping [ˈdʌmpiŋ]	n.	倾倒，倾销	9.1
dune [dju: n]	n.	沙丘	6.1
duration [djuəˈreiʃən]	n.	持续，持续期间	5.1
dynamic [daiˈnæmik]	adj.	动力的，动态的；有活力的，强有力的；不断变化的	3.1
dynamo [ˈdainəməu]	n.	发电机	22.2
earthfill	n.	填土，[地] 泥流	10.2
earthquake [ˈə: θkweik]	n.	地震，[喻] 在震荡，在变动	8.2
ebb [eb]	n.	退潮，落潮 v. 退潮，落潮；减少，衰落	11.3
ecological [ˌekəˈlɔdʒikəl]	adj.	生态学的，社会生态学的	24.2
ecosystem [ˈekəusistəm]	n.	生态系统	1.2
edible aquatic		可食用的水生生物	7.3
El Niño Southern oscillation		厄尔尼诺南方涛动	14.2
elaborate [ɪˈlæbəret]	adj.	精心制作的；详尽的；煞费苦心的	28.2
	vt.	精心制作；详细阐述；从简单成分合成	

	vi.	详细描述；变复杂	
electric current		电流	21.2
electromagnet [iˌlektrəu'mægnit]	n.	电磁石，电磁铁	21.2
elevation [ˌeli'veiʃən]	n.	海拔，上升，高地，正面图，提高，仰角	8.1
eliminate [ɪ'lɪmɪnet]	vt.	消除；排除	2.3
elimination [iˌlimi'neiʃən]	n.	排除，除去	22.3
ellipsoid [i'lipsɔid]	n.	椭球	6.2
elongated ['i:lɔŋgeitid]	adj.	加长的，拉长的，伸长的	21.1
embankment [em'bæŋkmənt]	n.	堤防，筑堤	10.2
embankment dam		堤坝	10.2
emergency [i'mə: dʒənsi]	n.	紧急情况，不测事件，非常时刻	11.1
empirical [em'pirikəl]	adj.	完全根据经验的，经验主义的；[化]实验式	18.1
empirical nature		实证性	20.1
encapsulate [in'kæpsəˌlet]	vt.	概括，压缩；封装；把…包于胶囊	3.1
end dumping		立堵法截流	19.1
engineering ['endʒiˌniəriŋ]	n.	工程技术，工程	2.1
enormous [i'nɔ: məs]	adj.	巨大的，庞大的，[古]极恶的，凶暴的	4.4
entrepreneurial[ˌɑntrəprə'njʊrɪəl]	adj.	企业家的，创业者的；中间商的	16.1
ephemeral [i'femərəl]	adj.	生命短暂的，朝生暮死的	7.2
equatorial [ˌekwə'tɔ: riəl]	n.	赤道仪	8.3
	adj.	赤道的，近赤道的	
equilibrium [ˌi: kwi'libriəm]	n.	平衡	6.1
equitable ['ekwɪtəb(ə)l]	adj.	公平的，公正的；平衡法的	3.2
equivalent [i'kwivələnt]	n.	等价物，意义相同的词	2.2
	adj.	相等的，相当的	
eruption [i'rʌpʃən]	n.	爆发，火山灰；[医] 出疹	8.3
estuary ['ɛstjʊ(ə)ri]	n.	河口，潮汐河口，一条大河的潮汐河口	10.1
eutrophication [juˌtrɔfi'keiʃən]	n.	营养化，超营养作用	9.2
evaluation [iˌvælju'eiʃən]	n.	估价；评估	2.1
evaporate [i'væpəreit]	v.	蒸发，消失	1.2
evaporation [iˌvæpə'reiʃ(ə)n]	n.	蒸发，脱水	4.2
evapotranspiration[iˌvæpəuˌtrænspi'reiʃən]	n.	土壤中水分损失总量	4.4
evolution [i: və'lu: ʃən]	n.	进展，发展，演变，进化	18.2
exceed [ik'si: d]	n.	过量，过度	3.1
	v.	超越	
exciter [ik'saitə]	n.	励磁机；刺激物，兴奋剂	21.2
executive [ig'zekjutiv]	adj.	行政的	2.2
	n.	执行者，主管	

externality [ˌekstəːˈnæliti]	n.	外形，外部效应，外部经济效果	15.2
facet [ˈfæsit]	n.	（宝石等）刻面，小平面	3.2
	vt.	在……上面	
facilitate [fəˈsiliteit]	vt.	使变得(更)容易，使便利	5.2
fearful [ˈfiəful]	adj.	担心的，可怕的	9.1
feasibility [ˌfiː zəˈbiləti]	n.	可行性	2.3
feather edge		削边，薄边，羽毛边	20.2
feature [ˈfiːtʃə]	n.	特色，特征；容貌；特写或专题节目	5.2
	vi.	起重要作用	
federally [ˈfɛdərəli]	adv.	联邦地，联邦政府地，同盟地	16.1
feedback [ˈfiː dbæk]	n.	反馈，反馈信息	14.2
feedlot [ˈfiː dlɔt]	n.	饲育场	9.2
fertigation [ˌfəːtiˈgeiʃən]	n.	水肥灌溉	17.1
fertilizer [ˈfəː tilaizə]	n.	肥料	9.1
filter [ˈfiltə]	v.	过滤，渗透，走漏	4.4
	n.	筛选，滤波器，过滤器，滤色镜	
fine[ˈfain]	adj.	纤细的，尖细的，细小的	6.1
finite [ˈfainait]	adj.	有限的，有限的	1.1
flap [flæp]	n.	折叠板，活板	24.2
flatland [ˈflæt, lænd, -lənd]	n.	平原，平坦地	11.1
flatten [ˈflætn]	vt.	使变平	23.2
fleet [fliːt]	n.	舰队	26.1
float [fləut]	n.	漂流物，浮舟，漂浮	
	vi.	漂流，浮动，漂浮	7.3
	vt.	使漂浮，容纳，淹没	
flood plain		洪泛平原，洪泛区，洪积平原	7.1
flooding [ˈflʌdiŋ]	n.	泛滥，灌溉；溢流，变色；产后出血	4.1
flourish [ˈflɜːrɪʃ]	vt.	夸耀，挥舞	16.4
	n.	兴旺，茂盛，挥舞，炫耀，华饰	
	vi.	繁荣，兴旺，茂盛，活跃，处于旺盛时期	
fluctuating [ˈflʌktjueitiŋ]	adj.	变动的，上下摇动的	23.2
fluctuation [ˌflʌktjuˈeiʃən]	n.	波动，起伏	14.2
fluid [ˈfluːid]	adj.	流体的，流动的	4.1
fluoride [ˈfluː əraid]	n.	氟化物	13.3
food chain		食物链	9.2
forecaster [fɔːˈkɑːstə]	n.	预报员	8.3
foreshore way [fɔːˈʃædəu]	n.	林荫路	7.3
forfeit [ˈfɔrfət]	vt.	（因犯罪、失职、违约等）丧失（权利、名誉、生命等）	

	adj.	因受罚而丧失的；被没收的	
	n.	罚金；没收物；丧失的东西	
formulation [ˌfɔrmjəˈleʃən]	n.	构想，规划；公式化；简洁陈述	27.2
forthcoming [ˌforθˈkʌmiŋ]	adj.	即将到来的，现成的，唾手可得的	
		化石	3.1
fossil fuel		燃料	14.2
foundation [faunˈdeiʃən]	n.	地基，房基；建立，设立，创办，创建；基础，基本原理，根据；基金（会）；粉底霜	10.2
fraction [ˈfrækʃən]	n.	小部分，片断；分数	5.1
fragment [ˈfrægmənt]	n.	碎片，片段	3.2
fragmentation [ˌfrægmənˈteʃən]	n.	分裂	3.2
freeboard [ˈfri: bɔ: d]	n.	超高，干舷，吃水线以上的船身	20.1
frequency [ˈfri: kwənsi]	n.	频率	23.2
freshwater [ˈfreʃˌwɔ:tə(r)]	n.	淡水，湖水	1.1
friction [ˈfrikʃən]	n.	摩擦，摩擦力	23.3
frontal dumping		平堵法截流	19.1
frustrating [ˈfrʌstretɪŋ]	adj.	令人沮丧的	28.3
furnish [ˈfɝniʃ]	vt.	提供；陈设，布置	3.1
gantry crane		龙门起重机，高架移动起重机	12.3
garlic [ˈgɑ:lik]	n.	大蒜	18.1
gaseous [ˈgæsiəs]	adj.	气体的，含气体的	4.4
generator [ˈdʒenəreitə]	n.	发电机	12.1
generic [dʒəˈnɛrɪk]	adj.	类的；一般的；属的；非商标的	27.6
geographic [dʒi: əˈgræfik]	adj.	地理(学)的	7.1
geohydrology [ˈdʒi: əuhaiˈdrɔlədʒi]	n.	水文地质学	4.1
geologic [dʒiəˈlɔdʒik]	adj.	地质的，地质学的	7.2
geothermal energy		地热能	14.2
geyser [ˈgi: zə]	n.	天然热喷泉	14.2
Gezhouba		葛洲坝	26.1
glacial drift		浮冰	6.1
glacier [ˈglæsjə, ˈgleiʃə]	n.	冰川	1.1
global [ˈgləubəl]	adj.	全球的，全世界的；总的，完整的	1.4
gorge [gɔ: dʒ]	n.	峡谷，咽喉，胃，暴食	10.2
	v.	狼吞虎咽，塞饱	
granular [ˈgrænjulə]	adj.	粒状的，小粒的	9.3
granule [ˈgrænju: l]	n.	小粒，微粒	6.2
gravitational collapse		引力坍缩（物）	6.1
gravitational force		引力，重力，地心吸力	22.1

grease [gri: s]	n.	油脂，贿赂	13.4
	vt.	涂脂于，[俗] 贿赂	
greenhouse ['gri: nhaus]	n.	温室，暖房	3.2
grid [grid]	n.	电网	23.1
grinding ['graindiŋ]	n.	碾碎	6.2
	adj.	磨的，摩擦的，碾的	
grinding cereal		谷物研磨，切削麦片	7.3
ground water table		地下水水位	13.3
groundwater ['graʊnd, wɔ: tə, -, wɔtə]	n.	地下水	4.1
groundwater hydrology		地下水水文学	4.1
gully ['gʌli]	n.	冲沟，溪谷，集水沟，雨水口，檐槽	8.2
gutter ['gʌtə]	n.	水槽，檐槽，排水沟；贫民区；装订线	9.1
halt [hɔlt]	n.&v.	停止，暂停，中断	25.2
handle ['hændl]	v.	处理，买卖，操作	9.1
	n.	柄，把手	
harness ['ha: nis]	vt.	利用(河流、瀑布等)产生动力(尤指电力)	12.4
hastily ['heistili]	adv.	急速地；草率地	11.1
head [hed]	n.	水头，落差	8.4
headbay ['hedbei]	n.	[水]上闸首，上游池	12.1
headgate [hedgeit]	n.	水头阀门，闸门	23.3
headwaters ['hedwɔ:tə]	n.	源头，河源	7.2
highlight ['haɪlaɪt]	vt.	突出；强调；使显著	28.3
	n.	最精彩的部分；最重要的事情；加亮区	
hinder ['hində]	vt.	阻碍，打扰	12.3
holistic [ho'listik]	adj.	从整体着眼的，全面的	3.2
hollow ['hɔləu]	n.	洞，窟窿，山谷	10.2
	adj.	空的，虚伪的，空腹的，凹的	
	adv.	彻底	
	vt.	挖空，弄凹	
	vi.	形成空洞	
horizontal [ˌhɔri'zɔntl]	adj.	水平的，平的	10.2
hurricane ['hʌrikən]	n.	飓风	8.2
hydraulic [hai'drɔ: lik]	adj.	水力的，水压的	6.2
hydro ['haidrəu]	n.	水电厂	21.1
	adj.	水电的，相当于 hydroelectric	
hydroelectric ['haidrəui'lektrik]	adj.	水电的，水力发电的	1.1
hydroelectricity [ˌhaidrəuiˌlek'trisiti]	n.	水电	22.1
hydrograph ['həidrəgra:fs]	n.	过程线，水位线；水利图表	5.1
hydrologic cyclen		水文循环	4.1

hydrology [hai'drɔlədʒi]	n.	水文学，水文地理学	3.2
hydropower ['haidrəupauə]	n.	水力发电；水力，水电	7.3
hydrosphere ['haidrəsfɪə]	n.	水圈	4.1
hydrostatic pressure		静水压力	10.3
hydrostorage ['haidrəu'stɔ: ridʒ]	n.	水存储	23.2
hyetograph ['haiitəugrɑ:f]	n.	时间雨量曲线，雨量记录表，雨量分布图	5.2
hypotheses [haɪ'pɑθəsiz]	n.	假定；臆测	14.2
Ice Age		冰河世纪	14.2
iceberg ['aisbəg]	n.	冰山，冷冰冰的人	1.1
icecap ['ais, kæp]	n.	冰盖，冰帽	4.2
igneous ['igniəs]	adj.	火的，火成的	14.2
illustrate ['ɪləstreɪt]	vt.	阐明，举例说明；图解	3.2
immature [ˌimə'tjuə]	adj.	不成熟的，未完全发展的	24.2
immediately [i'mi: diətli]	adv.	立即，马上；直接地，紧接着地	12.1
immense [i'mens]	adj.	极广大的，无边的	24.1
immune to		对……免疫，不受……影响	22.3
impermeable [im'pə: miəbl]	adj.	不渗透的	12.2
implicit [im'plisit]	adj.	无疑问的；含蓄的；内含的	5.2
impound [im'paund]	vt.	关在栏中，拘留，扣押，没收	10.2
impoundment [im'paundmənt]	n.	蓄水，积水，被坝所围住的水	12.5
impulse ['impʌls]	n.	推动	21.1
impurity [im'pjuəriti]	n.	杂质，不纯	9.3
inappropriate [ˌinə'prəupriit]	adj.	不适当的，不相称的	22.4
inclined [in'klaind]	adj.	倾斜的	12.1
inclinometer [ˌɪnkli'nɔmitə]	n.	测斜仪	20.1
incorrect [ˌinkə'rekt]	adj.	不正确的	5.1
incur [in'kə:]	v.	招致，带来	24.2
indicator ['indikeitə]	n.	污染指标，指示器，指示剂；[计算机]指示符	13.1
induce [in'dju: s]	vt.	感应，引起	3.2
inevitably [in'evitəbli]	adv.	不可避免地	2.1
inexhaustible [ˌinig'zɔ:stəbl]	adj.	无穷无尽的	24.1
inexorably [in'eksərəbəli]	adv.	不为所动的，坚决不变的	14.2
infiltrate [ɪn'filtret]	vt.	使潜入；使渗入	4.4
	vi.	渗入	
	n.	渗透物	
infiltration [ˌinfil'treiʃən]	n.	渗入，渗透	4.2
inflow ['infləu]	n.	入流，流入物	1.3
infrastructure ['infrəˌstrʌktʃə]	n.	下部构造，基础建设	17.3

inhabit [in'hæbit]	vt.	居住于，存在于，占据，栖息	8.1
innumerable [i'nju: mərəbl]	adj.	无数的，数不清的	26.1
installation[ˌinstə'leiʃən]	n.	(整套)装置，设备	12.1
instantaneous [ˌinstən'teiniəs]	adj.	瞬间的，即刻的	5.2
instrumental [ˌɪnstrə'mɛntl]	adj.	乐器的，有帮助的，仪器的，器械的	16.3
	n.	器乐曲，工具字，工具格	
intake ['inteik]	n.	入口，进口	12.3
intangible [ɪn'tæn(d)ʒɪb(ə)l]	adj.	无形的，触摸不到的，难以理解的	16.2
integrate ['intigreit]	v.	(使)成为一体，(使)合并	5.2
integrated ['intəˌgretid]	adj.	无种族界限的；和谐的 完整的，完全的，综合的	3.2
intensification [in, tensifi'keiʃən]	n.	增强，加剧	14.2
intensive [in'tensiv]	adj.	加强的，密集的；精工细作的，集约的	15.3
interaction [ˌintər'ækʃən]	n.	相互作用，相互影响	8.4
intermittently [intə'mitəntli]	adv.	间歇地	22.4
interrupt ['ɪntə'rʌpt]	vi.	打断；打扰	28.1
	vt.	中断；打断；插嘴；妨碍	
	n.	中断	
intimidating [ɪn'tɪmɪdetɪŋ]	adj.	吓人的	28.3
inundate ['inəndeit]	v.	淹没，浸水，泛滥	8.4
inundation [ˌinən'deiʃən]	n.	淹没	11.1
invertebrate [in'və: tibrit]	n.	无脊椎动物，无骨气的人	9.2
	adj.	无脊椎的，无骨气的	
investigate [in'vestigeit]	v.	调查，调查研究	23.3
iron ['aiən]	n.	铁	13.3
irrevocable [i'revəkəbəl]	adj.	不能取消的不能唤回的，不能变更的	2.3
isotope ['aisətəup]	n.	同位素	14.2
issue ['isju:]	n.	出版，发行，（报刊等）期，论点，问题，结果，（水、血等）流出	2.3
	vi.	发行，造成…结果，进行辩护	
	vt.	使流出，放出，发行（钞票等），发布（命令），出版（书等），发给	
jack [dʒæk]	n.	起重器，千斤顶，(J-)杰克(男子名，也指各种男性工人)，插孔，插座	7.3
jeopardize ['dʒɛpəˌdaiz]	vt.	危及，损害	3.1
jet [dʒet]	n.	喷嘴，喷射器，喷射流体	21.1
kayak ['kaiæk]	n.	皮船	7.3
kilowatt ['kɪləˌwɑt]	n.	千瓦	12.5
kinetic energy		动能	21.1

knot [nɔt]	n.	速度单位（每小时 1 海里，大约合每小时 1.85km）	24.2
lagoon [lə'ɡu:n]	n.	泻湖	9.3
land mass		大陆块，大陆板块	8.1
landslide ['lændslaid]	n.	[地] 崩塌，山崩，地滑，塌方	6.1
lateral ['lætərəl]	n.	侧面	17.2
	adj.	侧面的，横向的	
layout ['leiaut]	n.	规划，设计，布局	19.2
['li:tʃɪŋ]	n.	沥滤	
	v.	浸出，淋洗；滤取；滤去	
legislation [ˌledʒis'leʃən]	adj.	有立法权的	3.2
	n.	法律，法规；立法，制定法律	
lesser ['lesə]	adj.	较小的，更少的，次要的	9.4
levee ['levi]	n.	防洪堤，码头，大堤	
	v.	筑防洪堤于	8.1
lexical ['leksɪk(ə)l]	adj.	词汇的	29.1
lift [lift]	n.	无数的，数不清的	12.3
lift thickness	eral	铺层厚度	20.1
lignite ['liɡnait]	n.	褐煤	20.1
likelihood ['laiklihud]	n.	可能，可能性	8.4
liner ['lainə]	n.	衬垫，衬里	21.2
lining ['lainiŋ]	n.	内层，衬套	12.2
liquefy ['likwifai]	v.	液化，溶解	11.4
lithosphere ['lɪθəsfɪə]	n.	岩石圈	4.1
liver ['livə]	n.	肝脏	9.4
load balancing		负载平衡，负荷平衡	23.1
lumberjack ['lʌmbədʒæk]	n.	伐木工	7.3
magnet ['mæɡnit]	n.	磁体，磁铁	12.4
magnitude ['mæɡnitju: d]	n.	重要性，重大；巨大，广大	2.1
magnitude ['mæɡnɪtud]	n.	大小；量级；[地震]震级；重要；光度	5.1
mainstream ['meinstri:m]	n.	主流，干流	3.2
management ['mænidʒmənt]	n.	管理，处理	2.1
mandate ['mændet]	n.	授权，正式命令	3.2
mandatory ['mændətɔri]	adj.	强制的；托管的；命令的	27.6
	n.	受托者（等于 mandatary）	
manganese [ˌmæŋɡə'ni: z]	n.	锰	13.3
marine [mə'ri: n]	adj.	海的，海产的，航海的，船舶的，海运的	6.1
	n.	舰队，水兵，海运业	
maritime life		海生物	8.1

marsh [mɑ:ʃ]	n.	沼泽，湿地	11.3
masonry [ˈmeisnri]	n.	石工术，石匠职业	18.2
mass [mæs]	adj.	大规模的	7.1
	v.	集中	
master [ˈmæstər]	adj.	精通的；主要的	2.1
	n.	主人；专家；硕士；主管；少爷；	
		原件；……桅船	
	v.	精通，掌握；控制；战胜，克服；制作……母版	
mat [mæt]	n.	席子，垫子	11.1
meander [miˈændə]	vi.	蜿蜒而行	7.1
Mediterranean [ˌmedɪtəˈreini: ən]	n.	地中海海岸	11.1
medium [ˈmidiəm]	adj.&n.	中等的，中号的，媒介，手段	1.2
megawatt [ˈmegəˌwɑt]	n.	百万瓦特，兆瓦	12.5
Memphis [ˈmemfis]		孟菲斯（埃及尼罗河畔城市，美国城市）	10.1
Mesopotamian Civilization		美索不达米亚文明	11.1
meteorologist [ˌmi: tjəˈrɔlədʒist]	n.	气象学家	4.2
meteorology [ˌmi: tjəˈrɔlədʒi]	n.	气象学，气象状态	4.1
meticulous [məˈtɪkjələs]	adj.	一丝不苟的；小心翼翼的；拘泥小节的	28.3
metropolis [miˈtrɔpəlis]	n.	主要都市，都会，大城市	26.1
microorganism [maikrəʊˈɔ: gəniz(ə)]	n.	微生物	13.4
military [ˈmilitəri]	adj.&n.	军事（用）的[the～] 军队，武装力量	11.1
mine [main]	n.	矿，矿山，矿井	1.3
misaligned [ˌmɪsəˈlaɪnd]	adj.	不对齐的，方向偏离的；不重合的	17.2
Mississippi River		密西西比河	7.2
mobility [mouˈbiliti]	n.	流动性	6.2
modification [ˌmɔdifiˈkeiʃən]	n.	修改，修饰	2.1
modulation [ˌmɔdjuˈleiʃən]	n.	（声音之）抑扬；变调	14.2
moist [mɔist]	adj.	潮湿的，湿润的	4.4
moisture [ˈmɔistʃə]	n.	潮湿，水分，湿气	1.1
monitor [ˈmɔnitə]	v.	监测	9.1
monitoring point (section)		监测点位（断面）	13.2
moraine [mɔˈrein]	n.	冰碛	6.1
motion [ˈməuʃən]	n.	运动，动作	4.1
motivation [ˌmotəˈveʃən]	n.	动机；积极性；推动	28.1
mount [maunt]	vt.	装上，设置，安放	14.2
Mount Pinatubo		皮纳图博火山	14.2
mountain slide		山体滑坡	6.1
mounte[ˈmaʊntɚ]	v.	安装	14.2
movement monument		位移标志	20.1

mulch [mʌltʃ]	n.	覆盖物	17.1
multi-faceted [ˈmʌltiˌfæsitid]	adj.	多方面的；多才多艺的	3.2
multiple [ˈmʌltipl]	n.	倍数	5.1
	adj.	多样的，多重的	
multiple arch dam		连拱坝	10.2
multiply [ˈmʌltiplai]	v.	（使）相乘，(使)增加	5.1
multipurpose [ˈmʌltiˈpə: pəs]	adj.	综合利用，多种用途的，多目标的	26.1
municipal [mju:ˈnisipəl]	adj.	市政的，市立的，地方性的，地方自治的	9.3
navigation [ˌnævəˈgeʃən]	n.	航行（学）；导航，领航	3.2
neap tide		小潮	24.2
needlelike [ˈni: dllaik]	n.	针状（物）	6.2
net [net]	adj.	净余的，净的	5.1
New Orleans		新奥尔良	11.1
nitrate [ˈnaitreit]	n.	[化] 硝酸盐，硝酸钾	9.2
nitrogen [ˈnaitrədʒən]	n.	氮	9.3
nominal diameter		等容粒径	6.2
nominalization [ˌnɔminəlaiˈzeiʃən]	n.	名词化；名物化	29.2
nonstationarity [ˈnɔnsteiʃəˈnæriti]	n.	非平稳性	3.2
notoriously [nəuˈtɔ: riəsli]	adv.	声名狼藉地	26.1
Novarupta		诺瓦拉普塔火山	14.2
noxious [ˈnɑkʃəs]	adj.	有毒的，有害的	25.3
nozzle [ˈnɔzl]	n.	管口，喷嘴	17.1
nuclear power plant		核电站	9.3
nuclear power		核动力，核电	9.3
numerous [ˈnju:mərəs]	adj.	众多的，许多的	5.2
Nurek Dam		努列克坝	10.1
nutrient [ˈnju: triənt]	n.	养分，滋养物	9.1
	adj.	营养的，滋养的	
objective [əbˈdʒektɪv]	n.	目的；目标	18.1
	adj.	客观的；目标的；宾格的	
obligation [ɔbliga(ɑ)sjɔ̃]	n.	责任，义务；债务，债券；必须，需要	27.2
observe [əbˈzə: v]	vt.	注意到；观察；评论；遵守，奉行	3.2
occur [əˈkə:]	vi.	发生，出现，存在；被想起(到)	2.2
oceanographer [ˌəuʃiəˈnɔgrəfə]	n.	海洋学家	4.2
oceanography [ˌəuʃiəˈnɔgrəfi]	n.	海洋学	4.1
off-peak		非高峰（期）的	23.1
off-shore		离岸的，在近海处的	9.3
olive [ˈɔliv]	n.	橄榄	18.1
one-way	adj.	单行道的，单程的	26.1

Word	POS	Meaning	Section
ongoing ['ɔn,gəuiŋ]	adj.	正在进行的	23.3
optimization ['ɔptimaiziʒən]	n.	最优化	18.1
organic [ɔ:'gænik]	adj.	有机(体)的	6.1
organic content		有机质成分	20.1
organism ['ɔ: gənizəm]	n.	生物体，有机体	1.1
outsourcing ['aʊt,sɔrsɪŋ]	n.	外包；外购；外部采办	27.6
outstanding [aut'stændiŋ]	adj.	突出的，显著的	27.6
overall ['əuvərɔ: l]	adj.	总体的，全部的，全体的，一切在内的	2.1
	adv.	总的来说，全部地	
overcautious [əuvə'kɔ: ʃəs]	adj.	过分小心，保守	5.2
overflow ['əuvəfləu]	n.	泛滥，溢出；溢流口 v. 淹没，泛滥；充满；溢出	8.1
overhead [,ovɚ'hɛd]	adv.	在头顶上；在空中；在高处	17.1
	n.	天花板；[会计] 经常费用；间接费用；吊脚架空层	
overstressing ['əuvə'stresiŋ]	n.	大应力，过度（超限）应力，超负载	20.1
overtop ['əuvə'tɔp]	n.	溢过顶部	11.4
overtopping ['əuvə'tɔping]	n.	漫顶；漫溢	11.4
overwhelm ['əuvə'welm]	vt.	淹没，覆没，打击，制服，压倒	9.1
oxbow['ɑks,boʊ]	n.	由河流中的马蹄形弯曲形成的环；牛轭的U形项圈	7.1
oxidation [ɔksi'deiʃən]	n.	[化] 氧化	9.3
ozone ['əuzəun]	n.	臭氧	14.2
paradox ['pærədɔks]	n.	似乎矛盾却正确的说法；自相矛盾的人（物）	14.2
parcel ['pa:sl]	n.	包裹	18.1
	v.	打包	
participle ['pa: ti,sipəl]	n.	分词	29.2
particle ['pa: tikl]	n.	微粒，颗粒，粒子	1.1
particular [pə'tikjulə]	n.	细节，详细	1.4
	adj.	特定的，特殊的，特别的，详细的，精确的，挑剔的	
passageway ['pæsidʒwei]	n.	过道，出入口	12.3
pasture ['pæstʃɚ]	n.	牧场	9.2
	vt.	放牧；吃草	
pathogenic [,pæθə'dʒenik]	adj.	致病的，病原的，发病的	9.3
pathway ['pɑ:θwei]	n.	路，径	4.2
pattern ['pætən]	n.	模型，模式；花样，图案	1.2
	vt.	仿制	

peak discharges		最大下泄流量，最大泄水量；最大排出量	5.1
peneplain ['pi:neplein]	n.	准平原	7.1
penstock ['penstɔk]	n.	压力水管，水道，水渠	12.1
percentage [pə'sentidʒ]	n.	百分率	8.4
percolate['pɚkəlet]	vt.	过滤；渗出；浸透	4.2
	vi.	使渗出；使过滤	
	n.	滤过液；渗出液	
percolation [ˌpɚkə'leʃən]	n.	渗透，过滤	17.1
periodic [ˌpiəri'ɔdik]	adj.	周期的，定期的	2.1
permanganate [pə:'mæŋgəneit]	n.	[化] 高锰酸	13.1
permeability ['pə: miəbiliti]	n.	渗透性	4.1
permeable ['pə:miəbəl]	adj.	可渗透的	7.2
perpendicular axes		垂直轴线	6.2
pertinent ['pɚtnənt]	adj.	相关的，相干的；中肯的；切题的	27.2
petroleum [pi'trəuliəm]	n.	石油	9.3
phosphorus ['fɔsfərəs]	n.	磷	9.3
photovoltaic [ˌfəutəuvɔl'teiik]	adj.	光电的，光伏发电的	23.2
piezometer [ˌpaiə'zɔmitəs]	n.	压力计，压强计，测压计	20.1
pitch [pitʃ]	n.	斜度，螺距	21.1
pivot ['pɪvət]	n.	枢轴，中心点，旋转运动	16.3
	vt.	以……为中心旋转，把……置于枢轴上	
	vi.	在枢轴上转动，随……转移	
	adj.	枢轴的，关键的	
	n.	（Pivot）人名，（德）皮福特，（法）皮沃	
planning ['plænɪŋ]	n.	规划	2.1
plot [plɔt]	v.	绘图；密谋	5.2
	n.	故事情节；密谋；小块土地	
plough [plau]	v.	犁，耕	6.1
	n.	犁，犁形工具	
polar ['pəulə]	adj.	[天]两极的，极性的	1.1
	n.	极性	
policy ['pɔləsi]	n.	政策，方针；保险单	14.1
pollutant [pə'lu: tənt]	n.	污染物质	9.3
pollution extent		污染程度	13.1
popularity [ˌpɔpju'læriti]	n.	普及，流行	24.2
pore [pɔ:]	n.	毛孔，细孔	11.4
	vi.	（over）仔细阅读，审视	
porosity ['pɔ: rəs]	n.	多孔，空隙	6.2
post-tension		后张力	10.2

potassium [pə'tæsjəm]	n.	[化] 钾（19 号元素，符号 K）	13.1
potential energy		势能	21.1
power line		电力线，输电线	12.4
power plant		发电厂	9.3
power supply		电源，电力供应，供电	21.3
pragmatic [præg'mætɪk]	adj.	实际的，实用主义的，国事的	16.1
precede [prɪ'sid]	vi.	领先，在前面	27.3
	vt.	领先，在……之前；优于，高于	
precipitation [priˌsipi'teiʃən]	n.	降（雨、雪、雹）；降落；急躁，仓促；沉淀，沉淀作用	3.1
precisely [pri'saisli]	adv.	精确地；刻板地	5.2
precision [pri'siʒən]	n.	精确（性），精密（度）	17.2
predict [pri'dikt]	v.	预知，预言，预告	4.5
prehistoric [prihɪ'stɔrɪk]	adj.	史前的，陈旧的	16.4
preliminary [pri'liminəri]	n.	[pl.] 初步做法	5.2
	adj.	预备的，初步的	18.1
prerequisite ['pri:'rekwizit]	n.	先决条件	27.2
prescribed [prɪ'skraɪbd]	adj.	规定的，指定的	10.3
pressure trapezoid		梯形（扬）压力	
pressurized ['preʃəraizd]	adj.	加压的，受压的	15.1
prevailing [prɪ'veɪlɪŋ]	adj.	流行的，一般的，最普通的，占优势的，盛行很广的	
	v.	盛行，流行（prevail 的现在分词形式），获胜	3.2
proceed [prə'si: d]	v.	着手进行，继续进行	2.1
procrastinate [pro'kræstɪnet]	v.	耽搁，延迟	28.3
procurement[prə'kjʊrmənt]	n.	采购；获得，取得	27.5
	adj.	高架的；在头上的；在头顶上的	
professional [prə'fɛʃənl]	adj.	专业的；职业的；职业性的	29.1
	n.	专业人员；职业运动员	
prolonged [prə'lɔŋd]	adj.	持续很久的，长时期的，长时间的	8.3
prominent ['prɔminənt]	adj.	杰出的，突出的；凸起的	11.1
propeller [prə'pelə]	n.	螺旋桨；推进器	12.3
propeller blade		螺旋桨叶片	12.3
prototype ['prəutətaip]	n.	原型，样机	6.2
protozoan [ˌprəutəu'zəuən]	n.	[动] 原生动物	9.2
provincial [prə'vinʃəl]	adj.	省的，地方的	13.1
	n.	乡下人，地方人民	
pumped storage		抽水蓄能	12.5

purification [ˌpjuərifiˈkeiʃən]	n.	净化	4.5
purpose-built [ˈpə: pəsbilt]	adj.	为特定目的建造的	17.2
quiescent distilled water		静水，蒸馏过的纯水	6.2
rafting [ˈrɑ:ftiŋ]	n.	筏运	26.2
rainstorm [ˈreinstɔ:m]	n.	暴风雨	13.4
rainwater [ˈreinwɔ: tə(r)]	n.	雨水	4.4
rapid [ˈræpid]	n.	急流	7.3
realistic [riəˈlistik]	adj.	现实的；实际可行的；现实主义的	19.1
reap [rip]	vt.	收获，获得；收割	27.6
	vi.	收割，收获	
	n.	(Reap)人名；(英)里普	
recharge [ˈri: tʃɑ:dʒ]	n.	（水）补给	7.1
recognizable [ˈrekəgnaizəbl]	adj.	可辨认可认知的，可承认的	8.4
reconnaissance [riˈkɔnəsəns]	n.	侦察，搜查；勘察队，勘测	2.3
recreational [ˌrekriˈeiʃənəl, -kri: -]	adj.	娱乐的，休养的	13.4
rectangular section		矩形截面	21.1
rectilinear[ˈrɛktəˈlɪnɪɚ]	adj.	直线的	17.2
recycle [ˈri:ˈsaikl]	v.	使再循环，再制	4.5
regional [ˈri: dʒənəl]	adj.	整个地区的，地方的，地域性的	1.2
rehabilitation [ri: həˌbiliˈteiʃən]	n.	复原，康复	18.1
relative motion		相对运动	24.1
relevant [ˈreləvənt]	adj.	有关的，切题的	16.3
relocatee [ˈri: ləuˈkeiti:]	n.	迁至新址者	26.1
render [ˈrendə]	vt.	造就，使得	19.2
	vi.	给予补偿	
renewable [riˈnju: əbl]	adj.	可更新的，可恢复的	1.2
renewable energy		可再生能源	22.1
replenishment [riˈpleniʃmənt]	n.	补给，补充	8.1
reproductive [ˈri: prəˈdʌktiv]	adj.	再生的，复制的，生殖的	9.4
reserve [riˈzə: v]	n.	储备，保留，备用	23.2
reservoir [ˈrezər.vwɑ]	n.	水库，蓄水池	23.1
residue [ˈrezidju:]	n.	残余，渣滓，滤渣，残数，剩余物	9.3
resistive [riˈzistiv]	adj.	抗（耐、防）……的；电阻的；抵抗的，有抵抗力的，有耐力的	20.1
response [riˈspɔns]	n.	响应，反应	14.2
restriction [riˈstrikʃən]	n.	限制，限定，约束	5.2
retrieve [rɪˈtriv]	vt.	检索；恢复；重新得到	15.1
	vi.	找回猎物	
	n.	检索；恢复，取回	

revenue [ˈrevənju:]	n.	收入，国家的收入，税收	22.3
reversible [riˈvə: səbl]	adj.	可逆的	23.1
revetment [riˈvə:t mənt]	n.	护岸	11.1
revise [riˈvaiz]	v.	校订，修正，校正	2.1
	n.	校订，修正，再校稿	
ridge [ridʒ]	n.	脊，山脊，垄，埂，脊状突起	11.2
rill [ril]	n.	小河，小溪	7.1
	vi.	小河般地流	
rim [rim]	n.	边，轮缘	21.1
rockfill [ˈrɔkfil]	adj.	用石头填充的	19.2
rotation [rəuˈteiʃn]	n.	旋转	12.4
rotor [ˈrəutə]	n.	转子，回转轴，转动体	17.1
runner [ˈrʌnə]	n.	叶轮，转子	12.3
runoff [ˈrʌnwei]	n.	径流，水流	1.3
rush [rʌʃ]	v.	冲；仓促从事；突袭	11.3
	n.	冲；匆忙；繁忙时刻	
sabotage [ˈsæbəta: ʒ]	n.	阴谋破坏	22.4
saltwater [ˈsɔ:ltˈwɔ:tə]	adj.	盐水的，海产的	1.1
sanitation [ˌsænəˈteʃən]	n.	卫生系统	3.1
saturated [ˈsætʃəreitid]	adj.	饱和的	8.3
saturated soil		饱和土壤，饱水土壤	8.3
scatter [ˈskætə]	v.	分散，散开，驱散	20.1
scenario [səˈnɛrioʊ]	n.	方案，前景	25.2
scour [ˈskauə]	vt.	冲刷	22.4
screen [skri: n]	n.	筛子，掩蔽物	2.3
scroll case		蜗壳	12.3
scrutinize [ˈskrutənaɪz]	vt.	细阅；作详细检查	28.3
	n.	仔细或彻底检查	
	vi.	详细检查；细看	
sea level		海拔，海平面	8.1
sea wave		海浪	8.2
seal [si: l]	n.&vt.	封，密封	12.2
sealed [si: ld]	adj.	密封的，未知的	21.1
section [ˈsekʃən]	n.	部分，部件，节，项，区，地域，截面	26.2
sedimentation diameter		沉积直径，泥沙粒径	6.2
seep [si:p]	v.	渗出，渗流，漏	10.2
seepage [ˈsi: pidʒ]	n.	泄漏，渗水量	4.2
seepage analysis		渗流分析	20.1
segregation [ˌsegriˈgeiʃən]	n.	分离，隔离	20.2

seismic ['saizmik]	adj.	地震的，由地震引起的；震撼世界的	20.1
semi-impervious		半透水的	20.1
seniority [ˌsiniˈɔrəti]	n.	长辈，老资格，前任者的特权	16.3
separate ['sepəreit]	v.	使分离；区分；分居	4.1
	adj.	分离的，各自的	
series ['siəri: z]	n.	一系列，连续；丛书，连续剧	2.3
settlement gage		沉陷量测仪（计）	20.1
sevenfold ['sevn'fold]	adj.	七倍的	3.1
	adv.	由七部分组成地	
sewage ['sju: idʒ]	n.	脏水，污水	8.1
shaft [ʃa: ft]	n.	轴，杆状物	12.4
shallow water		浅水	8.1
shear strength		抗剪强度	20.1
shellfish ['ʃelfiʃ]	n.	水生贝壳类动物，贝，甲壳类动物	9.4
shoal [ʃəul]	n.	浅滩，沙洲	26.1
shrink [ʃriŋk]	vi.	起皱，收缩；畏缩	14.2
	vt.	使起皱，使收缩	
shrub [ʃrʌb]	n.	灌木，灌木丛	9.1
S-hydrograph		S 曲线	5.2
sieve diameter		筛孔直径	6.2
sieve mesh		筛分网，筛分孔	6.2
sift [sɪft]	vt.	筛选；撒；过滤；详查	9.2
	vi.	筛；详查；撒下；细究	
siltation [sliteiʃən]	n.	沉积作用，淤积	22.4
silting ['siltiŋ]	n.	淤积；淤塞；充填	18.2
silver iodide		碘化银微粒	4.5
sink [siŋk]	n.	水槽	9.3
	v.	陷入，衰退；下沉；降低；掘	
sinuous ['sinjuəs]	adj.	蜿蜒的	7.1
siphon ['saifən]	vt.	虹吸	12.2
sludge [slʌdʒ]	n.	软泥，淤泥，矿泥，煤泥	9.3
sluice [slu: s]	n.	水闸，蓄水	18.2
	v.	开闸放水，流出，冲洗，奔泻	
small hydro		小水电	25.1
snowfall ['snəufɔ: l]	n.	降雪，降雪量	4.2
snowfield ['snəufi: ld]	n.	雪原，雪地	4.4
snowpack ['snəupæk]	n.	积雪	7.1
snows melt		融雪	8.3
soak [səuk]	v.	浸，泡，浸透	4.2

Word	POS	Meaning	Section
	n.	浸透	
sodium ['səudiəm]	n.	[化] 钠	9.4
soluble ['sɔljubl]	adj.	可溶的	20.1
spatial analysis		空间分析	18.1
spawn [spɔ: n]	vi.	产卵	7.3
specialist ['speʃəlist]	n.	专家，专门医师	4.2
specification ['spɛsəfə'keʃən]	n.	规格；说明书；详述	2.3
specify ['spesifai]	vt.	明确说明，具体指定	2.2
sphere [sfiə]	n.	球(体)；范围，领域	4.1
spillway ['spilwei]	n.	溢洪道，泄洪道	10.2
spillway crest		溢洪道顶，溢流堰顶	10.2
spin [spin]	v.	旋转	12.5
spiral ['spaiərəl]	adj.	螺旋形的	12.3
split [splɪt]	n.&v.	分歧，分裂，使分裂，分开	6.2
sponge [spʌndʒ]	n.	海绵，海绵状物，（医）棉球，纱布	
	v.	用海绵等洗涤、擦拭或用海绵吸收（液体），依赖某人生活，……	8.3
sponsor ['spɔnsə]	vt.	发起，主办，赞助	9.1
	n.	主办人，发起人，保证人	
spontaneous [spɔn'teiniəs]	adj.	自动的，自发的	20.1
spray [spreɪ]	n.	喷雾，喷雾器；水沫	17.1
	vt.	喷射	
	vi.	喷	
spread [spred]	v.	展开；散布，涂	1.3
	n.	传播；幅度，范围	
stainless ['steɪnlɪs]	adj.	不锈的；纯洁的，未被玷污的；无瑕疵的	29.3
standard Proter optimum water content		标准普氏最优含水量	20.1
statistics [stə'tistiks]		统计数据，统计学，统计资料，统计数字	8.4
stator ['steitə]	n.	定子，固定片	21.2
steady ['stedɪ]	adj.	稳定的；不变的；沉着的	5.1
	vt.	使稳定；稳固；使坚定	
steep [stip]	adj.	陡峭的，不合理的；夸大的；急剧升降的	6.1
	n.	峭壁；浸渍	
	vt.	泡；浸；使……充满	
	vi.	泡；沉浸	
sterile ['sterail]	adj.	贫瘠的，不育的，不结果的	26.1
stimulus ['stimjuləs]	n.	刺激物，刺激，促进因素	26.2
stipulate ['stɪpjulet]	v.	规定；保证	27.3
	adj.	有托叶的	

storage ['stɔridʒ]	n.	蓄水量；储藏量，库存量；蓄电；储藏，[电脑]存储，存储器	4.5
storm [stɔ:m]	n.	暴雨(雪)	5.1
	vt.	猛攻	
	vi.	怒气冲冲地走	
strain indicator		应变指示仪	20.1
stream [stri:m]	n.	流，水流，人潮	20.2
	v.	使流出，流动	
subcontractors [ˌsʌbkən'træktə]	n.	转包商，次承包商；分承包方	27.2
subdivide [ˌsʌbdɪ'vaɪd,'sʌbdɪˌvaɪd]	n.	细分，把……再分	5.2
Subirrigation[sʌb'irigeiʃən]	n.	地下灌溉	17.3
subjective [səb'dʒektɪv]	adj.	主观的；个人的；自觉的	2.1
subsequent ['sʌbsɪkwənt]	adj.	后来的，随后的	2.1
subservient [sʌb'sə: viənt]	adj.	屈从的，有帮助的，有用的，奉承的	9.1
subside [səb'said]	v.	下沉，沉淀，平息，减退，衰减	8.1
substantiation [sʌbs'tænʃieiʃən]	n.	实体化，证明，证实	14.2
substitute ['sʌbstɪtut]	n.	代用品，代替者	2.1
	v.	替代	
succumb [sə'kʌm]	vi.	屈服；死；被压垮	28.3
suction ['sʌkʃən]	n.	吸入，吸力，抽气，抽气机，抽水泵	21.1
suffice [sə'fais]	vi.	足够	6.2
suffocate ['sʌfəkeit]	vt.	使窒息，噎住，闷熄；	9.2
	vi.	被闷死，窒息，受阻	
sulfate ['sʌlfeit]	n.	[化] 硫酸盐	13.3
	v.	以硫酸或硫酸盐处理，使变为硫酸盐	
surface ['sə: fis]	n.	表面，平面	14.2
	adj.	表面的，肤浅的	
surface runoff		地表径流，地面径流	4.4
surface water hydrology		地表水文学	4.1
surge [sə: dʒ]	n.	电流急冲，电涌；突然增大	21.3
sustaining [səs'teiniŋ]	adj.	维持的，支持的，持续的	4.5
swift [swift]	adv.	迅速地，敏捷地	8.2
	adj.	迅速的，快的，敏捷的，立刻的	
	n.	[鸟] 雨燕，（梳棉机等的）大滚筒	
synchronous ['sɪŋkrənəs]	adj.	同步的，同时发生的	25.2
syntactical [sɪn'tæktɪkəl]	adj.	句法的	29.2
tailrace ['teilreis]	n.	（水轮，涡轮的）尾水，放水路	12.3
tailwater [teil'wɔ: tə]	n.	尾水	21.1
talus ['teiləs]	n.	碎石堆	6.1

tap [tæp]	n.	轻打，活栓，水龙头	1.1
	vt.	轻敲，开发，分接，使流出	
	vi.	轻叩，轻拍，轻声走	
tarnish ['tɑ:niʃ]	vt.	玷污；（使）失去光泽，(使)变灰暗	26.2
tectonics [tek'tɒniks]	n.	地质学	7.2
telecommunications ['tɛləkə'mjʊnə'keʃənz]	n.	通信行业：服务类型变更，缴纳话费	27.6
		安装，装置；就职	
temporal and spatial		时间的和空间的	5.2
temporary ['tempərəri]	adj.	暂时的，临时的	11.1
tendency ['tendənsi]	n.	趋势，潮流；倾向；癖性；天分	2.2
tense [tens]	adj.	紧张的；拉紧的	5.2
	v.	变得紧张；使拉紧	
	n.	时态	
tensiometer [ˌtensɪ'ɒmɪtə]	n.	张力计；表面张力计；土壤湿度计	15.3
tentatively ['tentətivli]	adv.	试探（性）地，试验（性）地	19.1
terminology [ˌtɜ:mɪ'nɒlədʒɪ]	n.	术语，术语学	6.2
terrace ['terəs]	n.	阶梯；看台；排屋；地坪，草坪；[pl.] 梯田	11.1
terrain ['terein]	n.	地形	8.2
terrigenous [te'ridʒinəs]	adj.	陆源的	6.1
terrorism ['terərizəm]	n.	恐怖主义，恐怖行动	22.4
thawing ['θɔ:iŋ]	n.	熔化，融化	10.2
theoretical [ˌθiə'retikəl]	adj.	理论(上)的	19.1
Three Gorges		三峡	22.3
threshold ['θrɛʃhold]	n.	门槛；开始	3.1
thrust [θrʌst]	n.	推力，插，戳，刺，猛推	10.2
	vt.	冲，插入，挤进，刺，戳	
	vi.	插入，刺，戳，延伸，强行推进	
tidal wave		潮汐波，浪潮	8.2
tide [taid]	n.	潮，潮汐；潮流，趋势	11.3
tide by tide		一浪接着一浪	11.3
tile [tail]	n.	瓦，瓷砖	11.1
	vt.	铺瓦于，贴瓷砖于	
time-lag		时间滞后	14.2
topographic [ˌtɔpə'græfik]	adj.	地质的，地形学的	20.1
topographical [ˌtɔpə'græfikl]	adj.	地形学的，地形的	19.2
topography [tə'pɔgrəfi]	n.	地形学，地形	10.2
torrential [tə'renʃəl]	adj.	汹涌的，奔流的，猛烈的	7.1
total ['təutl]	n.	总数	1.1

	adj.	全部的，完全的	
	v.	合计	
toxicity [tɔk'sisiti]	n.	毒性	9.3
transcend [træn'sɛnd]	vt.	胜过，超越	16.4
transformer [træns'fɔ: mə]	n.	变压器	12.4
transition zone		移行区，过渡区	20.1
transmission [trænz'miʃən]	n.	播送，发射；传送，传递，传染	14.2
transpiration [ˌtrænspi'reiʃən]	n.	[生]（叶面的）水汽散发；蒸腾作用；[物] 流逸	4.4
transpire [træns'paiə]	v.	蒸腾，散发，使……蒸发，排出；泄露，为人所知；发生	4.2
transverse ['trænzvə: s]	n.	横向物，横轴，横断面；[数] 横截轴；[解] 横肌	20.1
	adj.	横向的，横断的，横切的；[数] 横截的	
traverse ['trævə: s]	vt.	横渡，横越	6.1
treatment ['tri: tmənt]	n.	处理	4.2
triaxial [trai'æksiəl]	adj.	三轴的	6.2
tribal ['traɪbl]	adj.	部落的，种族的	16.3
	n.	（Tribal）人名，（法）特里巴尔	
tributary ['tribjutəri]	adj.	支流的，进贡的	11.1
trigger ['trigə]	vt.	触发，引起	13.2
	n.	扳机；引起反应的行动	14.2
tropical ['trɔpikl]	adj.	热带的	8.2
tsunami [tsju:'nɑ:mi]	n.	海啸	8.2
tunnel ['tʌnl]	v.	挖（地道），开（隧道）	12.1
	n.	隧道，地道	
turbidity [tə:'biditi]	n.	混浊，混乱	12.2
turbine ['tə: bain]	n.	涡轮，轮机，透平	12.1
turbulence ['tə: bjuləns]	n.	湍流，紊流，（液体或气体的）紊乱	22.4
turf [tə: f]	n.	草皮	11.1
unambiguous ['ʌnæm'bigjuəs]	adj.	不引起歧义的，清楚的，清晰的	14.2
underground ['ʌndəgraund]	adj.	地下的，秘密的；	4.4
	n.	地下，地铁，地道，秘密活动	
underlying rock		下垫岩石	4.4
underrate [ˌʌndə'reit]	v.	低估	19.2
uneven ['ʌn'i: vən]	adj.	不平坦的，不平均的，不均匀的	1.3
uniformly ['ju: nifɔ:mli]	adv.	一致地	5.1
unravel [ʌn'rævəl]	vt.	解开	5.2
upland ['ʌplənd]	n.	高地，丘陵地，丘阜	7.3

adj. 高地的，山地的

upper and lower 上、下限 5.2

upriver ['ʌp'rivə] n. 向上游 7.1

upstream ['ʌp'stri:m] adj. 向上游，逆流地 2.2

uptake ['ʌpteik] n. 摄取，领会 14.2

uranium [juə'reiniəm] n. 铀 9.3

usage ['ju: sidʒ] n. 使用，用法；惯用法 14.1

utilitarian [jʊˌtɪlɪ'teərɪən] adj. 功利的，功利主义的，实利的 16.2

n. 功利主义者

utilities [ju:'tilitiz] n. 公共事业，电力公司 12.5

vagueness ['veignis] n. 模糊，含糊，暧昧，茫然 7.1

Vaksh River 瓦赫什河 10.1

valley['væli] n. 山谷，流域，溪谷 6.1

vane [vein] n. （风车、螺旋桨等的）翼，叶片 21.1

variant ['vɛəriənt] adj. 不同的 1.2

n. 变量

variation [ˌveəri'eiʃən] n. 变化，变动；变体，变种；变奏（曲） 5.2

vegetation [ˌvedʒi'teiʃən] n. 植物，草木 4.4

vendor ['vɛndɚ] n. 卖主；小贩；供应商；[贸易] 自动售货机 25.2

versatile ['və: sətail] adj. 通用的，万能的，多才多艺的，多面手的 18.2

versus ['və: səs] prep. 与……相对 5.2

vertical ['və: tikəl] adj. 垂直的 10.2

n. 垂直线

via ['vaiə] prep. 经，通过，经由 7.3

viable ['vaiəbl] adj. 可行的 24.2

viscous ['viskəs] adj. 黏性的，黏滞的 24.1

vital ['vaitl] adj. 极其重要的，生死攸关的，充满生机的 11.1

vocabulary [və'kæbjəlɛri] n. 词汇；词表；词汇量 29.1

volcanism ['vɔlkənizəm] n. 火山活动，火山作用 14.2

volcano [vɔl'keinəu] n. 火山 8.2

voltage ['vəultidʒ] n. 电压，伏特数 12.4

volume ['vɔlju:m] n. 体积，容量；卷，册，书卷；音量，响度 4.5

wastewater ['weistwɔ: tə] n. 废水（污水） 9.3

Water Diversion Project 南水北调工程 13.1

water vapor 水汽 1.1

waterborne ['wɔ: təbɔ: n] adj. 水传播的，水上的，水运的 9.1

watercourse ['wɔ: təkɔ: s] n. 水道，河道，航道 7.1

watermill ['wɔ: tə, mil] n. 水磨坊 7.3

watershed ['wɔ:təʃed] n. 转折点，重要关头；流域，分水岭， 4.5

		分水线	
waterwheel ['wɔ: təwi: l]	n.	水车，吊水机	21.1
wedge [wedʒ]	vt.	把……楔牢，塞入	6.1
	n.	楔（子）	
weir [wɪr]	n.	堰，拦河坝，导流坝	11.3
whitewater kayaking		激流皮艇	7.3
wicket ['wikit]	n.	导叶，小门	21.1
willow ['wiləu]	n.	柳，柳树	11.1
windmill ['windmil]	n.	风车	24.2
withdraw [wið'drɔ:]	vt.	收回，撤回	1.2
	vi.	离开，脱离	
Yangtze (=Yangtze River, Changjiang)		扬子江，长江	13.1
zooplankton [ˌzəuə'plæŋktən]	n.	浮游动物	12.2

Appendix 4 Abbreviations

A.D.	Anno Domini	公元 [拉丁文]
AC	alternating current	交流（电）
AGU	American Geophysical Union	美国地球物理协会
B.C.	Before Christ	公元前
BOD	Biochemical Oxygen Demand	生化需氧量
CIMIS	California Irrigation Management Information System	加利福尼亚灌溉管理信息系统
DC	direct current	直流（电）
DDT	dichloro-diphenyl-trichloroethane	滴滴涕，二氯二苯三氯乙烷
DOE	Department of Energy	能源部[美]
EST	English for Science and Technology	科技英语
EU	European Union	欧盟
FAO	Food and Agriculture Organization of the United Nations	联合国粮农组织
GDP	Gross Domestic Product	国内生产总值
GCIS	Genil-Cabra Irrigation Scheme	赫尼尔-卡布拉灌区灌溉计划
GW	gigawatt	千兆瓦
i.e.	拉丁文 id est，相当于 that is，	in other words 即，也就是
INFOHYDRO	Hydrological Information Referral Service	水文信息采集与获取
IPCC	Intergovernmental Panel on Climate Change	政府间气候变化小组
IUH	Instantaneous Unit Hydrograph	瞬时单位线
kW · h	kilowatt-hour	千瓦 · 时
LEPA	Low Energy Precision Application	法律实施计划局
MW	megawatt	兆瓦
PCBs	polychlorinated biphenyls	多氯联苯
PWS	Public Water System	公共用水系统
SBD	Standard Bidding Document	标准招标文件
SSTI	Subsurface Textile Irrigation	地下织物灌溉
UH	Unit Hydrograph	单位线
UN	United Nations	联合国
UNCED	United Nations Conference on Environment and Development	联合国环境与发展大会

UNEP	the United Nations Environmental Program	联合国环境计划
USBR	United States Bureau of Reclamation	美国垦务局
WCED	World Commission on Environment and Development	世界环境与发展委员会
WHO	World Health Organization	世界卫生组织
WMO	World Meteorological Organization	世界气象组织

Appendix 5 Website of Speciality Orgnizations

（相关专业网址）

1. 国际水协会：www.iwa-network.org or http://iwaponline.com
2. 世界水理事会：www.worldwatercouncil.org
3. 世界水周：www.worldwaterweek.org
4. 联合国水资源组织：www.unwater.org
5. 国际水资源协会：www.iwra.org
6. 国际水电协会：www.hydropower.org or www.hydro.org
7. 世界气象组织：http://public.wmo.int
8. 国际水文科学协会：https://iahs.info
9. 国际水文环境工程协会：www.iahr.org
10. 国家地下水协会：www.ngwa.org
11. 世界水土保持协会：www.waswac.org
12. 国际水文地质学家协会：https://iah.org
13. 国际水利与环境工程学会：www.iahr.org.cn
14. 国际灌排委员会：www.icid.org
15. 水技术协会（AWT）：www.awt.org
16. 世界健康水协会：www.whwanet.com
17. 国际瓶装水协会：www.bottledwater.org
18. 美国水资源协会：www.awra.org
19. 美国水质协会：www.wqa.org
20. 美国水行业协会：www.awwa.org
21. 美国国家海洋与大气管理局：www.noaa.gov
22. 美国水工程协会：www.iwa-network.org
23. 美国气象学会：www.ametsoc.org
24. 中国人民共和国水利部：www.mwr.gov.cn
25. 中国水利学会：www.ches.org.cn
26. 中国水网：http://www.h2o-china.com
27. 中国水协会：http://shuixiehui.cailiao.com
28. 中国水利水电研究协会：www.cwhra.org.cn
29. 中国水利工程协会：www.cweun.org
30. 中国水力发电工程学会：www.hydropower.org.cn
31. 中国水土保持学会：www.sbxh.org
32. 中国水利水电勘测设计协会：http://xh.giwp.org.cn

33. 中国净水网：www.zgjsw.com.cn
34. 中国海水淡化与水再利用学会：www.chinawatertech.org
35. 中国水利企业协会：www.cwec.org.cn
36. 中国气象服务协会：www.chinamsa.org
37. 中国城镇供水排水协会：www.cuwa.org.cn
38. 中国水产学会：www.csfish.org.cn
39. 中国水博览会：www.waterexpo.cn
40. 中国环境保护产业协会：www.caepi.org.cn
41. 中国环境科学学会：www.chinacses.org
42. 水环境联合会：www.wef.org
43. 中国环境科学学会：www.chinacses.org
44. 水环境联合会：www.wef.org
45. 中国土木工程学会: www.cces.net.cn
46. 中国岩石力学与工程学会：www.csrme.com
47. 北京水利学会：www.bjslxh.org.cn
48. 上海市水利学会：www.shhes.org.cn
49. 浙江省水利学会：www.zjwater.com
50. 辽宁省水利学会：www.lnslxh.org
51. 深圳市水利学会：www.slxh.com.cn
52. 上海市水利工程协会：www.swea.com.cn
53. 北京市水利工程协会：www.bwea.org.cn
54. 福建省水利工程协会：www.fjwea.org.cn
55. 湖南省水利工程协会：www.hnslxh.com
56. 贵州省水利工程学会：www.gzwea.com
57. 天津市水利工程协会：www.tjsw.gov.cn
58. 深圳市水利工程协会：www.szslxh.com
59. 湖北省水土保持协会：http://hubeihuaao.com.cn
60. 浙江省水利水电工程管理协会：www.zwema.org
61. 吉林省地下水协会：www.jldxs.cn
62. 北京地质学会：www.bjdzxh.org
63. 上海市地质协会：www.shdzxh.com
64. 湖南省地质学会：www.hnsdzxh.org.cn
65. 北京气象学会：www.bjqxxh.org
66. 天津市气象学会：www.tjqxxh.com.cn
67. 广州市气象学会：www.gz323.org.cn
68. 安徽气象学会：www.ahms1961.org

References

[1] 刘景植. 水利水电类专业英语 [M]. 武汉：武汉大学出版社，2001.
[2] 迟道才，周振民. 水利专业英语 [M]. 北京：中国水利水电出版社，2006.
[3] 刘双芹，胡震云，王薇薇，等. 水利水电工程英语 [M]. 北京：中国水利水电出版社，2007.
[4] 陈天照. 水利水电工程实用英语 [M]. 北京：中国电力出版社，2006.
[5] 李立功. 水利电力英语 [M]. 南京：河海大学出版社，1989.
[6] 李继清，朱永强，张玉山. 水利专业英语. 北京：中国水利水电出版社，2011
[7] 景志华，罗南华，孙石. 实用电力英语丛书：水力发电分册 [M]. 北京：中国电力出版社，2007.
[8] 朱永强，尹忠东. 电力专业英语阅读与翻译 [M]. 北京：中国水利水电出版社，2007.
[9] 高等学校外语专业教学指导委员会英语组. 高等学校英语专业英语教学大纲 [M]. 北京：外语教学与研究出版社，2000.
[10] 李俊梅. 重构专业英语教学模式的探讨 [J]. 郑州航空工业管理学院学报(社会科学版)，2007，26(5)：131-133.
[11] 李继清. 水利水电专业英语教学模式探究，华北电力大学学报（社科版），2012，（supp）: 258-260
[12] 刘延锋. 关于水文专业英语的教学实践与思考 [J]. 科技信息，2008（12）：11-12.
[13] 赵华. 专业英语教学中存在的问题及对策 [J]. 教育探索，2008，204(6):70-71.
[14] 亢树森. 科技英语写作与翻译教程 [M]. 西安：科学技术出版社，1998.
[15] 高文霞. 科技英语翻译模式和技巧[D]. 上海：上海大学硕士学位论文，2007.6.
[16] 刘然，包兰宇，景志华. 电力专业英语[M].北京：中国电力出版社，2006.
[17] 戴文进. 电气工程及自动化专业英语[M].北京：电子工业出版社，2004.
[18] 彭霞媚. 浅议电力专业英语的特点及翻译[J].华北电力大学学报（社会科学版），2009（1）.
[19] 李长栓. 非文学翻译理论与实践. 北京：中国对外翻译出版公司，2004.
[20] William D. Stevenson J. Elements of power System Analysis (Reading Materials for Students) [M]. Department of Electrical Engineering, 1985.
[21] Prabha Kundur. Power System Stability and Control. Power System Engineering series by Electric Power Research Institute [M]. Beijing: China Electric Power Press, 2001.
[22] Daily G C, English P R. Socio economic equity, sustainability, and earth capacity [J]. Ecological Application, 1996.
[23] Changjiang Water Resources Commission. The Three Gorges of the Yangtze River—A Guide for Tourists [M]. Beijing: China Water Power Press, 1997.
[24] Ren Decun. A General Introduction of the Yellow River—To Learn the Yellow River From Here [M]. The Yellow River Publishing House, 1999.
[25] Daniel Hillel. Advances in Irrigation [M]. London: Academic Press, 1982.
[26] Newmark P. A Textbook of Translation. Shanghai：Shanghai Foreign Language Education Press，2001.

[27] http://en.wik ipedia.org/wiki/Dike_(construction).

[28] http://en.wik ipedia.org/ wiki/River.

[29] http://en.wik ipedia.org/ wiki/ Climate_change.

[30] http://www.mwr.gov.cn/sj/tjgb/szygb/201707/t20170711_955305.html.